Advances in Food Mycology

Advances in Experimental Medicine and Biology

Recent Volumes in this Series

Volume 563
UPDATES IN PATHOLOGY
Edited by David C. Chhieng and Gene P. Siegal

Volume 564
GLYCOBIOLOGY AND MEDICINE: PROCEEDINGS OF THE 7TH JENNER GLYCOBIOLOGY AND MEDICINE SYMPOSIUM
Edited by John S. Axford

Volume 565
SLIDING FILAMENT MECHANISM IN MUSCLE CONTRACTION: FIFTY YEARS OF RESEARCH
Edited by Haruo Sugi

Volume 566
OXYGEN TRANSPORT TO ISSUE XXVI
Edited by Paul Okunieff, Jacqueline Williams, and Yuhchyau Chen

Volume 567
THE GROWTH HORMONE-INSULIN-LIKE GROWTH FACTOR AXIS DURING DEVELOPMENT
Edited by Isabel Varela-Nieto and Julie A. Chowen

Volume 568
HOT TOPICS IN INFECTION AND IMMUNITY IN CHILDREN II
Edited by Andrew J. Pollard and Adam Finn

Volume 569
EARLY NUTRITION AND ITS LATER CONSEQUENCES: NEW OPPORTUNITIES
Edited by Berthold Koletzko, Peter Dodds, Hans Akerbloom, and Margaret Ashwell

Volume 570
GENOME INSTABILITY IN CANCER DEVELOPMENT
Edited by Erich A. Nigg

Volume 571
ADVANCES IN MYCOLOGY
Edited by J.I. Pitts, A.D. Hocking, and U. Thrane

A Continuation Order Plan is available for this series. A continuation order will bring delivery of each new volume immediately upon publication. Volumes are billed only upon actual shipment. For further information please contact the publisher.

Advances in Food Mycology

Edited by

A.D. Hocking
Food Science Australia
North Ryde, Australia

J.I. Pitt
Food Science Australia
North Ryde, Australia

R.A. Samson
Centraalbureau voor Schimmelcultures
Utrecht, Netherlands

and

U. Thrane
Technical University of Denmark
Lyngby, Denmark

Springer

A.D. Hocking
Food Science Australia
PO Box 52, North Ryde
NSW 1670
Australia
Ailsa.Hocking@csiro.au

J.I. Pitt
Food Science Australia
PO Box 52, North Ryde
NSW 1670
Australia
John.Pitt@csiro.au

R.A. Samson
Centraalbureau voor Schimmelcultures
PO Box 85167
3508 AD Utrecht
Netherlands
Samson@cbs.knaw.nl

U. Thrane
Biocentrum-DTU
Technical University of Denmark
Building 221
DK-2800 Kgs. Lyngby
Denmark
ulf.thrane@biocentrum.dtu.dk

Library of Congress Control Number: 2005930810

ISBN-10: 0-387-28385-4 e-ISBN: 0-387-28391-9
ISBN-13: 978-0387-28385-2

Printed on acid-free paper.

Printed in the United States of America. (SPI/EB)

9 8 7 6 5 4 3 2 1

springeronline.com

FOREWORD

This book represents the Proceedings of the Fifth International Workshop on Food Mycology, which was held on the Danish island of Samsø from 15-19 October, 2003. This series of Workshops commenced in Boston, USA, in July 1984, from which the proceedings were published as *Methods for Mycological Examination of Food* (edited by A. D. King et al., published by Plenum Press, New York, 1986). The second Workshop was held in Baarn, the Netherlands, in August 1990, and the proceedings were published as *Modern Methods in Food Mycology* (edited by R. A. Samson et al., and published by Elsevier, Amsterdam, 1992). The Third Workshop was held in Copenhagen, Denmark, in 1994 and the Fourth near Uppsala, Sweden, in 1998. The proceedings of those two workshops were published as scientific papers in the *International Journal of Food Microbiology*.

International Workshops on Food Mycology are held under the auspices of the International Commission on Food Mycology, a Commission under the Mycology Division of the International Union of Microbiological Societies. Details of this Commission are given in the final chapter of this book.

This Fifth Workshop was organised by Ulf Thrane, Jens Frisvad, Per V. Nielsen and Birgitte Andersen from the Center for Microbial Biotechnology, Technical University of Denmark, Kgs. Lyngby,

Denmark. This Center, through numerous publications and both undergraduate teaching and graduate supervision has been highly influential in the world of food mycology for the past 20 years and more. Trine Bro and Lene Nordsmark from the Center also carried out the important tasks of providing secretarial help to the Organisers and solving the logistics of moving participants from Copenhagen to Samsø and back. Samsø provided an ideal setting for the Fifth Workshop, as the island is made up of rural agricultural communities, with old villages and rustic land and seascapes.

The Fifth Workshop was attended by some 35 participants, drawn from among food mycology and related disciplines around the world. The workshop was highly successful, with papers devoted to media and methods development in food mycology, as is usual with these workshops. Particular emphasis was placed on the fungi which produce mycotoxins, especially their ecology, and through ecology, potential control measures. Sessions were also devoted to yeasts, and the inactivation of fungal spores by the use of heat and high pressure. Nearly 40 scientific papers were presented over three days of the workshop, and these papers are the major contributions in these Proceedings.

The organisers especially wish to thank the sponsors of the Fifth Workshop: BCN Laboratories, Knoxville, Tennessee, USA; the Danish ECB5 Foundation, Copenhagen; Novozymes A/S, Bagsværd, Denmark; LMC Centre for Advanced Food Studies, Copenhagen; the Danish Research Agency STVF, Copenhagen, though Grant Number 26-03-0188; Eurofins Denmark A/S, Copenhagen and the Mycology Division of the International Union of Microbiological Societies, for their support which made this workshop possible.

A.D. Hocking
J.I. Pitt
R.A. Samson
U. Thrane

CONTENTS

CONTRIBUTORS

M. Lourdes Abarca, Departament de Sanitat i d'Anatomia Animals, Universitat Autónoma de Barcelona, 08193 Bellaterra, Barcelona, Spain

David Aldred, Applied Mycology Group, Biotechnology Centre, Cranfield University, Silsoe, Bedford MK45 4DT, UK

Stella M. Alzamora, Departament de Industrias, Facultad de Ciencias Exactas y Naturales, Universidad de Buenos Aires, Ciudad Universitaria 1428, Buenos Aires, Argentina

Birgitte Andersen, Center for Microbial Biotechnology, BioCentrum-DTU, Technical University of Denmark, DK-2800 Kgs. Lyngby, Denmark

Marta Bau, Departament de Sanitat i d'Anatomia Animals, Universitat Autónoma de Barcelona, 08193 Bellaterra, Barcelona, Spain

John Banks, Central Science Laboratory, Sand Hutton, York YO41 1LZ, UK

Mariam Begum, Food Science Australia, CSIRO, P.O. Box 52, North Ryde, NSW 1670, Australia

Ai Lin Beh, Food Science and Technology, School of Chemical Engineering and Industrial Chemistry, University of New South Wales, Sydney, NSW 2052, Australia

Thomas Börjesson, Svenska Lantmännen, Östra hamnen, SE-531 87 Lidköping, Sweden

M. Rosa Bragulat, Departament de Sanitat i d'Anatomia Animals, Universitat Autónoma de Barcelona, 08193 Bellaterra, Barcelona, Spain

Lloyd B. Bullerman, Department of Food Science and Technology, University of Nebraska, Lincoln, NE 68583-0919, USA

F. Javier Cabañes, Departament de Sanitat i d'Anatomia Animals, Universitat Autónoma de Barcelona, 08193 Bellaterra, Barcelona, Spain

Gözde Dalkiliç, Department of Food Engineering, Istanbul Technical University, Istanbul, 34469 Maslak, Turkey

Dwayne Deibert, Department of Food Science and Technology, University of Nebraska, Lincoln, NE 68583-0919, USA

Jan Dijksterhuis, Department of Applied Research and Services, Centraalbureau voor Schimmelcultures, Fungal Biodiversity Centre, Uppsalalaan 8, 3584 CT, Utrecht, Netherlands

Alan Dobson, Microbiology Department, University College Cork, Cork, Ireland

Robert W. Emmett, Department of Primary Industries, PO Box 905, Mildura, Vic. 3502, Australia

Corrado Fanelli, Laboratorio di Micologia, Univerisità "La Sapienza", Largo Cristina di Svezia 24, I-00165 Roma, Italy

Virginia E. Fernández Pinto, Laboratorio de Microbiología de Alimentos, Departamento de Química Orgánica, Area Bromatología, Facultad de Ciencias Exactas y Naturales, Universidad de Buenos Aires, Ciudad Universitaria, Pabellón II, 3° Piso, 1428, Buenos Aires, Argentina

Graham H. Fleet, Food Science and Technology, School of Chemical Engineering and Industrial Chemistry, University of New South Wales, Sydney, NSW 2052, Australia

Jens C. Frisvad, BioCentrum-DTU, Building 221, Technical University of Denmark, 2800 Lyngby, Denmark

Marketa Giesova, Department of Dairy and Fat Technology, Institute of Chemical Technology, Prague, Czech Republic

Funda K. Güler Istanbul Technical University, Dept. of Food Engineering Istanbul, Turkey, 34469 Maslak

Auli Haikara, VTT Biotechnology, PO Box 1500, FIN-02044 Espoo, Finland

Yousef Hassan, Department of Food Science and Technology, University of Nebraska, Lincoln, NE 68583-0919, USA

Gillian M. Heard, Food Science and Technology, School of Chemical Engineering and Industrial Chemistry, University of New South Wales, Sydney, NSW 2052, Australia

Dilek Heperkan, Istanbul Technical University, Dept. of Food Engineering Istanbul, 34469 Maslak, Turkey
Ailsa D. Hocking, Food Science Australia, CSIRO, PO Box 52, North Ryde, NSW 1670, Australia
Stephen Holmes, ADGEN Ltd, Nellies Gate, Auchincruive, Ayr KA6 5HW, UK
Russell Hope, Applied Mycology Group, Biotechnology Centre, Cranfield University, Silsoe, Bedford MK45 4DT, UK
Jos Houbraken, Centraalbureau voor Schimmelcultures, PO Box 85167, 3508 AD, Utrecht, The Netherlands
Beatriz T. Iamanaka, Food Technology Institute, ITAL C.P 139 CEP13.073-001 Campinas-SP, Brazil
Nils Jonsson, Swedish Institute of Agricultural and Environmental Engineering, PO Box 7033, SE-750 07 Uppsala, Sweden
Benozir A. Kazi, Department of Primary Industries, PO Box 905, Mildura, Vic. 3502, Australia
Reyna León-Cruz, Ingeniería Química y Alimentos, Universidad de las Américas, Puebla, Cholula 72820, Mexico
Su-lin L. Leong, Food Science Australia, CSIRO, PO Box 52, North Ryde, NSW 1670, Australia
Aurelio López-Malo, Ingeniería Química y Alimentos, Universidad de las Américas, Puebla, Cholula 72820, Mexico
Naresh Magan, Applied Mycology Group, Biotechnology Centre, Cranfield University, Barton Road, Silsoe, Bedford MK45 4DT, UK
Sonia Marín, Food Technology Department, Lleida University, 25198 Lleida, Spain
Hilary C. Menezes, Food Engineering Faculty (FEA), Unicamp, Campinas-SP, Brazil
Burçak E. Meriç, Istanbul Technical University, Dept. of Food Engineering Istanbul, 34469 Maslak, Turkey
Santiago Minués, Institut Català de la Vinya i el Vi (INCAVI), Generalitat de Catalunya, Vilafranca del Penedés, Barcelona, Spain
Kristian F. Nielsen, BioCentrum-DTU, Building 221, Technical University of Denmark, 2800 Lyngby, Denmark
Juhani Olkku, Oy Panimolaboratorio-Bryggerilaboratorium AB, P.O. Box 16, FIN-02150 Espoo, Finland
Monica Olsen, National Food Administration, PO Box 622, SE-751 26 Uppsala, Sweden
Enrique Palou, Ingeniería Química y Alimentos, Universidad de las Américas, Puebla, Cholula 72820, Mexico

Andrea Patriarca, Laboratorio de Microbiología de Alimentos, Departamento de Química Orgánica, Area Bromatología, Facultad de Ciencias Exactas y Naturales, Universidad de Buenos Aires, Ciudad Universitaria, Pabellón II, 3° Piso, 1428, Buenos Aires, Argentina

Sven-Johan Persson, Akron maskiner, SE-531 04 Järpås, Sweden

John I. Pitt, Food Science Australia, CSIRO, PO Box 52, North Ryde, NSW 1670, Australia

C. Prakitchaiwattana, Food Science and Technology, School of Chemical Engineering and Industrial Chemistry, University of New South Wales, Sydney, NSW 2052, Australia

Antonio J. Ramos, Food Technology Department, Lleida University, 25198 Lleida, Spain

Aldo Rizzo, National Veterinary and Food Res. Inst., PO Box 45, FIN-00581, Helsinki, Finland

Dojin Ryu, Department of Food Science and Technology, University of Nebraska, Lincoln, NE 68583-0919, USA

Robert A. Samson, Department of Applied Research and Services, Centraalbureau voor Schimmelcultures, Fungal Biodiversity Centre, Uppsalalaan 8, 3584 CT, Utrecht, Netherlands

Vicente Sanchis, Food Technology Department, Lleida University, 25198 Lleida, Spain

Eileen S. Scott, School of Agriculture and Wine, University of Adelaide, PMB 1, Glen Osmond, SA 5064, Australia

Gülçin Şişmanoğlu, Department of Food Engineering, Istanbul Technical University, Istanbul, 34469 Maslak, Turkey

Cindy Stewart, National Center for Food Safety and Technology, 6502 S. Archer Rd, Summit-Argo, IL 60501, USA

Marta H. Taniwaki, Food Technology Institute (ITAL), C.P 139 CEP13.073-001 Campinas-SP, Brazil

Ulf Thrane, Center for Microbial Biotechnology, BioCentrum-DTU, Technical University of Denmark, DK-2800 Kgs. Lyngby, Denmark

Graciela Vaamonde, Laboratorio de Microbiología de Alimentos, Departamento de Química Orgánica, Area Bromatología, Facultad de Ciencias Exactas y Naturales, Universidad de Buenos Aires, Ciudad Universitaria, Pabellón II, 3° Piso, 1428, Buenos Aires, Argentina

E. Vicente Food Technology Institute, ITAL C.P 139 CEP13.073-001 Campinas-SP, Brazil

Section 1.
Understanding the fungi producing important mycotoxins

IMPORTANT MYCOTOXINS AND THE FUNGI WHICH PRODUCE THEM

Jens C. Frisvad, Ulf Thrane,[*] *Robert A. Samson*[†] *and John I. Pitt*[‡]

1. INTRODUCTION

The assessment of the relationship between species and mycotoxins production has proven to be very difficult. The modern literature is cluttered with examples of species purported to make particular mycotoxins, but where the association is incorrect. In some cases, mycotoxins have even been named based on an erroneous association with a particular species: verruculogen, viridicatumtoxin and rubratoxin come to mind. As time has gone on, and more and more compounds have been described, lists of species-mycotoxin associations have become so large, and the inaccuracies in them so widespread in acceptance, that determining true associations has become very difficult. It does not need to be emphasised how important it is that these associations be known accurately. The possible presence of mycotoxigenic fungi in foods, and rational decisions on the status of foods suspected to contain mycotoxins, are ever present problems in the food industry around the world.

In defining mycotoxins, we exclude fungal metabolites which are active against bacteria, protozoa, and lower animals including insects.

* J. C. Frisvad and U. Thrane, Center for Microbial Biotechnology, BioCentrum-DTU, Technical University of Denmark, Building 221, DK-2800 Kgs. Lyngby, Denmark. Correspondence to jcf@biocentrum.dtu.dk

† R. A. Samson, Centraalbureau voor Schimmelcultures, PO Box 85167, 3508 AD Utrecht, Netherlands

‡ J. I. Pitt, Food Science Australia, CSIRO, PO Box 52, North Ryde, NSW 1670, Australia

Furthermore we exclude Basidiomycete toxins, because these are ingested by eating fruiting bodies, a problem different from the ingestion of toxins produced by microfungi. The definition of microfungi is not rigorous, but understood here to refer principally to Ascomycetous fungi, including those with no sexual stage. Lower fungi, from the subkingdom Zygomycotina, i.e. genera such as *Rhizopus* and *Mucor*, are not excluded, but compounds of sufficient toxicity to be termed mycotoxins have not been found in these genera, except perhaps for rhizonin A and B from *Rhizopus microsporus* (Jennessen et al., 2005).

This paper sets out to provide an up to date authoritative list of mycotoxins which are known to have caused, or we believe have the potential to cause, disease in humans or vertebrate animals, and the fungal species which have been shown to produce them.

We believe that all of the important and known mycotoxins produced by *Aspergillus, Fusarium* and *Penicillium* species have been included in this list. However, it is possible that other species will be found which are capable of producing known toxins, or other toxins of consequence will arise. It is also important to note that there are many errors in the literature concerning the mycotoxins and the fungi which produce them (Frisvad et al., 2006).

Many other toxic chemicals, known to be produced by species from these genera, have been excluded from this list for one reason or another. The very toxic chemicals, the janthitrems, have been excluded from this list because the species which make them, including *P. janthinellum*, normally do not grow to a significant extent in foods. On the other hand *Penicillium tularense* has recently been demonstrated to produce janthitrems in tomatoes (Andersen and Frisvad, 2004), so maybe these mycotoxins may occur sporadically. Other compounds which occur quite commonly in foods, including mycophenolic acid (Lafont et al., 1979, Lopez-Diaz et al., 1996; Overy and Frisvad, 2005), are of such low acute toxicity to vertebrate animals that their involvement in human or animal diseases appears unlikely. On the other hand mycophenolic acid has been reported to be strongly immunosuppressive (Bentley, 2000), so this fungal metabolite could pave the way for bacterial infections. Toxic low molecular weight compounds that may not be considered mycotoxins in a strict sense include aflatrem, botryodiploidin, brefeldin A, chetomin, chetocins, emestrin, emodin, engleromycin, fusarin C, lolitrems, paspalicine, paspaline, paspalinine, paspalitrems, paxilline, territrems, tryptoquivalins, tryptoquivalons, verruculotoxin, verticillins, and viridicatumtoxin which are among the fungal secondary metabolites listed as mycotoxins by Betina (1989).

Future research may show that some of these are more important for human and domestic animals health than currently indicated.

For convenience the list below has been set out by genus, but it should be kept in mind that some mycotoxins are common to both *Aspergillus* and *Penicillium* species (Samson, 2001). The list below sets out to be encyclopaedic, but at the same time we have indicated, where possible, which species producing a particular toxin are more likely to occur in foods and which are probably of little consequence.

2. *ASPERGILLUS* TOXINS

2.1. Aflatoxins

Aflatoxins are potent carcinogens (Class 1; JECFA, 1997) affecting man and all tested animal species including birds and fish. Four compounds are commonly produced in foods: aflatoxins B_1, B_2, G_1 and G_2, named for the colour of their fluorescence under ultra violet light, and their relative position on TLC plates.

Major sources. Aspergillus flavus is the most common species producing aflatoxins, occurring in most kinds of foods in tropical countries. This species has a special affinity with three crops, maize, peanuts and cottonseed, and usually produces only B aflatoxins. Only about 40% of known isolates produce aflatoxin.

Aspergillus parasiticus occurs commonly in peanuts, but is quite rare in other foods. It is also restricted geographically, and is rare in Southeast Asia (Pitt.et al., 1993). *A. parasiticus* produces both B and G aflatoxins, and virtually all known isolates are toxigenic.

Minor sources. Table 1 shows the species which are known to be capable of producing aflatoxin in culture, and some details concerning their appearance and their occurrence. Note that most of the minor species are known from only a very few isolates, and their occurrence in foodstuffs or feedstuffs is at most rare. On the other hand *A. nomius*, *A. toxicarius*, and *A. parvisclerotigenus* may be more common than expected, because it is very difficult to distinguish between those species and isolates may easily have been identified as *A. flavus* or *A. parasiticus*.

2.2. Cyclopiazonic acid (see also *Penicillium*)

Cyclopiazonic acid (CPA) (Holzapfel, 1968) is a potent mycotoxin that produces focal necrosis in most vertebrate inner organs in high

Table 1. Morphology and mycotoxin production characteristic of species in *Aspergillus* that are known aflatoxin producers [a]

Species	Heads	Conidia	Sclerotia	Known occurrence	Mycotoxins
Aspergillus flavus	Mostly biseriate	Spherical to ellipsoidal, smooth to finely rough walls	Large, spherical	Ubiquitous in tropics and subtropics	B aflatoxins (40% of isolates); CPA[b], ca 50%
A. parasiticus	Rarely biseriate	Spherical, rough walls	Large, spherical (uncommon)	USA, South America, Australia	B and G aflatoxins (nearly 100%)
A. nomius	Mostly biseriate	Spherical to ellipsoidal, smooth to finely rough walls	Small, bullet shaped	USA, Thailand	B and G aflatoxins (usually)
A. bombycis	Mostly biseriate	Spherical to subspheroidal, roughened	Not reported	Japan, Indonesia (silkworms only)	B and G aflatoxins
A. pseudotamarii	Biseriate	Spherical to subspheroidal, very rough walls	Large, spherical	Japan, Argentina	B aflatoxins, CPA
A. toxicarius	Rarely biseriate	Sphaerical, rough walled	Large, sphaerical	USA, Uganda	B and G aflatoxins
A. parvisclerotigenus	Mostly biseriate	Spherical, rough walled	Small, sphaerical	USA, Argentina, Japan, Nigeria	B and G aflatoxins, CPA
A. ochraceoroseus	Biseriate	Subspheroidal to ellipsoidal, smooth walled	Not reported	Ivory Coast (soil)	B aflatoxins, sterigmatocystin
A. rambellii	Biseriate	Ellipsoidal, smooth walled	Not reported	Ivory Coast (soil)	B aflatoxins, sterigmatocystin
Emericella astellata	Biseriate	Spheroidal, rugulose walls	Ascomata and hülle cells	Ecuador (leaves of *Ilex*)	B aflatoxins, Sterigmatocystin
Emericella venezuelensis	Biseriate	Sphaeroidal, rugulose walls	Ascomata and hülle cells	Venezuela (mangrove)	B aflatoxins, Sterigmatocystin

[a] From Kurtzman et al., 1987; Klich and Pitt, 1988; Pitt and Hocking, 1997; Klich et al., 2000; Ito et al., 2001; Peterson et al., 2001; Frisvad et al., 2004a; Frisvad and Samson, 2004a; Frisvad et al., 2005a.

[b] CPA, cyclopiazonic acid.

concentrations and affects the ducts or organs originating from ducts. It was originally believed that aflatoxins were responsible for all the toxic effects of *Aspergillus flavus* contaminated peanuts to turkeys in Turkey X disease, but it was later shown that cyclopiazonic acid had an additional severe effect on the muscles and bones of the turkeys (Jand et al., 2005).

Major sources. *Aspergillus flavus* and the domesticated form *A. oryzae* often produce large amounts of CPA. *A. flavus* is common on oil seeds, nuts, peanuts and cereals, but may also produce aflatoxin on dried fruits (Pitt and Hocking, 1997).

Minor sources. Other producers of CPA in *Aspergillus* include *A. tamarii*, *A. pseudotamarii*, *A. parvisclerotigenus*, but the role of these fungi concerning CPA production in foods or feeds is not clear.

2.3. Cytochalasin E

Cytochalasin E is a very toxic metabolite of *Aspergillus clavatus*. It may occur in malting barley (Lopez-Diaz and Flannigan, 1997)

Major source. Aspergillus clavatus.

Minor source. Rosellinia necatrix is not found in foods.

2.4. Gliotoxin

Gliotoxin is strongly immunosuppressive, but is probably only a potential problem in animal feeds (Betina, 1989).

Major sources. Aspergillus fumigatus has been found in animal feeds.

Minor sources. Gliocladium virens, P. lilacinoechinulatum and few other soil-borne species also produce gliotoxin.

2.5. β-Nitropropionic Acid (BNP)

β-nitropropionic acid has been reported to be involved in sugar cane poisoning of children, but may potentially also cause other intoxications, as producers are widespread (Burdock et al., 2001). Furthermore BNP has been found in miso, shoyu and katsuobushi and it can be produced by *A. oryzae* when artificially inoculated on cheese, peanuts etc. Unfortunately *A. flavus* has not been tested for the production of BNP, but BNP production by *A. oryzae* on peanuts indicates that *A. flavus* may be able to produce this mycotoxin in combination with aflatoxin B_1, cyclopiazonic acid and kojic acid. The possible synergistic effect of these mycotoxins on mammals is unknown.

Major sources. The BNP producing fungi from sugar cane are *Arthrinium phaeospermum* and *Art. sacchari*, but other species such as *Art. terminalis*, *Art. saccharicola*, *Art. aureum* and *Art. sereanis* also produce BNP (Burdock et al., 2001).

A. flavus may be an important producer of this mycotoxin in foods, but there are no surveys that include analytical determination of BNP alongside cyclopiazonic acid and aflatoxin B_1. *A. oryzae* and *A. sojae* can produce BNP in miso and shoyu, but it is probably more important that their wild-type forms, *A. flavus* and *A. parasiticus* respectively, may produce BNP in foods. More research is needed in this area.

Minor sources. Penicillium atrovenetum is another authenticated producer of BNP, but this fungus is only found in soil.

Incorrect sources. Penicillium cyclopium, *P. chrysogenum*, *Aspergillus wentii*, *Eurotium* spp., and *A. candidus* have been reported as producers of BNP (Burdock et al., 2001), but these identifications are doubtful.

2.6. Ochratoxin A (see also *Penicillium*)

Ochratoxin A (OA) is a nephrotoxin, affecting all tested animal species, though effects in man have been difficult to establish unequivocally. It is listed as a probable human carcinogen (Class 2B) (JECFA, 2001). Links between OA and Balkan Endemic Nephropathy have long been sought, but not established (JECFA, 2001).

Major sources. Aspergillus ochraceus (van der Merwe et al., 1965), occurring in stored cereals (Pitt and Hocking, 1997) and coffee (Taniwaki et al., 2003). *A. ochraceus* has been shown to consist of two species (Varga et al., 2000a, b; Frisvad et al., 2004b). The second and new species producing large amounts of ochratoxin A consistently, has been described as *A. westerdijkiae*. Actually the original producer of ochratoxin A from *Andropogon sorghum* in South Africa, NRRL 3174, has been designated as the type culture of *A. westerdijkiae* (Frisvad et al., 2004b). This is interesting as *A. westerdijkiae* is both a better and more consistent ochratoxin producer than *A. ochraceus*, and it may be also more prevalent in coffee than *A. ochraceus*. The ex type culture of *A. ochraceus* CBS 108.08 only produces trace amounts of ochratoxin A.

Aspergillus carbonarius (Horie, 1995) is a major OA producer. It occurs in grapes, producing OA in grape products, including grape juice, wines and dried vine fruits (IARC, 2002; Leong et al., 2004) and sometimes on coffee beans (Taniwaki et al., 2003; Abarca et al., 2004). *Aspergillus niger* is an extremely common species, but only few strains appear to be producers of OA, so this species may be of much less importance than *A. carbonarius* in grapes, wine and green coffee beans (Abarca et al., 1994; Taniwaki et al., 2003; Leong et al., 2004). It may be of major importance, however, as *A. niger* NRRL 337, referred to as the "food fungus", produces large amounts of OA in pure culture. This fungus is used for fermentation of potato peel waste etc. and used for animal feed (Schuster et al., 2002).

Petromyces alliaceus (Lai et al., 1970), produces large amounts of ochratoxin A in pure culture, and OA produced by this fungus has been found in figs in California (Bayman et al., 2002). *Aspergillus steynii*, from the *Aspergillus* section *Circumdati*, is also a very efficient producer of OA, and has been found in green coffee beans, mouldy soy beans and rice (Frisvad et al., 2004b). As with *A. westerdijkiae*, *A. steynii* may have been identified as *A. ochraceus* earlier, so the relative abundance of these three species is difficult to evaluate at present.

Penicillium verrucosum is the major producer of ochratoxin A in stored cereals (Frisvad, 1985; Pitt, 1987; Lund and Frisvad, 2003). *Penicillium nordicum* (Larsen et al., 2001) is the main OA producer found in meat products such as salami and ham. Both OA producing *Penicillium* species have been found on cheese also, but have only been reported to be of high occurrence on Swiss hard cheeses (as *P. casei*, Staub, 1911). The ex type culture of *P. casei* is a *P. verrucosum* (Larsen et al., 2001).

Minor sources. Several Aspergilli can produce ochratoxin A in large amounts, but they appear to be relatively rare. In *Aspergillus* section *Circumdati* (formerly the *Aspergillus ochraceus* group), the following species can produce ochratoxin A: *Aspergillus cretensis*, *A. flocculosus*, *A. pseudoelegans*, *A. roseoglobulosus*, *A. sclerotiorum*, *A. sulphureus* and *Neopetromyces muricatus* (Frisvad et al., 2004b). According to Ciegler (1972) and Hesseltine et al. (1972) *A. melleus*, *A. ostianus*, *A. persii* and *A. petrakii* may produce trace amounts of OA, but this has not been confirmed since publication of those papers. Strains of these species reported to produce large amounts of OA were reidentified by Frisvad et al. (2004b). In *Aspergillus* section *Flavi*, *Petromyces albertensis* produces ochratoxin A. In *Aspergillus* section *Nigri*, *A. lacticoffeatus* and *A. sclerotioniger* produce ochratoxin A (Samson et al., 2004).

2.7. Sterigmatocystin

Sterigmatocystin is a possible carcinogen. However, its low solubility in water or gastric juices limits its potential to cause human illness (Pitt and Hocking, 1997).

Major sources. The major source of sterigmatocystin in foods is *Aspergillus versicolor*. This fungus is common on cheese, but may also occur on other substrates (Pitt and Hocking, 1997).

Minor sources. A large number of species are able to produce sterigmatocystin, including *Chaetomium* spp., *Emericella* spp., *Monocillium nordinii* and *Humicola fuscoatra* (Joshi et al., 2002). These species are unlikely to contaminate foods.

2.8. Verruculogen and Fumitremorgins

Verrucologen is an extremely toxic tremorgenic mycotoxin, but it is unlikely to be founding significant levels in foods. *Neosartorya fisheri* may be present in heat treated foods, but *N. glabra* and allied species are much more common in foods, and the latter species do not produce verrucologen.

Major sources. *Aspergillus fumigatus* and *Neosartorya fischeri* are the major *Aspergillus* species producing verruculogen but these species are uncommon in foods. These species produce many other toxic compounds including gliotoxin, fumigaclavins, and tryptoquivalins (Cole et al., 1977; Cole and Cox, 1981; Panaccione and Coyle, 2005).

Minor sources. Aspergillus caespitosus, *Penicillium mononematosum* and *P. brasilianum* are efficient producers of verrucologen and fumitremorgins, but are very rare in foods and feeds.

3. *FUSARIUM* TOXINS

3.1. Antibiotic Y

Antibiotic Y has significant antibiotic properties towards phytopathogenic bacteria but low cell toxicity (Golinski et al., 1986). However, this compound, which originally was named lateropyrone (Bushnell et al., 1984), has not been studied in detail. Producers of antibiotic Y are widespread and common in agricultural products, so the natural occurrence of antibiotic Y may be of importance. Natural occurrence in cherries, apples and wheat grains has been reported (Andersen and Thrane, 2005).

Major sources. The main producer is *Fusarium avenaceum* which occurs frequently in cereal grain, fruit and vegetables. Another consistent producer is *F. tricinctum*, which also is very frequently found on cereal grains in temperate climates.

Minor sources. F. lateritium is known as a plant pathogen, but also causes spoilage in fruits and has been reported from apples and cherries in which antibiotic Y was detected (Andersen and Thrane, 2006). In warmer climates *F. chlamydosporum* is a potential producer of antibiotic Y in cereal grain and other seeds.

3.2. Butenolide

Butenolide is a collective name for compounds with a given ring structure; however in *Fusarium* mycotoxicology butenolide is a synonym for 4-acetamido-2-buten-4-olide, which has been associated with cattle diseases (fescue foot) since the mid 1960s (Yates et al., 1969). The toxicology has been thoroughly discussed by Marasas et al. (1984). There have been no reports of butenolide in foods, but it may be an important toxin due to the reported synergistic effect with enniatin B (Hershenhorn et al., 1992).

Major sources. The original reported producer of butenolide is *F. sporotrichioides* [reported as *F. nivale,* see Marasas et al. (1984) for details] and other frequent producers of butenolide in cereals are *F. graminearum* and *F. culmorum*.

Minor sources. Other potential producers of butenolide are *F. avenaceum*, *F. poae* and *F. tricinctum* which are frequently found in cereal grains together with *F. crookwellense*, *F. sambucinum* and *F. venenatum*. The latter three species also can be found in potatoes and other root vegetables.

3.3. Culmorin

Culmorin has a low toxicity in several biological assays (Pedersen and Miller, 1999) but a synergistic effect with deoxynivalenol towards caterpillars has been demonstrated (Dowd et al., 1989). Culmorin and hydroxyculmorins have been detected in cereals (Ghebremeskel and Langseth, 2000). These samples also contained deoxynivalenol and acetyl-deoxynivalenol.

Major sources. F. culmorum and *F. graminearum*, found in cereals, are the major producers of culmorin. The less widely distributed species *F. poae* and *F. langsethiae* are also consistent producers of culmorin and derivatives (Thrane et al., 2004).

Minor sources. Other species producing culmorin are *F. crookwellense* and *F. sporotrichioides*, also found in cereals.

3.4. Cyclic Peptides

The two groups of cyclic peptides, beauvericin and enniatins, are structurally related and they show antibiotic and ionophoric activities (Kamyar et al., 2004). Both groups of cyclic peptides have been detected in agricultural products (Jestoi et al., 2004).

3.4.1. Beauvericin

Beauvericin was originally found in entomopathogenic fungi such as *Beauveria bassiana* and *Isaria fumosorosea* (formerly *Paecilomyces fumosoroseus*; Luangsa-Ard et al., 2005) but has also been detected in several *Fusarium* species occurring on food (Logrieco et al., 1998).

Major sources. Fusarium subglutinans, F. proliferatum and *F. oxysporum* are consistent producers of beauvericin and have often been found to produce high quantities under laboratory conditions. These species are often found on maize and fruits.

Minor sources. Several species of the *Gibberella fujikuroi* complex have been reported to produce beauvericin in low amounts, including *F. nygamai, F. dlaminii* and *F. verticillioides* from cereals and fruits. The systematics of these Fusaria has developed dramatically during the last years, so a lot of species specific information of toxin production is not available.

F. avenaceum, F. poae and *F. sporotrichioides* on cereal grain, fruits and vegetables are known to produce beauvericin in low amounts (Morrison et al., 2002; Thrane et al., 2004). In addition, *F. sambucinum* and a few strains of *F. acuminatum, F.equiseti* and *F. longipes* from agricultural products have also been reported low producers of beauvericin (Logrieco et al., 1998).

3.4.2. Enniatins

Enniatins are a group of more than 15 related compounds produced by several *Fusarium* species, but also from *Halosarpeia* sp. and *Verticillium hemipterigenum*; however these are not of food origin.

Major sources. Fusarium avenaceum is the most important enniatin producer in cereals and other agricultural food plants, because this species is a very frequent and consistent producer of enniatin B (Morrison et al., 2002). *Fusarium sambucinum* is a consistent producer of enniatin B and diacetoxyscirpenol and causes dry rot in potatoes; however the role of these toxins has not been examined.

Minor sources. F. langsethiae, F. poae and *F. sporotrichioides*, mainly occur on cereal grain, *F. lateritium* from fruits and *F. acuminatum* from herbs.

3.5. Fumonisins

Since the discovery of fumonisins in the late 1980s much attention has been paid to these highly toxic compounds. Several reviews on the chemistry, toxicology and mycology have been published (Marasas et al., 2001; Weidenbörner, 2001).

Major sources. F. verticillioides (formerly known as *F. moniliforme*; Seifert et al., 2003) and *F. proliferatum* are the main sources of fumonisins in maize. These species and fumonisins in maize and to a lesser extent other cereal crops have been reported from all over the world in numerous papers and book chapters.

Minor sources. Other fumonisin producing species are *Fusarium nygamai, F. napiforme, F. thapsinum, F. anthophilum* and *F. dlamini* from millet, sorghum and rice. Some strains of these species have also been isolated from soil debris.

3.6. Fusaproliferin

Fusaproliferin is a recent discovered mycotoxin which shows teratogenic and pathological effects in cell assays (Bryden et al., 2001). Fusaproliferin has been detected in natural samples together with beauvericin and fumonisin (Munkvold et al., 1998). Nothing is known about a possible synergistic effect in such toxin combinations.

Major sources. Fusarium proliferatum and *F. subglutinans* are the major sources in maize and other cereal grains. The fungi and fusaproliferin have been detected in Europe, North America and South Africa (Wu et al., 2003).

Minor sources. A few strains of *F. globosum, F. guttiforme, F. pseudocircinatum, F. pseudonygamai* and *F. verticillioides* have been found to produce fusaproliferin, however the systematics in this section of *Fusarium* has developed dramatically within recent years so specific information on the toxin production by recently described species is unknown.

3.7. Moniliformin

Moniliformin is cytotoxic, inhibits protein synthesis and enzymes, causes chromosome damages and induces heart failure in mammals and poultry (Bryden et al., 2001). Moniliformin has been found world wide in cereal samples

Major sources. In maize *F. proliferatum* and *F. subglutinans* are the main producers of moniliformin, whereas *F. avenaceum* and *F. tricinctum* are the key sources in cereals grown in temperate climates.

Minor sources. In sorghum, millet and rice *F. napiforme, F. nygamai, F. verticillioides* and *F. thapsinum* may be responsible for moniliformin production. Some strains of *F. oxysporum* produce a significant amount of moniliformin under laboratory condition; however there is no detailed information on a possible production in vegetables and fruits. An overview of other minor sources has been published (Schütt et al., 1998).

3.8. Trichothecenes

More than 200 trichothecenes have been identified and the non-macrocyclic trichothecenes are among the most important mycotoxins. Trichothecenes are haematotoxic and immunosuppressive. In animals, vomiting, feed refusal and diarrhoea are typical symptoms. Skin oedema in humans has also been observed. An EU working group on has reported on trichothecenes in food (Schothorst and van Egmond, 2004).

3.8.1. Deoxynivalenol (DON) and Acetylated Derivatives (3ADON, 15ADON)

Deoxynivalenol (DON) and its acetylated derivatives (3ADON, 15ADON) are by far the most important trichothecenes. Numerous reports on world-wide occurrence have been published and several international symposia and workshops have focussed on DON (Larsen et al., 2004).

Major sources. Fusarium graminearum and *F. culmorum* are consistent producers of DON, especially in cereals. Within both species strains have been grouped into those that produce DON and its derivatives, and those that produce nivalenol and furarenon X as their major metabolites. Intermediates have also been found (Nielsen and Thrane, 2001). Recently, *F. graminearum* has been divided into nine phylogenetic species (O'Donnell et al., 2004); however in the present context this species concept will not be used as a correlation to existing mycotoxicological literature is impossible at this stage.

Minor sources. Production of DON by *F. pseudograminearum* has been reported, but this species is restricted to warmer climates.

3.8.2. Nivalenol (NIV) and Fusarenon X (FX, 4ANIV)

Nivalenol (NIV) and fusarenon X (FX, 4ANIV) occur in the same commodities as DON and are in many cases covered by the same surveys due to the high degree of similiarity. NIV is often detected in much lower concentrations than DON, but is considered to be more toxic.

Major sources. Fusarium graminearum is a well known producer of NIV and FX in cereals. In temperate climates *F. poae,* which is a consistent producer of NIV (Thrane et al., 2004), may be responsible for NIV in cereals.

Minor sources. Strains of *F. culmorum* that produce NIV are less commonly isolated than those that produce DON producers. *F. equiseti* and *F. crookwellense* found in some cereal samples and in vegetables may also produce NIV. In potatoes *F. venenatum* strains that produce NIV have been detected (Nielsen and Thrane, 2001).

3.8.3. T-2 toxin

T-2 toxin is one of the most toxic trichothecenes, whereas the derivative HT-2 toxin is less toxic. Due to structural similarity these toxins are often included in the same analytical method.

Major sources. Fusarium sporotrichioides and *F. langsethiae*, frequently isolated from cereals in Europe, are consistent producers of T-2 and HT-2 (Thrane et al., 2004).

Minor sources. Only a few T-2 and HT-2 producing strains of *F. poae* and *F. sambucinum* have been found (Nielsen and Thrane, 2001; Thrane et al., 2004).

3.8.4. Diacetoxyscirpenol (DAS)

Diacetoxyscirpenol (DAS) and monoacetylated derivatives (MAS) are a fourth group of important trichothecenes in food.

Major sources. Fusarium venenatum isolates often produce high levels of DAS and this species is frequently isolated from cereals and potatoes (Nielsen and Thrane, 2001). *F. poae* isolates also often produce high levels of DAS.

Minor sources. Fusarium equiseti isolates can produce DAS and MAS in high amounts, but this species is infrequently isolated from cereals and vegetables. *F. sporotrichioides* and *F. langsethiae* also produce DAS and MAS; however at lower levels (Thrane et al., 2004). *F. sambucinum* isolates produce DAS and MAS and are a probable cause of DAS in potatoes (Ellner, 2002).

3.9. Zearalenone

Zearalenone causes hyperoestrogenism in swine and possible effects in humans have also been reported. Derivatives of zearalenone have been used as growth promoters in livestock; however this is now banned in European Union (Launay et al., 2004). The toxicity of zearalenone and its derivatives have been reviewed recently (Hagler et al., 2001).

Major sources. Fusarium graminearum and *F. culmorum* are the most pronounced producers of zearalenone and several derivatives. They occur frequently in cereals all over the world. Recently, *F. graminearum* has been divided into nine phylogenetic species (O'Donnell et al., 2004); however in the present context this species concept will not be used as a correlation to existing mycotoxicological literature is impossible at this stage.

Minor sources. Under laboratory conditions *Fusarium equiseti* produces a number of zearalenone derivatives in high amounts, but little is known about production under natural conditions. *F. crookwellense* also produces zearalenone.

4. *PENICILLIUM* TOXINS

4.1. Chaetoglobosins

The chaetoglobosins are toxic compounds that may be involved in mycotoxicosis. They are produced by common food-borne Penicillia and have been found to occur naturally (Andersen et al., 2004).

Major sources. Penicillium expansum and *P. discolor* are major sources of the chaetoglobosins. Both species cause spoilage in fruits and vegetables, and the latter species also occurs on cheese (Frisvad and Samson, 2004b).

Minor sources. Chaetomium globosum and *P. marinum* are probably not of significance in foods.

4.2. Citreoviridin

Citreoviridin was reported as a cause of acute cardiac beriberi (Ueno, 1974), but a more in depth toxicological evaluation of this metabolite is needed. It has been associated with yellow rice disease, but this disease has also been associated with *P. islandicum* and its toxic metabolites cyclic peptides cyclochlorotine and islanditoxin, and anthraquinones luteoskyrin and rugulosin (Enomoto and Ueno, 1974).

Major sources. Eupenicillium cinnamopurpureum has been found in cereals in USA and in Slovakia (Labuda and Tancinova, 2003) and is an efficient producer of citreoviridin. *P. citreonigrum* may be of some importance in yellowed rice.

Minor sources. P. smithii, P. miczynskii and *P. manginii* (Frisvad and Filtenborg, 1990) have most often been recovered from soil and only rarely from foods. *Aspergillus terreus* has occasionally been reported from foods, but is primarily a soil-borne fungus.

4.3. Citrinin

Citrinin is a nephrotoxin, but probably of less importance than ochratoxin A (Reddy and Berndt, 1991), however, producers of citrinin are widespread and common in foods. Citrinin has been found in cereals, peanuts and meat products (Reddy and Berndt, 1991).

Major sources. P. citrinum is an efficient and consistent producer of citrinin and has been found in foods world-wide (Pitt and Hocking, 1997). *P. verrucosum* is predominantly cereal-borne in Europe and often produces citrinin as well as ochratoxin A (Frisvad et al., 2005b). *P. expansum*, common in fruits and other foods, sometimes produces citrinin. *P. radicicola* is commonly found in onions, carrots and potatoes (Overy and Frisvad, 2003).

Minor sources. Aspergillus terreus, A. carneus, P. odoratum and *P. westlingii* have been reported as producers of citrinin, but are not likely to occur often in foods.

4.4. Cyclopiazonic acid (see also *Aspergillus*)

Major sources. Penicillium commune and its domesticated form *P. camemberti*, and the closely related species *P. palitans*, are common on cheese and meat products and may produce cyclopiazonic acid in these products (Frisvad et al., 2004c). *P. griseofulvum* is also a major producer of cyclopiazonic acid, and may occur in long stored cereals and cereal products such as pasta (Pitt and Hocking, 1997).

Minor sources. P. dipodomyicola occurs in the environs of the kangaroo rat in the USA, but has also been reported from rice in Australia and in a chicken feed mixture in Slovakia (Frisvad and Samson, 2004b).

4.5. Mycophenolic acid

Despite having a low acute toxicity, mycophenolic acid may be a very important indirect mycotoxin as it highly immunosuppressive, perhaps influencing the course of bacterial and fungal infections (Bentley, 2000).

Major sources. Penicillium brevicompactum is a ubiquitous species and may produce mycophenolic acid in foods, e.g. ginger (Overy and Frisvad, 2005). Two other major species producing mycophenolic acid are *P. roqueforti* and *P. carneum*. Another important producer is *Byssochlamys nivea* (Puel et al., 2005). Mycophenolic acid has been found to occur naturally in blue cheeses (Lafont et al., 1979).

Minor sources. The soil-borne species *Penicillium fagi* also produces mycophenolic acid (Frisvad and Filtenborg, 1990, as *P. raciborskii*). *Septoria nodorum* (Devys et al., 1980) is another source but is unimportant as a food contaminant.

4.6. Ochratoxin A (see also *Aspergillus*)

Major sources. Penicillium verrucosum (Frisvad, 1985; Pitt, 1987) is the major producer of ochratoxin A in cool climate stored cereals (Lund and Frisvad, 2003).

Penicillium nordicum (Larsen et al., 2001) is the main OA producer found in manufactured meat products such as salami and ham. Both OA producing *Penicillium* species have been found on cheese also, but have only been reported to be of high occurrence on Swiss hard cheeses (as *P. casei* Staub, 1911). The ex type culture of *P. casei* is a *P. verrucosum* (Larsen et al., 2001).

4.7. Patulin

Patulin is generally very toxic for both prokaryotes and eukaryotes, but the toxicity for humans has not been conclusively demonstrated. Several countries in Europe and the USA have now set limits on the level of patulin in apple juice.

Major sources. Penicillium expansum is by far the most important source of patulin. *P. expansum* is the major species causing spoilage of apples and pears, and is the major source of patulin in apple juice and other apple and pear products.

Byssochlamys nivea may be present in pasteurised fruit juices and may produce patulin and mycophenolic acid (Puel et al., 2005).

Penicillium griseofulvum is a very efficient producer of high levels of patulin in pure culture, and it may potentially produce patulin in cereals, pasta and similar products.

P. carneum may produce patulin in beer, wine, meat products and rye-bread as it has been found in those substrates (Frisvad and Samson, 2004b), but there are no reports yet on patulin production by this species in those foods. *P. carneum* also produces mycophenolic acid, roquefortine C and penitrem A (Frisvad et al., 2004c). *P. paneum* occurs in rye-bread (Frisvad and Samson, 2004b), but again actual production of patulin in this product has not been reported.

P. sclerotigenum is common in yams and has the ability to produce patulin in laboratory cultures.

Minor sources. The coprophilous fungi *P. concentricum*, *P. clavigerum*, *P. coprobium*, *P. formosanum*, *P. glandicola, P. vulpinum, Aspergillus clavatus, A. longivesica* and *A. giganteus* are very efficient producers of patulin in the laboratory, but only *A. clavatus* may play any role in human health, as it may be present in beer malt (Lopez-Diaz and Flannigan, 1997). *Aspergillus terreus, Penicillium novae-zeelandiae, P. marinum, P. melinii* and other soil-borne fungi may produce patulin in pure culture, but are less likely to occur in any foods.

4.8. Penicillic acid

Penicillic acid (Alsberg and Black, 1911) and dehydropenicillic acid (Obana et al., 1995) are small toxic polyketides, but their major role in mycotoxicology may be in their possible synergistic toxic effect with OA (Lindenfelser at al., 1973; Stoev et al., 2001) and possible additive or synergistic effect with the naphtoquinones hepatotoxins xanthomegnin, viomellein and vioxanthin.

Major sources. Penicillic acid is likely to co-occur with OA, xanthomegnin, viomellein and vioxanthin produced by members of *Aspergillus* section *Circumdati* and *Penicillium* series *Viridicata* (which often co-occur with *P. verrucosum*). The *Aspergillus* species often occur in coffee and the Penicillia are common in cereals. The major sources of penicillic acid are *P. aurantiogriseum*, *P. cyclopium*, *P. melanoconidium* and *P. polonicum* (Frisvad and Samson, 2004b) and all members of *Aspergillus* section *Circumdati* (Frisvad and Samson, 2000). Penicillic acid is produced by *P. tulipae* and *P. radicicola*, which are occasionally found on onions, carrots and potatoes (Overy and Frisvad, 2003).

Minor sources. Penicillic acid has been found in one strain of *P. carneum* (Frisvad and Samson, 2004b).

4.9. Penitrem A

Penitrem A is a highly toxic tremorgenic indol-terpene. It has primarily been implicated in animal mycotoxicoses (Rundberget and Wilkins, 2002), but has also been suspected to cause tremors in humans (Cole et al., 1983; Lewis et al., 2005).

Major sources. Penicillium crustosum is the most important producer of penitrem A (Pitt, 1979). This species is of world-wide distribution and often found in foods. This mycotoxins is produced by all isolates of *P. crustosum* examined (Pitt, 1979; Sonjak et al., 2005). *P. melanoconidium* is common in cereals (Frisvad and Samson, 2004b), but it is not known whether this species can produce penitrem A in infected cereals.

Minor sources. P. glandicola, P. clavigerum, and *P. janczewskii* are further producers of penitrem A (Ciegler and Pitt, 1970; Frisvad and Samson, 2004b; Frisvad and Filtenborg, 1990), but have been recovered from foods only sporadically.

4.10. PR toxin

PR toxin is a mycotoxin that is acutely toxic and can damage DNA and proteins (Moule et al., 1980; Arnold et al., 1987). It is unstable in cheese (Teuber and Engel, 1983), but it may be produced in silage and other substrates.

Major sources. Penicillium roqueforti is the major source of PR toxin. It has been reported also from *P. chrysogenum* (Frisvad and Samson, 2004b).

4.11. Roquefortine C

The status of roquefortine C as a mycotoxin has been questioned, but it is a very widespread fungal metabolite, and is produced by a large number of species. The acute toxicity of roquefortine C is not very high (Cole and Cox, 1981), but it has been reported as a neurotoxin.

Major sources. Penicillium albocoremium, P. atramentosum, P. allii, P. carneum, P. chrysogenum, P. crustosum, P. expansum, P. griseofulvum, P. hirsutum, P. hordei, P. melanoconidium, P. paneum, P. radicicola, P. roqueforti, P. sclerotigenum, P. tulipae and *P. venetum* are all producers that have been found in foods, but the natural occurrence of roquefortine C has been reported only rarely.

Minor sources. P. concentricum, P. confertum, P. coprobium, P. coprophilum, P. flavigenum, P. glandicola, P. marinum, P. persicinum and *P. vulpinum* are less likely food contaminants.

4.12. Rubratoxin

Rubratoxin is a potent hepatotoxin (Engelhardt and Carlton, 1991) and is of particular interest as it has been implicated in severe liver damage in three Canadian boys, who drank rhubarb wine contaminated with *Penicillium crateriforme*. One of the boys needed to have the liver transplanted (Richer et al., 1997).

Major producers. P. crateriforme is the only known major producer of rubratoxin A and B (Frisvad, 1989).

4.13. Secalonic Acid D

The toxicological data on secalonic acid D and F are somewhat equivocal (Reddy and Reddy, 1991), so the significance of this metabolite in human and animal health is somewhat uncertain.

Major sources. Claviceps purpurea, Penicillium oxalicum, Phoma terrestris and *Aspergillus aculeatus* produce large amounts of secalonic acid D and F in pure culture. Secalonic acid D has been found to occur in grain dust in USA (Palmgren, 1985; Reddy and Reddy, 1991).

4.14. Verrucosidin

Verrucosidin is a of the mycotoxin from species in *Penicillium* series *Viridicata* that has been claimed to cause mycotoxicosis in animals (Burka et al., 1983).

Major sources. Penicillium polonicum, P. aurantiogriseum and *P. melanoconidium* are the major known sources of verrucosidin (Frisvad and Samson, 2004b).

4.15. Xanthomegnin, Viomellein and Vioxanthin

These toxins have been reported to cause experimental mycotoxicosis in pigs and they apparently are more toxic to the liver than to kidneys in mammals (Zimmerman et al., 1979). They have been found to be naturally occurring in cereals (Hald et al., 1983; Scudamore et al., 1986).

Major sources. P. cyclopium, P. freii, P. melanoconidium, P. tricolor and *P. viridicatum* are common in cereals. *A. ochraceus, A. westerdijkiae* and possibly *A. steynii* are common in green coffee beans and are occasionally found in grapes and on rice.

Minor sources. *P. janthinellum* and *P. mariaecrucis* are soil-borne species producing these hepatotoxins (Frisvad and Filtenborg, 1990).

5. TOXINS FROM OTHER GENERA

5.1. *Claviceps* Toxins

Ergot alkaloids are common in sclerotia of *Claviceps*, which are produced on cereals, especially in whole rye. These sclerotia are often removed before milling of the rye, and outbreaks of ergotism rarely occur now.

Major sources. Claviceps purpurea and *C. paspali* are the major sources of ergot alkaloids (Blum, 1995). Several Penicillia and Aspergilli can produce clavinet type alkaloids also, but their possible role in mycotoxicology is unknown.

5.2. *Alternaria* Toxins

Tenuazonic acid is regarded as the most toxic of the secondary metabolites from *Alternaria* (Blaney, 1991). It is also produced by a *Phoma* species.

Major sources. Phoma sorghina appears to be the most important producer of tenuazonic acid. It has been associated with onyalai, a haematological disease (Steyn and Rabie, 1976). Species in the *Alternaria tenuissima* complex often produce tenuazonic acid, but it has not been found in isolates of *A. alternata sensu stricto. A. citri, A. japonica, A. kikuchiana, A. longipes, A. mali, A. oryzae,* and *A. solani* have also been reported to produce tenuazonic acid (Sivanesan, 1991).

Many other metabolites have been found in *Alternaria*, and some can occur naturally in tomatoes, apples and other fruits (Sivanesan, 1991; Andersen and Frisvad, 2004). The toxicity of such compounds, including alternariols, is not well examined.

5.3. *Phoma* and *Phomopsis* Toxins

Lupinosis toxin (phomopsin) is produced when *Phomopsis leptostromiformis* grows on lupin plants (*Lupinus* species) and lupin seeds

(Culvenor et al., 1977). It is a hepatotoxin which has caused widespread disease in sheep grazing lupins in Australia, South Africa and parts of Europe (Marasas, 1974; Culvenor et al., 1977). As lupin seed is used for human food in South Asia, quality control of phomopsin is important.

5.4. *Pithomyces* Toxins

Sporidesmin is produced by *Pithomyces chartarum* and causes facial eczema in sheep (Atherton et al., 1974). However, this is a disease of pasture only.

5.5. *Stachybotrys* Toxins

Stachybotrys and *Memnoniella* spp. are primarily of importance for indoor air, but stachybotrytoxicosis was one of the first equine mycotoxicosis to be reported (Rodrick and Eppley, 1974). *Stachybotrys chartarum* and *S. chlorohalonata* are the two important fungi producing cyclic trichothecenes (satratoxins) and toxic atranones (Andersen et al., 2003; Jarvis, 2003).

5.6. *Monascus* Toxins

Monascus ruber is used in the production of red rice in the Orient, and is a source of red food colouring. However, it has been repeatedly reported to produce citrinin (Blanc et al., 1995).

6. DISCUSSION

A large number of filamentous fungi are able to produce secondary metabolites that are toxic to vertebrate animals, i.e. mycotoxins. Only a fraction of these fungi can produce mycotoxins in food or feeds, and among those, pathogenic field fungi and deteriorating storage fungi are the most significant. When misidentified fungi are excluded, only a few fungal species are highly toxigenic, and producing their toxins in sufficiently large amounts to cause public alarm. The most important among these are trichothecenes, fumonisins, aflatoxin, ochratoxin A and zearalenone (Miller, 1995), because their fungal producers are widespread and can grow and produce their toxins on many kinds of foods. Other mycotoxins are important, but may only occur on a single type of crop and cause mycotoxicosis in one kind of animal. Phomopsin is an example of this.

Correct identification and knowledge of the associated mycobiota of the different foods and feeds will assist in determining which mycotoxins to look for. There have been examples of mycotoxins analysis for aflatoxin, trichothecenes, zearalenone, fumonisin and ochratoxin A in silage, where the dominant mycobiota is *P. roqueforti, P. paneum, Monascus ruber* and *Byssochlamys nivea*. In that particular case patulin, mycophenolic acid, PR toxin, and citrinin would be more relevant mycotoxins to analyse for. Rapid methods may are effective to secure healthy foods and feeds, but such methods should be based on mycological and ecological knowledge. We hope that our compilation of mycotoxin producers will help in deciding the most appropriate mycotoxin analyses of foods and feeds. New mycotoxins and new mycotoxin producers will no doubt appear, but we believe that the most important ones are listed here.

7. ACKNOWLEDGEMENTS

Jens Frisvad and Ulf Thrane thank LMC and the Technical Research Council for financial support.

8. REFERENCES

Abarca, M. L., Bragulat, M. R., Castella, G., and Cabañes, F. J., 1994, Ochratoxin A production by strains of *Aspergillus niger* var. *niger*, *Appl. Environ. Microbiol.* **60**:2650-2652.

Abarca, M. L., Accensi, F., Cano, J., and Cabañes, F. J., 2004, Taxonomy and significance of black Aspergilli, *Antonie van Leeuwenhoek* **86**:33-49.

Alsberg, C. L. and Black, O. F., 1913, Contributions to the study of maize deterioration; biochemical and toxicological investigations of *Penicillium puberulum* and *Penicillium stoloniferum*, *Bull. Bur. Anim. Ind. U.S. Dept. Agric.* **270**:1-47.

Andersen, B., and Frisvad, J. C., 2004, Natural occurrence of fungi and fungal metabolites in moldy tomatoes, *J. Agric. Food Chem.* **52**:7507-7513.

Andersen, B., and Thrane, U., 2006, Food-borne fungi in fruit and cereals and their production of mycotoxins, in: *Advances of Food Mycology*, A. D. Hocking, J. I. Pitt, R. A. Samson and U. Thrane, eds, Springer, New York. pp. 137-152.

Andersen, B., Nielsen, K. F., Thrane, U., Szaro, T., Taylor, J. W., and Jarvis, B. B., 2003, Molecular and phenotypic descriptions of *Stachybotrys chlorohalonata* sp. nov. and two chemotypes of *Stachybotrys chartarum* found in water-damaged buildings, *Mycologia* **95**:1227-1238.

Andersen, B., Smedsgaard, J., and Frisvad, J. C., 2004, *Penicillium expansum*: consistent production of patulin, chaetoglobosins and other secondary metabolites in culture and their natural occurrence in fruit products, *J. Agric. Food Chem.* **52**:2421-2429.

Arnold, D. L., Scott, P. M., McGuire, P. F., and Harwig, J., 1987, Acute toxicity studies on roquefortine C and PR-toxin, metabolites of *Penicillium roqueforti* in the mouse, *Food Cosmet. Toxicol.* **16**:369-371.

Atherton, L. G., Brewer, D., and Taylor, A., 1974, *Pithomyces chartarum*: a fungal parameter in the aetiology of some diseases of domestic animals, in: *Mycotoxins,* I. F. H. Purchase. ed., Elsevier, Amsterdam, pp. 29-68.

Bayman, P., Baker, J. L., Doster, M. A., Michailides, T. J., and Mahoney, N. E., 2002, Ochratoxin production by the *Aspergillus ochraceus* group and *Aspergillus alliaceus, Appl. Environ. Microbiol.* **68**:2326-2329.

Bentley, R., 2000, Mycophenolic acid: a one hundred year odyssey from antibiotic to immunosuppressant, *Chem. Rev.* **100**:3801-3825.

Betina, V., 1989, *Mycotoxins. Chemical, biological and environmental aspects,* Elsevier, Amsterdam, 438 pp.

Blanc, P. J., Loret, M. O., and Goma, G., 1995, Production of citrinin by various species of *Monascus*, *Biotechnol. Lett.* **17**:210-213.

Blaney, B. J., 1991, *Fusarium* and *Alternaria* toxins, in: *Fungi and mycotoxins in stored products*, B. R. Champ, E. Highley, A. D. Hocking and J. I. Pitt, eds, ACIAR Proceedings No. 36, Australian Centre for International Agricultural Research, Canberra, pp. 86-98.

Blum, M. S., 1995, *The toxic action of marine and terrestrial alkaloids*, Alaken, Inc., Fort Collins, CO.

Bryden,W. L., Logrieco,A., Abbas, H. K., Porter, J. K., Vesonder, R. F., Richard, J. L., and Cole, R. J., 2001, Other significant *Fusarium* mycotoxins, in: Fusarium. *Paul E. Nelson Memorial Symposium*. B. A. Summerell, J. F. Leslie, D. Backhouse, W. L. Bryden and L. W. Burgess, eds, APS Press, St. Paul, MN, pp. 360-392.

Burdock, G. A., Carabin, I. G., and Soni, M. G., 2001, Safety assessment of β-nitropropionic acid: a monograph in support of an acceptable daily intake in humans, *Food Chem.* **75**:1-27.

Burka, L. T., Ganguli, M., and Wilson, B. J., 1983, Verrucosidin, a tremorgen from *Penicillium verrucosum* var. *cyclopium*. *J. Chem. Soc. Chem. Commun.* **1983**:544-545.

Bushnell, G. W., Li, Y.-L., and Poulton, G. A., 1984, Pyrones. X. Lateropyrone, a new antibiotic from the fungus *Fusarium lateritium* Nees, *Can. J. Chem.* **62**:2101-2106.

Ciegler, A., 1972, Bioproduction of ochratoxin A and penicillic acid by members of the *Aspergillus ochraceus* group, *Can. J. Microbiol.* **18**:631-636.

Ciegler, A., and Pitt, J. I., 1970, Survey of the genus *Penicillium* for tremorgenic toxin production, *Mycopath. Mycol. Appl.* **42**:119-124.

Cole, R. J. and Cox, R. H., 1981, *Handbook of toxic fungal metabolites*, Academic Press, New York.

Cole, R. J., Dorner, J. W., Cox, R. H., 1983, Two classes of alkaloid mycotoxins produced by *Penicillium crustosum* Thom isolates from contaminated beer, *J. Agric Food Chem.* **31**:655-657.

Cole, R. J., Kirksey, J. W., Dorner, J. W., Wilson, D. M., Johnson, J. C., Jr., Johnson, A. N., Bedell, D. M., Springer, J. P., Chexal, K. K., Clardy, J. J., and Cox, R. H.,

1977, Mycotoxins produced by *Aspergillus fumigatus* species isolated from molded silage, *J. Agric. Food Chem.* **25**:826-830.

Devys, M., Bousquet, J. F., Kollman, A., and Barbier, M, 1980, Dihydroisocoumarines et acide mycophenolique du milieu de culture du champignon phytopathogene *Septoria nodorum*, *Phytochemistry* **19**: 2221-2222.

Dowd, P. F., Miller, J. D., and Greenhalgh, R., 1989, Toxicity and interactions of some *Fusarium graminearum* metabolites to caterpillars, *Mycologia* **81**:646-650.

Ellner, F. M., 2002, Mycotoxins in potato tubers infected by *Fusarium sambucinum*, *Mycotoxin Res.* **18**:57-61.

Engelhardt, J. A., and Carlton, W. W., 1991, Rubratoxins, in: *Mycotoxins and Phytoallexins*, R. P. Sharma and D. K. Salunkhe, eds, CRC Press, Boca Raton, Florida., pp. 259-289.

Enomoto, M., and Ueno, I., 1974, *Penicillium islandicum* (toxic yellowed rice) – luteoskyrin – islanditoxin cyclochlorotine, in: *Mycotoxins,* I. F. H. Purchase, ed., Elsevier, Amsterdam, pp. 303-326.

Frisvad, J. C., 1985, Profiles of primary and secondary metabolites of value in classification of *Penicillium viridicatum* and related species, in: *Advances in* Penicillium *and* Aspergillus *Systematics*, R. A. Samson and J. I. Pitt, eds, Plenum Press, New York, pp. 311-325.

Frisvad, J. C., 1989, The connection between Penicillia and Aspergilli and mycotoxins with special emphasis on misidentified isolates, *Arch. Environ. Contam. Toxicol.* **18**:452-467.

Frisvad, J. C., and Filtenborg, O., 1990, Revision of *Penicillium* subgenus *Furcatum* based on secondary metabolites and conventional characters, in: *Modern Concepts in* Penicillium *and* Aspergillus *Systematics*, R. A. Samson and J. I. Pitt, eds, Plenum Press, New York, pp. 159-170.

Frisvad, J. C., and Samson, R. A., 2000. *Neopetromyces* gen. nov. and an overview of teleomorphs of *Aspergillus* section *Circumdati*, *Stud. Mycol. (Baarn)* **45**:201-207.

Frisvad, J. C., and Samson, R. A., 2004a, *Emericella venezuelensis*, a new species with stellate ascospores producing sterigmatocystin and aflatoxin B_1, *System. Appl. Microbiol.* **27**:672-690.

Frisvad, J. C., and Samson, R. A., 2004b, Polyphasic taxonomy of *Penicillium* subgenus *Penicillium*. A guide to identification of food and air-borne terverticillate Penicilllia and their mycotoxins, *Stud. Mycol. (Utrecht)* **49**:1-173.

Frisvad, J. C., Samson, R. A., and Smedsgaard, J., 2004a, *Emericella astellata*, a new producer of aflatoxin B_1, B_2 and sterigmatocystin, *Lett. Appl. Microbiol.* **38**: 440-445.

Frisvad, J. C., Frank, J. M., Houbraken, J. A. M. P., Kuijpers, A. F. A., and Samson, R. A., 2004b.,New ochratoxin A producing species of *Aspergillus* section *Circumdati. Stud. Mycol. (Utrecht)* **50**:23-43.

Frisvad, J. C, Nielsen, K. F., and Samson, R. A., 2006, Recommendations concerning the chronic problem of misidentification of species associated with mycotoxigenic fungi in foods and feeds, in: *Advances in Food Mycology*, A. D. Hocking, J. I. Pitt, R. A. Samson and U. Thrane, eds, Springer, New York. pp. 33-46.

Frisvad, J. C., Smedsgaard, J., Larsen, T. O., and Samson, R. A., 2004c, Mycotoxins, drugs and other extrolites produced by species in *Penicillium* subgenus *Penicillium*. *Stud. Mycol. (Utrecht)* **49**:201-242.

Frisvad, J. C., Skouboe, P., and Samson, R. A., 2005a, Taxonomic comparison of three different groups of aflatoxin producers and a new efficient producer of aflatoxin B_1, sterigmatocystin and 3-O-methylsterigmatocystin, *Aspergillus rambellii* sp. nov., *System. Appl. Microbiol.* **28**:442-453.

Frisvad, J. C., Lund, F., and Elmholt, S., 2005b, Ochratoxin A producing *Penicillium verrucosum* isolates from cereals reveal large AFLP fingerprinting variability, *J. Appl. Microbiol.* **98**:684-692.

Ghebremeskel, M., and Langseth, W., 2000, The occurrence of culmorin and hydroxy-culmorins in cereals, *Mycopathologia* **152**:103-108.

Golinski, P., Wnuk, S., Chelkowski, J., Visconti, A., and Schollenberger, M,. 1986, Antibiotic Y: biosynthesis by *Fusarium avenaceum* (Corda ex Fries) Sacc., isolation, and some physiochemical and biological properties, *Appl. Environ. Microbiol.* **51**:743-745.

Hagler Jr, W. M., Towers, N. R., Mirocha, C. J., Eppley, R. M., and Bryden, W. L., 2001, Zearalenone: mycotoxin or mycoestrogen? in: Fusarium, *Paul E. Nelson Memorial Symposium.* B. A. Summerell, J. F. Leslie, D. Backhouse, W. L. Bryden and L. W. Burgess, eds, APS Press, St. Paul, MN, pp. 321-331.

Hald, B., Christensen, D. H., and Krogh, P., 1983, Natural occurrence of the mycotoxin viomellein in barley and the associated quinone-producing penicillia, *Appl. Environ. Mcirobiol.* **46**:1311-1317.

Hershenhorn, J., Park, S. H., Stierle, A., and Strobel, G. A., 1992, *Fusarium avenaceum* as a novel pathogen of spotted knapweed and its phytotoxins, acetamido-butenolide and enniatin B, *Plant Sci.* **86**:155-160.

Hesseltine, C. W., Vandegraft, E. E., Fennell, D. I., Smith, M., and Shotwell, O., 1972, Aspergilli as ochratoxin producers, *Mycologia* **64**:539-550.

Holzapfel, C. W., 1968, The isolation and structure of cyclopiazonic acid, a toxic product from *Penicillium cyclopium* Westling, *Tetrahedron* **24**:2101-2119.

Horie, Y., 1995, Productivity of ochratoxin A of *Aspergillus carbonarius* in *Aspergillus* section *Nigri*, *Nippon Kingakkai Kaiho* **36**:73-76.

IARC (International Agency for Research on Cancer), 2002, Some traditional herbal medicines, some mycotoxins, naphthalene and styrene, IARC Monographs in the Evaluation of Carcinogenic Risks to Humans, Vol. 82, IARC Press, Lyons, France.

Ito, Y., Peterson, S. W., Wicklow, D. T., and Goto, T., 2001, *Aspergillus pseudotamarii*, a new aflatoxin producing species in *Aspergillus* section *Flavi*, *Mycol. Res.* **105**: 233-239.

Jand, S. K., Kaur, P., and Sharma, N. S., 2005, Mycoses and mycotoxicosis in poultry, *Ind. J. Anim. Sci.* **75**: 465-476.

Jarvis, B. B., 2003, *Stachybotrys chartarum*: a fungus for our time, *Phytochemistry* **64**:53-60.

JECFA (Joint FAO/WHO Expert Committee on Food Additives), 2001, *Safety Evaluation of Certain Mycotoxins in Food.* Prepared by the Fifty-sixth meeting of the JECFA. FAO Food and Nutrition Paper 74, Food and Agriculture Organization of the United Nations, Rome, Italy.

Jennessen, J., Nielsen, K. F., Houbraken, J., Lyhne, E. K., Schnürer, J., Frisvad, J. C., and Samson, R. A., 2005, Secondary metabolite and mycotoxin production by the *Rhizopus microsporus* group, *J. Agric. Food Chem.* **53**:1833-1840.

Jestoi, M., Rokka, M., Yli-Mattila, T., Parikka, P., Rizzo, A., and Peltonen, K., 2004, Presence and concentrations of the *Fusarium*-related mycotoxins beauvericin, enniatins and moniliformin in Finnish grain samples, *Food Addit. Contam.* **21**:794-802.

Joshi, B. K., Gloer, J. B., and Wicklow, D. T., 2002, Bioactive natural products from a sclerotium-colonizing isolate of *Humicola fuscoatra*, *J. Nat. Prod.* **65**:1734-1737.

Kamyar, M., Rawnduzi, P., Studenik, C. R., Kouri, K., and Lemmens-Gruber, R., 2004, Investigation of the electrophysiological properties of enniatins, *Arch. Biochem. Biophys.* **429**:215-223.

Klich, M. A., and Pitt, J. I., 1988, Differentiation of *Aspergillus flavus* from *A. parasiticus* and other closely related species, *Trans. Br. Mycol. Soc.* **91**:99-108.

Klich, M., Mullaney, E. J., Daly, C. B., and Cary, J. W., 2000, Molecular and physiological aspects of aflatoxin and sterigmatocystin biosynthesis by *Aspergillus tamarii* and *A. ochraceoroseus, Appl. Microbiol. Biotechnol.* **53**:605-609.

Kurtzman, C. P., Horn, B. W., and Hesseltine, C. W., 1987, *Aspergillus nomius*, a new aflatoxin-producing species related to *Aspergillus flavus* and *Aspergillus parasiticus*, *Antonie van Leeuwenhoek* **53**:147-158.

Labuda, R., and Tancinova, D., 2003, *Eupenicillium ochrosalmoneum*, a rare species isolated from a pig feed mixture in Slovakia, *Biologia* **58**:1123-1126.

Lafont, O., Debeaupuis, J.-P., Gaillardin, M., and Payen, J., 1979, Production of mycophenolic acid by *Penicillium roqueforti* strains, *Appl. Environ. Microbiol.* **39**:365-368.

Lai, M., Semeniuk, G., and Hesseltine, C. W., 1970, Conditions for production of ochratoxin A by *Aspergillus* species in a synthetic medium, *Appl. Microbiol.* **19**: 542-544.

Larsen, J. C., Hunt, J., Perrin, I., and Ruckenbauer, P, 2004, Workshop on trichothecenes with a focus on DON: summary report, *Toxicol. Lett.* **153**:1-22.

Larsen, T. O., Svendsen, A., and Smedsgaard, J., 2001, Biochemical characterization of ochratoxin A-producing strains of the genus *Penicillium*, *Appl. Environ. Microbiol.* **67**:3630-3635.

Launay, F. M., Ribeiro, L., Alves, P., Vozikis,V., Tsitsamis, S., Alfredsson, G., Sterk, S. S., Blokland, M., Iitia, A., Lovgren, T., Tuomola, M., Gordon, A., and Kennedy, D. G., 2004, Prevalence of zeranol, taleranol and *Fusarium* spp. toxins in urine: implications for the control of zeranol abuse in the European Union, *Food Addit. Contam.* **21**:833-839.

Leong, S. L., Hocking, A. D., and Pitt, J. I., 2004, The occurrence of fruit rot fungi (*Aspergillus* section *Nigri*) on some drying varieties of irrigated grapes. *Aust. J. Grape Wine Res.* **10**:83-88.

Lewis, P. R., Donoghue, M. B., Hocking, A. D., Cook, L., and Granger, L. V., 2005, Tremor syndrome associated with a fungal toxin: sequelae of food contamination, *Med. J. Aust.* **82**:582-584.

Lindenfelser, L. A., Lillehoj, E. B., and Milburn, M. S., 1973, Ochratoxin and penicillic acid in tumorigenic and acute toxicity tests with mice, *Dev. Ind. Microbiol.* **14**:331-336.

Logrieco, A., Moretti, A., Castella, G., Kostecki, M., Golinski, P., Ritieni, A., and Chelkowski, J., 1998, Beauvericin production by *Fusarium* species, *Appl. Environ. Microbiol.* **64**:3084-3088.

Lopez-Diaz, T. M., and Flannigan, B., 1997, Mycotoxins of *Aspergillus clavatus*: toxicity of cytochalasin E, patulin, and extracts of contaminated barley malt, *J. Food Prot.* **60**:1381-1385.

Lopez-Diaz, T. M., Roman-Blanco, C., Garcia-Arias, M. T., Garcia-Fernández, M. C., and Garcia-López, M. L., 1996, Mycotoxins in two Spanish cheese varieties, *Int. J. Food Microbiol.* **30**:391-395.

Luangsa-Ard, J. J., Hywel-Jones, N. L., Manoch, L., and Samson, R. A., 2005, On the relationships of *Paecilomyces* sect. *Isarioidea* species, *Mycol. Res.* **109**:581-589.

Lund, F., and Frisvad, J. C., 2003, *Penicillium verrucosum* in wheat and barley indicates presence of ochratoxin A, *J. Appl. Microbiol.* **95**:1117-1123.

Marasas, W. F. O., Miller, J. D., Riley, R. T., and Visconti, A., 2001, Fumonisins occurrence, toxicology, metabolism and risk assessment, in: *Fusarium.* Paul E. Nelson Memorial Symposium, B. A. Summerell, J. F. Leslie, D. Backhouse, W. L. Bryden and L. W. Burgess, eds, APS Press, St. Paul, MN, pp. 332-359.

Marasas, W. F. O., Nelson, P. E., and Toussoun, T. A., 1984, *Toxigenic* Fusarium *species. Identity and mycotoxicology*, The Pennsylvania State University Press, University Park, PA and London, 328 pp.

Miller, J. D., 1995, Mycotoxins in grains: issues for stored product research, *J. Stored Prod. Res.* **31**:1-16.

Morrison, E., Kosiak, B., Ritieni, A., Aastveit, A. H., Uhlig, S., and Bernhoft, A., 2002, Mycotoxin production by *Fusarium avenaceum* strains isolated from Norwegian grain and the cytotoxicity of rice culture extracts to porcine epithelial cells, *J. Agric. Food Chem.* **50**:3070-3075.

Moule, Y., Moreau, S., and Aujard, C., 1980, Induction of cross-links between DNA and protein by PR-toxin, a mycotoxin from *Penicillium roquefortii, Mutat. Res.* **77**:79-89.

Munkvold, G., Stahr, H. M., Logrieco, A., Moretti, A., and Ritieni, A., 1998, Occurrence of fusaproliferin and beauvericin in *Fusarium*-contaminated livestock feed in Iowa, *Appl. Environ. Microbiol.* **64**:3923-3926.

Nielsen, K. F., and Thrane, U., 2001, Fast methods for screening of trichothecenes in fungal cultures using gas chromatography-tandem mass spectrometry, *J. Chromatogr. A* **929**:75-87.

Obana, H., Kumeda, Y., and Nishimune, T., 1995, *Aspergillus ochraceus* production of 5,6-dihydropenicillic acid in culture and foods, *J. Food Prot.* **58**:519-523.

O'Donnell, K., Ward, T.J., Geiser, D.M., Kistler, H.C., and Aoki, T., 2004, Genealogical concordance between the mating type locus and seven other nuclear genes supports formal recognition of nine phylogenetically distinct species within the *Fusarium graminearum* clade, *Fungal Genet. Biol.* **41**:600-623.

Overy, D. P., and Frisvad, J. C., 2003, New *Penicillium* species associated with bulbs and root vegetables, *System. Appl. Microbiol.* **26**:631-639.

Overy, D. P., and Frisvad, J. C., 2005, Mycotoxin production and postharvest storage rot of ginger (*Zingiber officinale*) by *Penicillium brevicompactum*, *J. Food Prot.* **68**:607-609.

Palmgren, M. S., 1985, Microbial and toxic constituents of grain dust and their health implications, in: *Trichothecenes and Other Mycotoxins*, J. Lacey, ed., John Wiley, Chichester, pp. 47-57.

Panaccione, D. G., and Coyle, C. M., 2005, Abundant respirable ergot alkaloids from the common airborne fungus *Aspergillus fumigatus*, *Appl. Environ. Microbiol.* **71**:3106-3111.

Pedersen, P. B., and Miller, J. D., 1999, The fungal metabolite culmorin and related compounds, *Natural Toxins* **7**:305-309.

Peterson, S. W., Ito, Y., Horn, B. W., and Goto, T., 2001. *Aspergillus bombycis*, a new aflatoxigenic species and genetic variation in its sibling species, *A. nomius*, *Mycologia* **93**:689-703.

Pitt, J. I., 1979, *Penicillium crustosum* and *P. simplicissimum*, the correct names for two common species producing tremorgenic mycotoxins, *Mycologia* **71**:1166-1177.

Pitt, J. I., 1987, *Penicillium viridicatum, P. verrucosum*, and the production of ochratoxin A, *Appl. Environ. Microbiol.* **53**:266-269.

Pitt, J. I., and Hocking, A. D., 1997, *Fungi and Food Spoilage*, 2nd edition, Blackie Academic and Professional, London, 596 pp.

Pitt, J. I., Hocking, A. D., Bhudhasamai, K., Miscamble, B. F., Wheeler, K. A., and Tanboon-Ek, P., 1993, The normal mycoflora of commodities from Thailand. 1. Nuts and oilseeds, *Int. J. Food Microbiol.* **20**:211-226.

Puel, O., Tadrist, S. Galtier, P., Oswald, I. P., and Delaforge, M., 2005, *Byssochlamys nivea* as a source of mycophenolic acid, *Appl. Environ. Microbiol.* **71**:550-553.

Reddy, R.,V., and Berndt, W. O., 1991, Citrinin, in: *Mycotoxins and Phytoalexins*, R. P. Sharma and D. K. Salunkhe, eds, CRC Press, Boca Raton, Florida., pp. 237-250.

Reddy, C. S. and Reddy, R. V. 1991. Secalonic acids, in: *Mycotoxins and Phytoalexins*, R. P. Sharma and D. K. Salunkhe, eds, CRC Press, Boca Raton, Florida., pp. 167-190.

Richer, L., Sigalet, D., Kneteman, N., Jones, A., Scott, R. B., Ashbourne, R., Sigler, L., Frisvad, J., and Smith, L., 1997, Fulminant hepatic failure following ingestion of moldy homemade rhubarb wine, *Gastenterology* **112**: A1366.

Rodrick, J. V., and Eppley, R. M., 1974, *Stachybotrys* and stachybotryotoxicosis, in: *Mycotoxins*, I. F. H. Purchase, ed., Elsevier, Amsterdam, pp. 181-197.

Rundberget, T., and Wilkins, A. L., 2002, Thomitrems A and E, two indole-alkaloid isoprenoids from *Penicillium crustosum* Thom, *Phytochemistry* **61**:979-985.

Samson, R. A., 2001, Current fungal taxonomy and mycotoxins, in: *Mycotoxins and phycotoxins in perspective at the turn of the century*, de Koe, W. J., Samson, R. A., van Egmond, H. P., Gilbert, J., and Sabino, M., eds, Proceedings of the X international IUPAC Symposium, Mycotoxins and Phycotoxins, Guaruja-Sao Paulo, Brazil-May 21-25, 2000, W. J. de Koe, Wageningen, pp. 343-350.

Samson, R. A., Houbraken, J. A. M. P., Kuijpers, A. F. A., Frank, J. M., and Frisvad, J. C., 2004, New ochratoxin A or sclerotium producing species in *Aspergillus* section *Nigri*, *Stud. Mycol. (Utrecht)* **50**:45-61.

Schothorst, R. C., and van Egmond, H. P., 2004, Report from SCOOP task 3.2.10, Collection of occurrence data of *Fusarium* toxins in food and assessment of dietary intake by the population of EU member states; Subtask: trichothecenes, *Toxicol. Lett.* **153**:133-143.

Schuster, E., Dunn-Coleman, N., Frisvad, J. C., and van Dijck, P. W. M., 2002, On the safety of *Aspergillus niger* -a review, *Appl. Microbiol. Biotechnol.* **59**:426-435.

Schütt, F., Nirenberg, H. I., and Deml, G., 1998, Moniliformin production in the genus *Fusarium*, *Mycotoxin Res.* **14**:35-40.

Scudamore, K. A., Atkin, P., and Buckle, A. E., 1986, Natural occurrence of the naphthoquinone mycotoxins, xanthomegnin, viomellein and vioxanthin in cereals and animal foodstuffs, *J. Stored Prod. Res.* **22**:81-84.

Seifert, K. A., Aoki, T., Baayen, R. P., Brayford, D., Burgess, L. W., Chulze, S., Gams, W., Geiser, D., de Gruyter, J., Leslie, J. F., Logrieco, A., Marasas, W. F. O., Nirenberg, H. I., O'Donnell, K., Rheeder, J. P., Samuels, G. J., Summerell, B. A., Thrane, U. and Waalwijk, C., 2003, The name *Fusarium moniliforme* should no longer be used, *Mycol. Res.* **107**:643-644.

Sivanesan, A., 1991, The taxonomy and biology of dematiaceous hyphomycetes and their mycotoxins, in: *Fungi and mycotoxins in stored products*, B. R. Champ, E. Highley, A. D. Hocking and J. I. Pitt, eds, ACIAR Proceedings No. 36, Australian Centre for International Agricultural Research, Canberra, pp. 47-64.

Sonjak, S., Frisvad, J. C. and Gunde-Cimerman, N., 2005, Comparison of secondary metabolite production by *Penicillium crustosum* strains, isolated from Arctic and other various ecological niches, *FEMS Microbiol. Ecol.* **53**: 51-60.

Staub, W., 1911, *Penicillium casei* n. sp. als Ursache die rotbraunen Rinderfarbung bei Emmenthaler Käsen, *Centrabl. f. Bakt. (II)* **31**:454.

Steyn, P. S., and Rabie, C. J., 1976, Characterisation of magnesium and calcium tenuazonate from *Phoma sorghina*, *Phytochemistry* **15**:1977-1979.

Stoev, S. D., Vitanov, S., Anguelov, G., Petkova-Bocharova, T., and Creppy, E. E., 2001, Experimental mycotoxic nephropathy in pigs provoked by a diet containing ochratoxin A and penicillic acid, *Vet. Res. Commun.* **25**:205-223.

Taniwaki, M. H., Pitt, J. I., Teixeira, A. A., and Iamanaka, B. T., 2003, The source of ochratoxin A in Brazilian coffee and its formation in relation to processing methods, *Int. J. Food Microbiol.* **82**:173-179.

Teuber, M., and Engel, G., 1983, Low risk of mycotoxin production in cheese, *Microbiol. Alim. Nutr.* **1**:193-197.

Thrane, U., Adler, A., Clasen, P.-E., Galvano, F., Langseth, W., Lew, H., Logrieco, A., Nielsen, K. F., and Ritieni, A., 2004, Diversity in metabolite production by *Fusarium langsethiae, Fusarium poae,* and *Fusarium sporotrichioides*, *Int. J. Food Microbiol.* **95**:257-266.

Ueno, Y., 1974, Citreoviridin from *Penicillium citreo-viride* Biourge, in: *Mycotoxins* I. F. H. Purchase, ed., Elsevier, Amsterdam, pp. 283-302.

Van der Merwe, K. J., Steyn, P. S., Fourie, L., Scott., D. B., and Theron, J. J., 1965, Ochratoxin A, a toxic metabolite produced by *Aspergillus ochraceus* Wilh., *Nature* **205**:1112-1113.

Varga, J., Tóth, B., Rigó, K, Téren, J, Hoekstra, R. F., and Kozakiewics, Z., 2000a, Phylogenetic analysis of *Aspergillus* section *Circumdati* based on sequences of the internal transcribed spacer regions of the 5.8 S rRNA gene, *Fungal Gen. Biol.* **30**:71-80.

Varga, J., Kevei, É., Tóth, B., Kozakiewicz, Z., and Hoekstra, R. F., 2000b, Molecular analysis of variability within the toxigenic *Aspergillus ochraceus* species, *Can. J. Microbiol.* **46**:593-599.

Weidenbörner, M., 2001, Food and fumonisins, *Eur. Food Res. Technol.* **212**:262-273.

Wu, X., Leslie, J. F., Thakur, R. A., and Smith, J. S., 2003, Purifiction of fusaproliferin form cultures of *Fusarium subglutinans* by preparative high-performance liquid chromatography, *J. Agric. Food Chem.* **51**: 383-388.

Yates, S. G., Tookey, H. L., Ellis, J. J., Tallent, W. H., and Wolff, I. A., 1969, Mycotoxins as a possible cause of fescue toxicity, *J. Agric. Food Chem.* **17**:437-442.

Zimmerman, J. L., Carlton, W. W., and Tuite, J., 1979, Mycotoxicosis produced by cultural products of an isolate of *Aspergillus ochraceus*. 1. Clinical observations and pathology, *Vet. Pathol.* **16**:583-592.

RECOMMENDATIONS CONCERNING THE CHRONIC PROBLEM OF MISIDENTIFICATION OF MYCOTOXIGENIC FUNGI ASSOCIATED WITH FOODS AND FEEDS

*Jens C. Frisvad, Kristian F. Nielsen and Robert A. Samson**

1. INTRODUCTION

Since the aflatoxins were first reported in 1961 from *Aspergillus flavus*, mycotoxins have often been named after the fungus which was first found to produce them. A long list of connections between fungal species and mycotoxins and antibiotics has been reported, but unfortunately many of the identifications, and hence the connection between mycotoxin name and the source of the toxin, are incorrect (Frisvad, 1989). The most famous example of such incorrect connections was Alexander Fleming's identification of the original penicillin producer as *Penicillium rubrum*. Fortunately, in this example, the substance was named after the genus *Penicillium*, rather than the species, as K. B. Raper re-identified the strain as *P. notatum*, which was subsequently determined to be a synonym of *P. chrysogenum* (Pitt, 1979b). Later, penicillin was found in other strains of *P. chrysogenum* (Raper and Thom, 1949).

The early aflatoxin literature is plagued with wrong reports of aflatoxin production by *Penicillium puberulum* (Hodges et al., 1964),

* J. C. Frisvad and K. F. Nielsen: BioCentrum-DTU, Building 221, Technical University of Denmark, 2800 Lyngby, Denmark; R. A. Samson: Centraalbureau voor Schimmelcultures, PO Box 85167, 3508 AD, Utrecht, Netherlands. Correspondence to: jcf@biocentrum.dtu.dk

Aspergillus ostianus (Scott et al., 1967), *Rhizopus* sp. (Kulik and Holaday, 1966), the bacterium *Streptomyces* (Mishra and Murthy, 1968) and several other taxa. The most famous of these reports was the paper of El-Hag and Morse (1976). They reported that *Aspergillus oryzae*, the domesticated species used in the manufacture of soy sauce and other Oriental fermented foods, produced aflatoxin. However, the culture of *A. oryzae* they used was quickly shown to be contaminated by an aflatoxin producing *A. parasiticus* (Fennell, 1976). Immediate correction of this error did not prevent Adebajo et al. (1992), El-Kady et al. (1994), Atalla et al. (2003) or Drusch and Ragab (2003) reporting that *A. oryzae* produces aflatoxin.

Often, publications reporting mycotoxin production are reviewed by people who have little or no understanding of mycological taxonomy. For example, "*P. patulinum*" and "*P. clavatus*" are mentioned in Drusch and Ragab (2003). In Bhatnagar et al. (2002), "*P. niger*" is mentioned as producing ochratoxin A. Each of these names is an incorrect combination of genus and species. Bhatnagar et al. (2002) give *P. viridicatum* as producing ochratoxin A in a table, while using *P. verruculosum* as the species name in the text, confusing it with *P. verrucosum*, the correct name for the producer of this toxin.

Such mistakes could have been avoided. This paper provides a set of recommendations to be followed to ensure correct reports of connections between mycotoxin production and fungal species.

2. EXAMPLES OF INCORRECT CITATIONS OF SOME FUNGI PRODUCING WELL KNOWN MYCOTOXINS

2.1. Aflatoxin

The known producers of aflatoxin are given in a separate paper in these Proceedings (Frisvad et al., 2006). The list of other species that have been (incorrectly) reported to produce aflatoxins includes *Aspergillus flavo-fuscus, A. glaucus, A. niger, A. oryzae, A. ostianus, A. sulphureus, A. tamarii, A. terreus, A. terricola, A. wentii, Emericella nidulans* (as *A. nidulans), Emer. rugulosa* (as *A. rugulosus*), *Eurotium chevalieri, Eur. repens, Eur. rubrum, Mucor mucedo, Penicillium citrinum, P. citromyces, P. digitatum, P. frequentans, P. expansum, P. glaucum, P. puberulum, P. variabile, Rhizopus* sp. and the bacterium *Streptomyces* sp. None of these species produce aflatoxins, and many of these names are not accepted as valid species in any case.

2.2. Sterigmatocystin

Fungi known to produce sterigmatocystin include *Aspergillus versicolor, Emericella nidulans,* several other *Emericella* species and some *Chaetomium* species. Although sterigmatocystin is a precursor of aflatoxins (Frisvad, 1989), only *Aspergillus ochraceoroseus* (Frisvad et al., 1999; Klich et al., 2000), and some *Emericella* species accumulate both sterigmatocystin and aflatoxin (Frisvad et al., 2004a; Frisvad and Samson, 2004a). Species in *Aspergillus* section *Flavi,* which includes the major aflatoxin producers, efficiently convert sterigmatocystin into 3-methoxysterigmatocystin and then into aflatoxins (Frisvad et al., 1999).

Many *Aspergillus* species have been reported to produce sterigmatocystin, incorrectly except for those cited above. Sterigmatocystin production by *Penicillium* species has not been reported, apart from an obscure reference to *Penicillium luteum* (Dean, 1963). However, Wilson et al. (2002) claimed that *P. camemberti, P. commune* and *P. griseofulvum* produce sterigmatocystin. Perhaps they mistook sterigmatocystin for cyclopiazonic acid. Three *Eurotium* species have been claimed to produce sterigmatocystin (Schroeder and Kelton, 1975), but this was based only on unconfirmed TLC assays. Unfortunately the strains used were not placed in a culture collection.

2.3. Ochratoxin A

Ochratoxin A is produced by four main species, *Aspergillus carbonarius, A. ochraceus, Petromyces alliaceus, Penicillium verrucosum*, and a few other related species as detailed elsewhere (Frisvad and Samson, 2004b; Samson and Frisvad, 2004; Frisvad et al., 2006). A very large number of species have been claimed to produce ochratoxin A, but not all will be detailed here. However, some of the names frequently cited in reviews will be mentioned. Of the Penicillia, *P. viridicatum* was the name cited for many years as the major ochratoxin A producer, but it was shown that *P. verrucosum* was the correct name for this fungus, the only species that produces ochratoxin A in cereals in Europe (Frisvad and Filtenborg,1983; Frisvad, 1985; Pitt. 1987). The closely related *P. nordicum*, which occurs on dried meat in Europe, was mentioned as producing ochratoxin A by Frisvad and Filtenborg (1983) and Land and Hult (1987), but not accepted as a separate species until the publication of Larsen et al. (2001).

P. verrucosum has been correctly cited as the main *Penicillium* species producing ochratoxin A for a number of years now, but in a series of recent reviews and papers *P. viridicatum* and *P. verruculosum* (no doubt mistaken for *P. verrucosum*) have been mentioned again (Mantle and McHugh, 1993; Bhatnagar et al., 2002; Czerwiecki et al., 2002a, b). In the latter two papers *P. chrysogenum, P. cyclopium, P. griseofulvum, P. solitum, Aspergillus flavus, A. versicolor* and *Eurotium glaucum* were listed as ochratoxin A producers. The strain of *P. solitum* reported by Mantle and McHugh (1993) to produce ochratoxin A were assigned more recently to *P. polonicum,* but neither species produces ochratoxin A (Lund and Frisvad, 1994; 2003). These isolates were contaminated by *P. verrucosum*. The reports by Czerwiecki et al. (2002 a, b) are more problematic in that the fungi have been discarded, so it will never be possible to check the results.

The following species were listed as ochratoxin A producers by Varga et al. (2001): *Aspergillus auricomus, A. fumigatus, A. glaucus, A. melleus, A. ostianus, A. petrakii, A. repens, A. sydowii, A. terreus, A. ustus, A. versicolor, A. wentii, Penicillium aurantiogriseum, P. canescens, P. chrysogenum, P. commune, P. corylophilum, P. cyaneum, P. expansum, P. fuscum, P. hirayamae, P. implicatum, P. janczewskii, P. melinii, P. miczynskii, P. montanense, P. purpurescens, P. purpurogenum, P. raistrickii, P. sclerotiorum, P. spinulosum,, P. simplicissimum, P. variabile* and *P. verruculosum*. None of these species produces ochratoxin A, and it seems clear that the authors have uncritically accepted lists from earlier reviews. In the recent *Handbook of Fungal Secondary Metabolites* (Cole and Schweikert, 2003a, b; Cole et al., 2003), only two of the species cited as producing ochratoxin A are correct: *A. ochraceus* and *A. sulphureus*. The others mentioned are not.

2.4. Citrinin

Citrinin is produced by a number of species in *Penicillium* and *Aspergillus*, notably *P. citrinum, P. expansum, P. verrucosum, A. carneus, A. niveus* and an *Aspergillus* species resembling *A. terreus* (Frisvad, 1989; Frisvad et al., 2004b), but not by *Aspergillus oryzae* or *P. camemberti*, as claimed by Bennett and Klich (2003). Critical checking of the original reports clearly did not occur. Many other species have been claimed to produce citrinin, including *A. ochraceus* (Mantle and McHugh, 1993), *A. wentii* (Abu-Seidah, 2002) and *Eurotium pseudoglaucum* (El-Kady et al., 1994), but either fungus or mycotoxin may have been misidentified in these cases.

2.5. Patulin

A number of species in different genera, notably *Penicillium, Aspergillus* and *Byssochlamys*, produce patulin. Among the most efficient producers of patulin are *Aspergillus clavatus, A. giganteus, A. terreus, Byssochlamys nivea, P. carneum, P. dipodomyicola, Penicillium expansum, P. griseofulvum, P. marinum, P. paneum* and several dung associated Penicillia (Frisvad, 1989; Frisvad et al., 2004b). It is not, however, produced by species in all of the 42 genera listed by Steiman et al. (1989) and Okele et al. (1993). These papers include erroneous statements that *Alternaria alternata, Fusarium culmorum, Mucor hiemalis, Trichothecium roseum* and many others produce patulin. The production of patulin by *Alternaria alternata* was later reported by Laidou et al. (2001), and mentioned in a review by Drusch and Ragab (2003). However patulin was not found in hundreds of analyses of *Alternaria* extracts (Montemurro and Visconti, 1992), or in extracts from more than 200 *Alternaria* cultures tested by us at the Technical University of Denmark (B. Andersen, personal communication).

2.6. Penitrem A

Many species have been claimed to produce penitrem A, but most have been misidentifications of *Penicillium crustosum* (Pitt, 1979; Frisvad, 1989). Names given to isolates that were in fact *P. crustosum* include *P. cyclopium, P. verrucosum* var. *cyclopium*, *P. verrucosum* var. *melanochlorum, P. viridicatum, P. commune, P. lanosum, P. lanosocoeruleum, P. granulatum, P. griseum, P. martensii, P. palitans* and *P. piceum* (Frisvad, 1989). Other species which do produce penitrem A include *P. carneum, P. melanoconidium, P. tulipae, P. janczewskii, P. glandicola* and *P. clavigerum* (Frisvad et al., 2004b). Only the first three of these species are likely to occur in foods.

2.7. Cyclopiazonic Acid

Cyclopiazonic acid is produced by *Aspergillus flavus, A. oryzae, A. tamarii, A. pseudotamarii, Penicillium camemberti, P. commune, P. dipodomyicola, P. griseofulvum* and *P. palitans* (Goto et al., 1996; Huang et al., 1994; Pitt et al., 1986; Polonelli et al., 1987; Frisvad et al., 2004b). Cyclopiazonic acid was originally isolated from and named after *P. cyclopium* CSIR 1082, but this fungus was reidentified as *P. griseofulvum* (Hermansen et al., 1984; Frisvad, 1989). Despite this, most reviews still cite *P. cyclopium* or *P. aurantiogriseum* [of which

Pitt (1979) considered *P. cyclopium* to be a synonym] as producers (Scott, 1994; Bhatnagar, 2002; Bennett and Klich, 2003). Scott (1994) drew an incorrect conclusion

> "α-cyclopiazonic acid is a metabolite of several *Penicillium* and *Aspergillus* species and is of Canadian interest from two viewpoints. First, one of the important producers (*P. aurantiogriseum*, formerly *P. cyclopium*, Pitt et al., 1986), commonly occurs in stored Canadian grains..."

Although *P. aurantiogriseum* no doubt occurs in cereal grains, it is not a producer of cyclopiazonic acid.

Another example of an error being cited repeatedly is the claimed production of cyclopiazonic acid by *Aspergillus versicolor* (Ohmomo et al., 1973; cited by Bhatnagar et al., 2002) even though Domsch et al. (1980) and Frisvad (1989) had stated that the isolate described by Ohmomo et al. (1973) was correctly identified as *A. oryzae*, a well-known producer of cyclopiazonic acid (Orth, 1977). *Penicillium hirsutum, P. viridicatum, P. chrysogenum, P. nalgiovense, Aspergillus nidulans* and *A. wentii* have also wrongly been claimed to produce cyclopiazonic acid (Cole et al., 2003; Abu-Seidah, 2003).

2.8. Xanthomegnin, Viomellein and Vioxanthin

Xanthomegnin, viomellein and vioxanthin are nephrotoxins produced by all members of *Aspergillus* section *Circumdati* (Frisvad and Samson, 2000), *Penicillium cyclopium, P. freii, P. melanoconidium, P. tricolor* and *P. viridicatum* (Lund and Frisvad, 1994), and by *P. janthinellum* and some other genera and species which do not occur in foods. Some of these *Penicillium* species occur in cereals, so these toxins have been found occurring naturally (Scudamore et al., 1986). These toxins are not produced, however, by *P. crustosum* as reported by Hald et al. (1983), by *P. oxalicum* as reported by Lee and Skau (1981) or by *A. nidulans, A. flavus, A. oryzae* or *A. terreus* as reported by Abu-Seidah (2003).

2.9. Penicillic Acid

Penicillic acid is associated with *Penicillium* series *Viridicata* and *Aspergillus* section *Circumdati* (Lund and Frisvad, 1994; Frisvad and Samson, 2000; Frisvad et al., 2004). Production reported by *P. roqueforti* (Moubasher et al., 1978; Olivigni and Bullman, 1978) is now considered to be due to the similar species *P. carneum* (Boysen et al., 1996).

2.10. Rubratoxins

Rubratoxins are hepatoxic mycotoxins known to be produced only by the rare species *Penicillium crateriforme* (Frisvad, 1989). Rubratoxins are not produced by *P. rubrum, P. purpurogenum* or *Aspergillus ochraceus* as reported by Moss et al. (1968), Natori et al. (1970) and Abu-Seibah (2003).

2.11. Trichothecenes

Trichothecenes are especially troublesome as it is only after the introduction of capillary gas chromatography coupled to mass spectrometry (MS) and more recently the introduction of liquid chromatography combined with atmospheric ionization MS that reliable methods have been available for these mycotoxins. Because immunochemical methods have been improved in recent years they also can now be considered valid. However results from TLC and HPLC based methods are dubious, unless combined with immunoaffinity cleanup, as many authors have neglected very time consuming but crucial clean-up steps.

Trichothecene have been reported to be produced by several *Fusarium* species as detailed elsewhere in these proceedings (Frisvad et al., 2006). Marasas et al. (1984) showed that *Fusarium nivale,* which gave nivalenol its name, does not produce trichothecenes. However, under its newer, correct name, *Microdochium nivale* was still incorrectly cited as a trichothecene producer in a recent review (Bhatnagar et al., 2002). It has even been claimed recently that *Aspergillus* species (*A. oryzae, A. terreus, A. parasiticus* and *A. versicolor*) produce nivalenol, deoxynivalenol and T-2 toxin (Atilla et al., 2003). *A. parasiticus* was claimed to produce very high amounts of deoxynivalenol and T-2 toxin after growth on wheat held at 80% relative humidity for 1-2 months. These data are totally implausible. Possibly the wheat was already contaminated with trichothecenes before use, but the high levels indicate that there may have been false positives as well.

3. RECOMMENDATIONS

To avoid incorrect reporting of fungal species producing particular mycotoxins, we recommend the following rules when working with mycotoxin producing fungi and the reporting of the results:

3.1. Ensure correct identification and purity of fungal isolates

- Fungal isolates from the particular substrate should be checked with the literature on the mycobiota of foods, e.g. Filtenborg et al. (1996), Pitt and Hocking (1997) or Samson et al. (2004), which correlate particular fungal species with particular food types or substrates. Unusual findings especially should be carefully checked. For example, *Aspergillus oryzae* is the domesticated form of *A. flavus* and *A. sojae* is the domesticated form of *A. parasiticus*, and these fungi are not expected to be isolated other than from production plants used for making Oriental foods or enzymes.
- Use typical cultures as reference for comparison, both for identification and mycotoxin production. Frisvad et al. (2000), lists typical cultures for each species of common foodborne *Penicillium* subgenus *Penicillium* species. Some effective mycotoxin producing cultures are listed in Table 1.
- Check the purity of cultures, as contaminated cultures are a very common problem. Check for contaminants by growing cultures on standard media such as CYA (Pitt and Hocking, 1997). Especially when fungi are grown on cereals or liquid cultures it is very difficult to assess if the culture is pure, and it necessary to streak them out on agar substrates where it is much easier to see if the culture is pure.

Table 1. Reference cultures for the production of the more common *Aspergillus* and *Penicillium* mycotoxins

Mycotoxin	Producing species and reference culture
Aflatoxins B_1 and B_2	*Aspergillus parasiticus* CBS[a] 100926
	Aspergillus flavus CBS 573.65
Aflatoxins $G_1$and G_2	*Aspergillus parasiticus* CBS 100926
Sterigmatocystin	*Aspergillus versicolor* CBS 563.90
Ochratoxin A	*Petromyces alliaceus* CBS 110.26
	Penicillium verrucosum CBS 223.71
Patulin	*Aspergillus clavatus* CBS 104.45
	Penicillium griseofulvum CBS 295.97
Cyclopiazonic acid	*Penicillium griseofulvum* CBS 295.97
Roquefortine C	*Penicillium griseofulvum* CBS 295.97
Citrinin	*Penicillium citrinum* CBS 252.55
	Penicillium verrucosum CBS 223.71
Penicillic acid	*Penicillium cyclopium* CBS 144.45
Penitrem A	*Penicillium crustosum* CBS 181.89
Verrucosidin	*Penicillium polonicum* CBS 101479
Xanthomegnin	*Penicillium cyclopium* CBS 144.45
Rubratoxin B	*Penicillium crateriforme* CBS 113161

[a]CBS = Culture collection of the Centraalbureau voor Schimmelcultures, Utrecht, Netherlands

- If unusual producers are found, check them carefully for purity and correct identity using the references cited above. A specialist taxonomist may be consulted.

3.2. Ensure that cultures are deposited in a recognised culture collection

- Deposit all interesting strains producing mycotoxins in international culture collections, and cite the culture collection numbers in any publications regarding the strains. This procedure should be mandatory for all microbial, biochemical and chemical journals.

3.3. Ensure substrate is sterile and not already contaminated with mycotoxins

- If natural substrates, such as cereals, are used for mycotoxin production, they should be sterilised before use (e.g. by autoclaving or by gamma-irradiation). Control assays should be carried out for all mycotoxins being studied on the material intended for use as the substrate. This will ensure false positives are not reported.
- Also check for interfering peaks. Natural substrates such as grains may contain interfering compounds, and the chemical composition of these matrices may change during fungal growth. In such matrices, highly selective cleanup procedures should be used and combined with highly selective analytical methods.

3.4. Ensure optimal conditions for mycotoxin production are used

- Use several media and growth conditions to ensure that the fungus can actually produce the mycotoxins. Four good media for mycotoxin production are listed in Table 2.

3.5. Ensure appropriate analytical and confirmatory procedures for mycotoxin extraction and identification

- Sample preparation methods are important and should be validated. Sample preparation is specific for the food matrix it is designed for. Use only validated analytical methods.

Table 2. Efficient media for mycotoxin production

Czapek Yeast Autolysate agar (CYA) (Pitt, 1979; Pitt and Hocking, 1997)		Yeast Extract Sucrose agar (YES) (Frisvad and Filtenborg, 1983)	
$NaNO_3$	3 g	Yeast extract (Difco)	20 g
K_2HPO_4	1 g	Sucrose	150 g
KCl	0.5 g	$MgSO_4 \cdot 7H_2O$	0.5 g
$MgSO_4 \cdot 7H_2O$	0.5 g	$ZnSO_4 \cdot 7H_2O$	0.01 g
$FeSO_4 \cdot 7H_2O$	0.01 g	$CuSO_4 \cdot 5H_2O$	0.005 g
$ZnSO_4 \cdot 7H_2O$	0.01 g	Agar	20 g
$CuSO_4 \cdot 5H_2O$	0.005 g	Distilled water	1 litre
Yeast extract (Difco)	5 g		
Sucrose	30 g		
Agar	20 g		
Distilled water	1 litre		
Rice powder Corn steep agar (RC) (Bullerman, 1974)		**Mercks Malt Extract (MME) agar (El-Banna and Leistner, 1988)**	
Rice powder	50 g	Malt extract	30 g
Corn steep liquid	40 g	Soy peptone	3 g
$ZnSO_4 \cdot 7H_2O$	0.01 g	$ZnSO_4 \cdot 7H_2O$	0.01 g
$CuSO_4 \cdot 5H_2O$	0.005 g	$CuSO_4 \cdot 5H_2O$	0.005 g
Agar	20 g	Agar	20 g
Distilled water	1 litre	Distilled water	1 litre
		pH 5.6	

- Use efficient extraction techniques, for example, fumonisins are very polar and penitrem A is very apolar. Extractions should be validated by recovery experiments.
- Use authenticated standards of the mycotoxins for comparison, ideally as internal and external standards.
- More than one separation technique should be use, combined with selective detection principles. Single UV, refractive index, evaporative light scattering, or flame ionisation detection are non-specific. Fluorescence and full UV spectra are specific to some compounds, while mass spectrometry and especially tandem mass spectrometry is very selective for most compounds when monitoring several ions. Generally four identification points should give a very specific detection, e.g. obtained by LC-MS/MS monitoring two fragmentation reactions.
- Use more than one discretionary test to secure correct identification of the mycotoxin, Combined these with derivatization or alternative clean-up procedures when finding unexpected results.

4. REFERENCES

Abu-Seidah, A. A., 2003, Secondary metabolites as co-markers in the taxonomy of Aspergilli, *Acta Microbiol. Pol.* **52**:15-23.

Adebajo, L. O., 1992, Spoilage moulds and aflatoxins from poultry feeds, *Nahrung* **36**:523-529.

Atalla, M. M., Hassanein, N. M., El-Beih, A. A., and Youssef, Y. A.-G., 2003, Mycotoxin production in wheat grains by different Aspergilli in relation to different humidities and storage periods, *Nahrung* **47**:6-10.

Bennett, J. W., and Klich, M. A., 2003, Mycotoxins, *Clin. Microbiol. Rev.* **16**:497-516.

Bhatnagar, D., Yu, J., and Ehrlich, K. C., 2002, Toxins of filamentous fungi, in: *Fungal Allergy and Pathogenicity,* M. J. Breitenbach, R. Crameri, and S. B. Lehrer, eds, *Chem. Immunol.* **81**:167-206.

Boysen, M., Skouboe, P., Frisvad, J.C., and Rossen, L., 1996, Reclassification of the *Penicillium roqueforti* group into three species on the basis of molecular genetic and biochemical profiles, *Microbiology* **142**: 541-549.

Bullerman, L. B., 1974, A screening medium and method to detect several mycotoxins in mold cultures, *J. Milk Food Technol.* **37**:1-3.

Cole, R. J., and Schweikert, M. A. 2003. *Handbook of Secondary Fungal Metabolites.* Vol. 1, Academic Press, New York.

Cole, R. J., and Schweikert, M. A. 2003. *Handbook of Secondary Fungal Metabolites.* Vol. 2, Academic Press, New York.

Cole, R. J., Jarvis, B. B., and Schweikert, M. A., 2003, *Handbook of Secondary Fungal Metabolites.* Vol. 3, Academic Press, New York.

Czerwiecki, L., Czajkowska, D., and Witkowska-Gwiazdowska, A., 2002a, On ochratoxin A and fungal flora in Polish cereals from conventional and ecological farms. Part 1: Occurrence of ochratoxin A and fungi in cereals in 1997, *Food Addit. Contam.* **19**:470-477.

Czerwiecki, L., Czajkowska, D., and Witkowska-Gwiazdowska, A., 2002b, On ochratoxin A and fungal flora in Polish cereals from conventional and ecological farms. Part 2: Occurrence of ochratoxin A and fungi in cereals in 1998, *Food Addit. Contam.* **19**:1051-1057.

Dean, F. M., 1963, *Naturally Occurring Oxygen Compounds*, Butterworth, London, p. 526.

Drusch, S., and Ragab, W., 2003, Mycotoxins in fruits, fruit juices, and dried fruits, *J. Food Prot.* **66**:1514-1527.

Domsch, K. H., Gams, W., and Anderson, T.-H., 1980, *Compendium of Soil Fungi,* Academic Press, London.

El-Banna, A. A., and Leistner, L., 1988, Production of penitrem A by *Penicillium crustosum* isolated from foodstuffs, *Int. J. Food Microbiol.* **7**:9-17.

El-Hag, N., and Morse, R. E., 1976, Aflatoxin production by a variant of *Aspergillus oryzae* (NRRL 1988) on cowpeas (*Vigna sinensis*), *Science* **192**:1345-1346.

El-Kady, I., El-Maraghy, S., and Zihri, A.-N., 1994, Mycotoxin producing potential of some isolates of *Aspergillus flavus* and *Eurotium* groups from meat products, *Microbiol. Res.* **149**:297-307.

Fennell, D. I., 1976, *Aspergillus oryzae* (NRRL strain 1988): a clarification, *Science* **194**:1188.

Filtenborg, O., Frisvad, J. C., and Thrane, U., 1996, Moulds in food spoilage, *Int. J. Food Microbiol.* **33**:85-102.

Frisvad, J. C. 1985. Classification of asymmetric Penicillia using expressions of differentiation, in *Advances in* Penicillium *and* Aspergillus *systematics*, R. A. Samson and J. I. Pitt, eds, Plenum Press, New York, pp. 327-333.

Frisvad, J. C. 1989. The connection between the Penicillia and Aspergilli and mycotoxins with special emphasis on misidentified isolates, *Arch. Environ. Contam. Toxicol.* **18**:452-467.

Frisvad, J. C., and Filtenborg, O., 1983, Classification of terverticillate Penicillia based on profiles of mycotoxins and other secondary metabolites, *Appl. Environ. Microbiol.* **46**:1301-1310.

Frisvad, J. C., and Samson, R. A., 2000, *Neopetromyces* gen. nov. and an overview of teleomorphs of *Aspergillus* subgenus *Circumdati, Stud. Mycol.* **45**:201-207.

Frisvad, J.C., and Samson, R.A., 2004a, *Emericella venezuelensis*, a new species with stellate ascospores producing sterigmatocystin and aflatoxin B_1, *System. Appl. Microbiol.* **27**:672-680.

Frisvad, J.C., and Samson, R.A., 2004b, New ochratoxin producing species of *Aspergillus* section *Circumdati, Stud. Mycol.* **50**:23-43.

Frisvad, J. C., Filtenborg, O., Lund, F., and Samson, R. A., 2000, The homogeneous species and series in subgenus *Penicillium* are related to mammal nutrition and excretion, in: *Integration of Modern Taxonomic Methods for* Penicillium *and* Aspergillus *Classification*. R. A. Samson and J. I. Pitt, eds, Harwood Academic Publishers, Amsterdam, pp. 265-283.

Frisvad, J. C., Houbraken, J., and Samson, R. A., 1999, *Aspergillus* species and aflatoxin production: a reappraisal, in: *Food Microbiology and Food Safety into the Next Millennium,* A. C. J. Tuijtelaars, R. A. Samson, F. M. Rombouts and S. Notermans, eds, Foundation Food Micro '99, Zeist, Netherlands. pp. 125-126.

Frisvad, J. C., Samson, R. A., and Smedsgaard, J., 2004a, *Emericella astellata*, a new producer of aflatoxin B_1, B_2, and sterigmatocystin, *Lett. Appl. Microbiol.* **38**:440-445.

Frisvad, J. C., Smedsgaard, J., Larsen, T. O., and Samson, R. A., 2004b, Mycotoxins, drugs and other extrolites produced by species in *Penicillium* subgenus *Penicillium, Stud. Mycol.* **49**:201-241.

Frisvad, J. C., Thrane, U., Samson, R. A. and pitt, J. I., 2006, Important mycotoxins, and fungi which produce them, in: *advances in Food Mycology*, A. D. Hocking, J. I. Pitt, R. A. Samson and U. Thrane, eds, Springer, New York, pp. 3–25.

Goto, T., Wicklow, D. T., and Ito, Y, 1996, Aflatoxin and cyclopiazonic acid production by a sclerotium-producing *Aspergillus tamarii* strain, *Appl. Environ. Microbiol.* **62**:4036-4038.

Hald, B., Christensen, D. H., and Krogh, P., 1983, Natural occurrence of the mycotoxin viomellein in barley and the associate quinone-producing Penicillia, *Appl. Environ. Microbiol.* **42**:446-449.

Hermansen, K., Frisvad, J. C., Emborg, C., and Hansen, J., 1984, Cyclopiazonic acid production by submerged cultures of *Penicillium* and *Aspergillus* strains, *FEMS Microbiol. Lett.* **21**:253-261.

Hodges, F. A., Zust, J. R., Smith, H. R., Nelson, A. A., Armbrecht, B. H., and Campbell, A. D., 1964, Mycotoxins: aflatoxin produced by *Penicillium puberulum*, *Science* **145**:1439.

Huang, X., Dorner, J. W., and Chu, F. S., 1994, Production of aflatoxin and cyclopiazonic acid by various Aspergilli: an ELISA approach, *Mycotox. Res.* **10**:101-106.

Ito, Y., Peterson, S. W., Wicklow, D. T., and Goto, T., 2001, *Aspergillus pseudotamarii*, a new aflatoxin producing species in *Aspergillus* section *Flavi, Mycol. Res.* **105**:233-239.

Klich, M., Mullaney, E. J., Daly, C. B., and Cary, J. W., 2000, Molecular and physiological aspects of aflatoxin and sterigmatocystin biosynthesis by *Aspergillus tamarii* and *A. ochraceoroseus*, *Appl. Microbiol. Biotechnol.* **53**:605-609.

Kulik, M. M. and Holaday, C. E., 1966, Aflatoxin: a metabolic product of several fungi, *Mycopath. Mycol. Appl.* **30**:137-140.

Laidou, I. A., Thanassoulopoulos, C. C., and Liakopoulou-Kyriakidis, M., 2001, Diffusion of patulin in the flesh of pears inoculated with four post-harvest pathogens, *J. Phytopathol.* **149**:457-461.

Land, C. J., and Hult, K., 1987, Mycotoxin production by some wood-associated *Penicillium* spp., *Lett. Appl. Microbiol.* **4**:41-44.

Larsen, T. O., Svendsen, A., and Smedsgaard, J., 2001, Biochemical characterization of ochratoxin A-producing strains of the genus *Penicillium*, *Appl. Environ. Microbiol.* **67**:3630-3635.

Lee, L. S., and Skau, D. B., 1981, Thin layer chromatographic analysis of mycotoxins: a review of recent literature, *J. Liquid Chromatogr.* **4**, Suppl. 1:43-62.

Lund, F., and Frisvad, J. C., 1994, Chemotaxonomy of *Penicillium aurantiogriseum* and related species, *Mycol. Res.* **98**:481-492.

Lund, F., and Frisvad, J. C., 2003, *Penicillium verrucosum* in cereals indicates production of ochratoxin A. *J. Appl. Microbiol.* **95**:1117-1123.

Mantle, P. G., and McHugh, K. M., 1993, Nephrotoxic fungi in foods from nephropathy households in Bulgaria, *Mycol. Res.* **97**:205-212.

Marasas, W. F. O., Nelson, P. E., and Toussoun, T. A., 1984, *Toxigenic* Fusarium *Species. Identity and Mycotoxicology,* Pennsylvania State University Press, University Park, Pennsylvania.

Mishra, S. K., and Murthy, H. S. R., 1968, An extra fungal source of aflatoxins, *Curr. Sci. (Mysore)* **37**:406.

Montemurro, N., and Visconti, A., 1992, *Alternaria* metabolites -chemical and biological data, in: Alternaria. *Biology, Plant Diseases and Metabolites,* J. Chelkowski and A. Visconti, eds, Elsevier, Amsterdam, pp. 449-557.

Moss, M. O., Robinson, F. V. and Wood, A. B., 1968, Rubratoxin B, a toxic metabolite of *Penicillium rubrum*, *Chemy Ind.* **1968**:587-588.

Moubasher, A. H., Abdel-Kader, M. I. A., and El-Kady, I. A., 1978, Toxigenic fungi isolated from Roquefort cheese, *Mycopathologia* **66**:187-190.

Natori, S., Sakaki, S., Kurata, M., Udagawa, S., Ichinoe, M., Saito, M., Umeda, M., and Ohtsubo, K., 1970, Production of rubratoxin B by *Penicillium purpurogenum, Appl. Microbiol.* **19**:613-617.

Ohmomo, S., Sugita, M., and Abe, M., 1973, Isolation of cyclopiazonic acid, cyclopiazonic acid imine and bissecodehydrocyclopiazonic acid from the cultures of *Aspergillus versicolor* (Vuill.) Tiraboschi, *J. Agric. Chem. Soc. Japan* **47**:57-93.

Okeke, B., Seigle-Murandi, F., Steiman, R., Benoit-Guyod, J.-L., and Kaouadjii, M., 1993, Identification of mycotoxin-producing fungal strains: a step in the isolation of compounds active against rice fungal diseases, *J. Agric. Food Chem.* **41**:1731-1735.

Olivigni, F. J., and Bullerman, L. B., 1978, Production of penicillic acid and patulin by an atypical *Penicillium roqueforti* isolate, *Appl. Microbiol.* **35**:435-438.

Orth, R., 1977, Mycotoxins of *Aspergillus oryzae* strains for use in the food industry as starters and enzyme producing molds, *Ann. Nutr. Aliment.* **31**:617-624.

Pitt, J. I., 1979a, *Penicillium crustosum* and *P. simplicissimum*, the correct names for two common species producing tremorgenic mycotoxins, *Mycologia* **71**:1166-1177.

Pitt, J. I., 1979b, *The Genus* Penicillium *and its Teleomorphic States* Eupenicillium *and* Talaromyces, Academic Press, London.

Pitt, J. I., 1987., *Penicillium viridicatum, P. verrucosum*, and the production of ochratoxin A, *Appl. Environ. Microbiol.* **53**:266-269.

Pitt, J. I., and Hocking, A. D., 1997, *Fungi and Food Spoilage*. 2nd edition, Blackie Academic and Professional, London.

Pitt, J. I., Cruickshank, R. H., and Leistner, L., 1986, *Penicillium commune, P. camembertii*, the origin of white cheese moulds, and the production of cyclopiazonic acid, *Food Microbiol.* **3**:363-371.

Polonelli, L., Morace, G., Rosa, R., Castagnola, M., and Frisvad, J. C., 1987, Antigenic characterization of *Penicillium camemberti* and related common cheese contaminants, *Appl. Environ. Microbiol.* **53**:872-878.

Raper, K. B. and Thom, C., 1949, *A Manual of the Penicillia,* Williams and Wilkins, Baltimore.

Samson, R. A., and Frisvad, J. C., 2004, New ochratoxin or sclerotium producing species in *Aspergillus* section *Nigri*, *Stud. Mycol.* **50**:45-61.

Samson, R. A., Hoekstra, E. S., and Frisvad, J. C., eds, 2004, *Introduction to Food- and Airborne Fungi*, 7th edition, Centraalbureau voor Schimmelcultures, Utrecht, Netherlands, 389 pp.

Schroeder, H. W., and Kelton, W. H., 1975, Production of sterigmatocystin by some species of the genus *Aspergillus* and its toxicity to chicken embryos, *Appl. Microbiol.* **30**:589-591.

Scott, P. M., 1994, *Penicillium* and *Aspergillus* toxins, in: *Mycotoxins in Grain. Compounds other than Aflatoxin*. J. D. Miller and H. L. Trenholm, H. L., eds, Eagan Press, St. Paul, Minnesota, pp. 261-285.

Scott, P. M., van Walbeek, W., and Forgacs, J., 1967, Formation of aflatoxins by *Aspergillus ostianus* Wehmer, *Appl. Microbiol.* **15**:945.

Scudamore, K. A., Atkin, P., and Buckle, A. E., 1986, Natural occurrence of the naphthoquinone mycotoxins, xanthomegnin, viomellein and vioxanthin in cereals and animal foodstuffs, *J. Stored Prod. Res.* **22**:81-84.

Steiman R., Seigle-Murandi, F., Sage, L., and Krivobok S., 1989, Production of patulin by micromycetes, *Mycopathologia* **105**:129-133.

Varga, J., Rigó, K., Réren, J., and Mesterházy, Á., 2001, Recent advances in ochratoxin research. I. Production, detection and occurrence of ochratoxins, *Cereal Res. Commun.* **29**:85-92.

Wilson, D. M., Mutabanhema, W., and Jurjevic, Z., 2002, Biology and ecology of mycotoxigenic *Aspergillus* species as related to economy and health concerns, in: *Mycotoxins and Food Safety,* J. W. DeVries, M. W. Trucksess and L. S. Jackson, eds, Kluwer Academic Publishers, Dordrech, Netherlands. pp. 3-17.

Section 2. Media and method development in food mycology

COMPARISON OF HYPHAL LENGTH, ERGOSTEROL, MYCELIUM DRY WEIGHT, AND COLONY DIAMETER FOR QUANTIFYING GROWTH OF FUNGI FROM FOODS

*M. H. Taniwaki, J. I. Pitt, A. D. Hocking and G. H. Fleet**

1. INTRODUCTION

Fungi are significant environmental microorganisms, as they are responsible for spoilage of foods, production of mycotoxins and in some cases desirable bioconversions. It is important therefore to have reliable, convenient methods for measuring fungal growth. However, the growth of fungi is not easy to quantify because, unlike bacteria and yeasts, fungi do not grow as single cells, but as hyphal filaments that cannot be quantified by the usual enumeration techniques. Fungal hyphae can penetrate solid substrates, such as foods, making their extraction difficult. In addition, fungi differentiate to produce spores, resulting in large increases in viable counts often with little relationship to biomass (Pitt, 1984).

A number of methods have been developed for quantifying fungal growth and their principles and applications comprehensively reviewed (Matcham et al.,1984; Hartog and Notermans, 1988; Williams, 1989; Newell, 1992; Samson et al., 1992; de Ruiter et al., 1993; Pitt and Hocking, 1997). The most frequently used method is

* M. H. Taniwaki, Instituto de Tecnologia de Alimentos, Campinas-Sp, Brazil; J. I. Pitt, A. D. Hocking, Food Science Australia, PO Box 52, North Ryde, NSW 2113, Australia; G. H. Fleet, Food Science and Technology, University of New South Wales, Sydney, NSW 2052, Australia. Correspondence to: mtaniwak@ital.sp.gov.br

the counting of viable propagules, i.e. colony forming units (CFU), a technique derived from food bacteriology. However, this method suffers from serious drawbacks. Viable counts usually reflect spore numbers rather than biomass (Pitt, 1984). When fungal growth consists predominantly of hyphae, i.e. in young colonies or inside food particles, viable counts will be low, but when sporulation occurs, counts often increase rapidly without any great increase in biomass. Some fungal genera, e.g. *Alternaria* and *Fusarium,* produce low numbers of spores in relation to hyphal growth, whereas others, e.g. *Penicillium*, produce very high numbers of spores. Consequently, viable counts are a poor indicator of the extent of fungal growth and appear to correlate poorly with other measures such as ergosterol (Saxena et al., 2001).

A second commonly used method is measurement of colony diameter (Brancato and Golding, 1953). When measured over several time intervals, colony diameters can be translated into growth rates, which are frequently linear over quite long periods (Pitt and Hocking, 1977) and have been widely used in water activity studies (e.g. Pitt and Hocking, 1977; Pitt and Miscamble, 1995) and to model growth (Gibson et al., 1994). However colony diameter as a measure of fungal biomass takes no account of colony density (Wells and Uota, 1970).

Estimation of mycelium dry weight is a third commonly used method to assess fungal growth or biomass. This is the method of choice for growth in liquid systems, such as fermentors, however, mycelium dry weight measurements lack sensitivity and are destructive (Deploey and Fergus, 1975). A fourth approach, measurement of hyphal length, was used by Schnürer (1993) to estimate the biomass of three fungi grown in pure culture. This technique has the advantage of actually measuring growth, but is particularly laborious. None of these methods can be used to estimate fungal biomass in foods.

Chemical assays have also been used to measure fungal growth. The two substances commonly assayed are chitin and ergosterol. The chitin assay is well documented (Ride and Drysdale, 1972), but has major disadvantages: it lacks sensitivity, is time consuming, and is subject to interference from insect fragments (Pitt and Hocking, 1997). Ergosterol is the dominant sterol in most fungi (Weete, 1974), and is not found to any significant extent in plants, animals or bacteria (Schwardorf and Muller, 1989). Thus, its measurement in environmental samples can be taken as an index of the presence of fungi (Seitz et al., 1977, 1979; Nylund and Wallander, 1992; Miller and

Young, 1997). The advantages of this method are high sensitivity, specificity and relatively short analysis time (Seitz et al., 1977; Schwardorf and Muller, 1989). However, the ergosterol assay has never been validated against more traditional methods, an essential step before it can be accepted as a reliable method for quantifying fungal growth in foods. In addition, the influences of such factors as medium composition, water activity and age of colony on the ergosterol content of mycelium have not been evaluated adequately. One study has attempted to compare ergosterol content with mycelial dry weight over a range of species: for nine aquatic fungi, only three showed correlations between these parameters (Bermingham et al., 1995). Studies comparing ergosterol content with mould viable counts have reported mixed results in grains (Schnürer and Jonsson, 1992) and pure cultures (Saxena et al., 2001).

This paper reports a comparison of the ergosterol, colony diameter, dry weight and hyphal length methods for quantifying the growth of several fungal species significant in foods. Studies were carried out in pure culture under a range of conditions.

2. MATERIALS AND METHODS

2.1. Fungi

Single isolates of nine food spoilage fungi, representing examples of heat resistant, xerophilic and toxigenic species commonly found in foods, were obtained from the FRR culture collection at Food Science Australia North Ryde, NSW, Australia (Table 1). These species were *Aspergillus flavus, Byssochlamys fulva, Byssochlamys nivea, Eurotium chevalieri, Fusarium oxysporum, Mucor plumbeus, Penicillium commune, Penicillium roqueforti* and *Xeromyces bisporus*.

2.2. Media

The following media were used: Czapek Yeast Extract Agar (CYA), Malt Extract Agar (MEA) and Potato Dextrose Agar (PDA) representative of high water activity (a_w) media, (all of *ca* 0.997 a_w); and Czapek Yeast Extract 20% Sucrose Agar (CY20S), 0.98 a_w and Malt Extract Yeast Extract 50% Glucose Agar (MY50G), 0.89 a_w, as reduced a_w media. PDA was from Oxoid Ltd, Basingstoke, UK, and the formulae for the others are given by Pitt and Hocking (1997).

Table 1. Origins of cultures used[a]

Species	Strain number	Source
Aspergillus flavus	FRR 2757	Peanut, Queensland, Australia, 1984
Byssochlamys fulva	FRR 3792	Strawberry puree, NSW, Australia, 1990
Byssochlamys nivea	FRR 4421	Strawberry, Brazil, 1993
Eurotium chevalieri	FRR 547	Animal feed, Queensland, Australia, 1970
Fusarium oxysporum	FRR 3414	Orange juice, NSW, Australia, 1987
Mucor plumbeus	FRR 2412	Apple juice, NSW, Australia, 1981
Penicillium commune	FRR 3932	Cheddar cheese, NSW, Australia, 1991
Penicillium roqueforti	FRR 2162	Cheddar cheese, USA, 1978
Xeromyces bisporus	FRR 2351	Dates, NSW, 1981

[a] FRR denotes the culture collection of Food Science Australia, North Ryde, NSW, Australia

2.3. Cultivation

Inocula were prepared from 5 to 7 day cultures grown on CYA, except for *B. nivea, E. chevalieri* and *X. bisporus* which were grown on MEA for 7 to 10 days, CY20S for 10 to 15 days and MY50G for 15 to 20 days, respectively. Cultures for growth estimates and assays were grown in 90 mm plastic Petri dishes, inoculated at a single central point. Each fungus was grown on several plates of each medium. Plates were incubated upright at 25°C.

2.4. Growth Measurement

Fungal growth was measured by the methods described below throughout the growth period, but at intervals which varied widely with species and medium. Measurements and assays were carried out in duplicate.

2.4.1. Colony Diameters

Colonies were measured from the reverse side in millimetres with a ruler. Only well formed, circular colonies were chosen for measurement.

2.4.2. Mycelium Dry Weight

A colony and surrounding agar were cut from a Petri dish, transferred to a beaker containing distilled water (100 ml), then heated in a steamer for 30 min to melt the agar. The mycelium, which remained intact, was rinsed once in distilled water and then transferred to a dried, weighed filter paper which was placed in an aluminium dish and dried in an oven at 80°C for 18 h. After cooling to room temperature in a desiccator, the filter papers and mycelium were weighed and the dry weight calculated by difference. The method was based on those of Paster et al. (1983) and Zill et al. (1988).

2.4.3. Hyphal Length

Hyphal lengths were estimated by direct microscopy using a haemocytometer and a modification of the method of Schnürer (1993). Colonies and associated agar were cut into pieces and homogenized with distilled water (3-100 ml, depending on the size of the colony), for about 30 s using an Ultra-Turrax homogeniser (Ystral GmbH, Dottingen, Germany). The suspension was then treated in a sonicator (Branson Sonic Power Company, Danbury, CT) at 100 watts for about 20 s to break up hyphal clumps. After dilution in distilled water, drops (0.5 ml) were placed in a haemocytometer and hyphal fragments counted. Hyphal lengths were measured using the intersection technique (Olson, 1950) at a magnification of 400 X. Colonies from two plates were measured separately and for each plate 10 fields were counted. Results were calculated from the means of the two plates.

2.4.4. Ergosterol Assay

A colony was excised from a Petri dish culture, transferred to a beaker of distilled water (100 ml) containing Tween 80 (0.05%) and steamed for 30 min to melt the agar. The intact mycelium was collected, rinsed with water and transferred to a round bottomed flask. Ergosterol was extracted from the mycelium by refluxing with 95% ethanol: water (100 ml, 50: 50 v/v) and potassium hydroxide (5 g) for 30 min (Zill et al., 1988). This crude extract was partitioned three times with n-hexane in a separating funnel. The combined hexane extracts were concentrated under vacuum to near dryness. The residue was redissolved in n-hexane (2 ml), then filtered through a polypropylene membrane, 0.45 µm pore size, 13 mm diameter (Activon, Sydney, NSW). The filtrate was dried under N_2 and redissolved in n-hexane for ergosterol quantification.

Ergosterol was assayed by high pressure liquid chromatography (HPLC) using a Millipore Waters system fitted with a LiChrosorb SI

60 column (Gold Pak, Activon). The column was eluted with nhexane: isopropanol (97: 3, v/v) at 1 ml/min and ergosterol was detected by absorption at 280 nm about 8-10 min after injection of the sample. Ergosterol was quantified by reference to an ergosterol standard calibration curve, prepared from a standard solution (2 mg/ml, Sigma Chemicals, St. Louis, MO). For five of the fungal

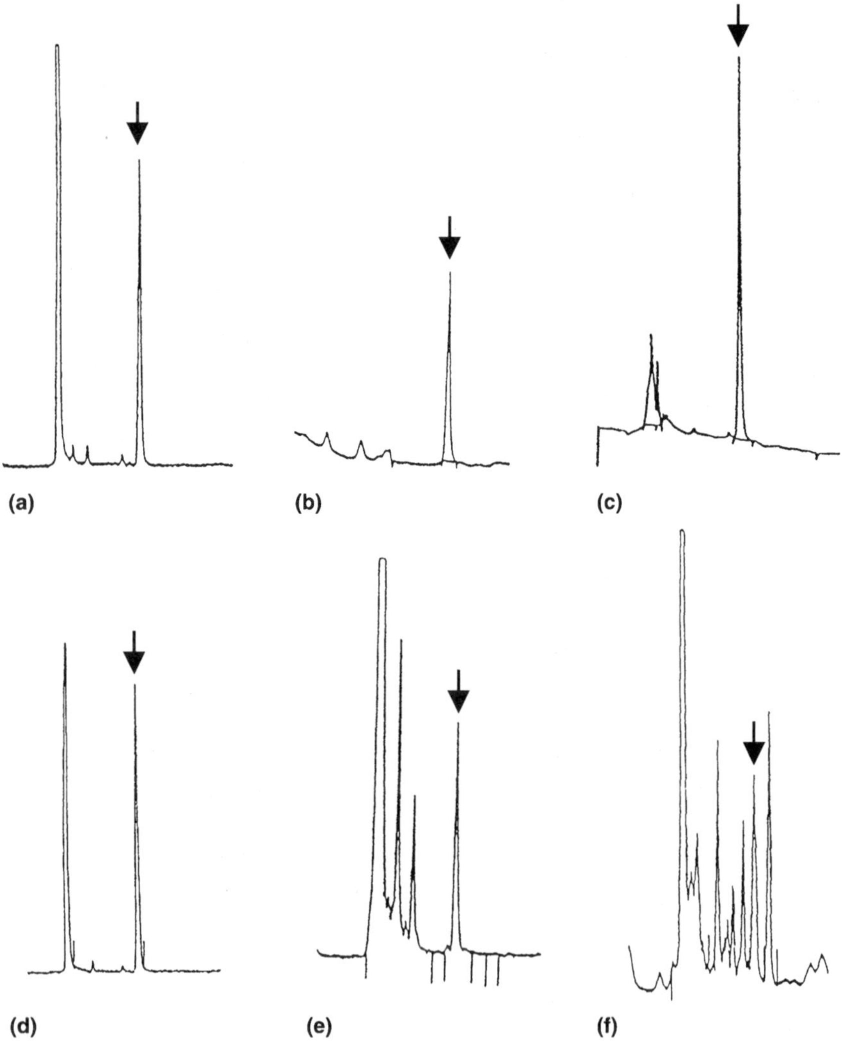

Figure 1. HPLC traces of ergosterol from six fungi: (a) *Penicillium commune*, (b) *Penicillium roqueforti*, (c) *Byssochlamys nivea*, (d) *Fusarium oxysporum*, (e) *Aspergillus flavus*, (f) *Eurotium chevalieri*. The ergosterol peak is indicated by an arrow.

species, well separated single peaks for ergosterol were obtained in HPLC traces of mycelial extracts (Figure 1). However, in extracts from *E. chevalieri* and *X. bisporus* the peak eluted close to interfering substances. In these cases a second filtration of the extract or addition of ergosterol standard was necessary to conclusively identify the ergosterol peaks.

3. RESULTS

3.1. Validation of ergosterol assay

A linear relationship was observed between peak heights measured on HPLC chromatograms and ergosterol concentration. The lower limit of detection was 0.01 μg of ergosterol. The coefficient of variation for ergosterol peaks detected by HPLC in extracts ranged between 1 and 14% provided that the ergosterol content was greater than 20 μg. Variation was greatest when the ergosterol content was less than 50 μg and least for high amounts (e.g. 500 μg) (Table 2). Ergosterol recoveries from spiked samples of *F. oxysporum* and *B. fulva* mycelium were 80-100% using the described method.

For most species, the ergosterol peak in the HPLC traces was clear and well separated from other peaks (Figure 1a-d). However, the ergosterol peaks for *Aspergillus flavus* (Figure 1e) and more especially for *Eurotium chevalieri* (Figure 1f) and *Xeromyces bisporus* (not shown) were close to peaks likely to be other sterols. Levels of ergosterol observed in these species were lower than expected.

Table 2. Reproducibility of ergosterol analysis (n=3) in colonies of *Fusarium oxysporum* and *Byssochlamys fulva* grown on Czapek yeast extract agar for various incubation periods

Species	Incubation time (d)	Ergosterol (μg)	Coefficient of variation (%)
Fusarium oxysporum	2	5.5, 7.5,11.1	34.9
	5	620.5, 654.8, 696.2	5.8
	6	884.2, 767.4, 835.0	7.0
Byssochlamys fulva	2	0.76, 0.42, 0.76	28.4
	4	23.5, 31.1, 27.7	13.9
	7	694.7, 682.0, 686.2	0.9

3.2. Growth of Fungi as Assessed by Colony Diameters

Most of the fungi grew on all of the media used, though with varying vigour, reflecting their water relations (Table 3). Best growth of most species occurred on CY20S, 0.98 a_w except for *B. fulva* and *F. oxysporum* which grew faster at 25°C on CYA, 0.997 a_w. Along with *M. plumbeus*, these species produced little or no growth on MY50G at 0.89 a_w, an a_w near their lower limit for growth (Pitt and Hocking, 1997). *A. flavus* and *P. commune* grew strongly at all a_w tested. Growth of *E. chevalieri,* a xerophilic species, was slow on CYA, and faster on CY20S and MY50G. *X. bisporus*, an extreme xerophile, grew only on MY50G.

3.3. Influence of Colony Age on Growth Parameters

The influence of colony age on some of the data obtained by the four methods used for measuring fungal growth is shown in Table 4, for four representative species. Ratios were calculated for colony diameter over hyphal length, mycelium dry weight and ergosterol content over hyphal length, and ergosterol over mycelium dry weight. The ratio of colony diameter over hyphal length showed a general downward trend, indicating a greater rate of hyphal extension than colony diameter increase as colonies aged. Colony diameters are therefore not a good measure of fungal biomass production in aging colonies. The other ratios were reasonably constant, indicating a general correspondence between mycelial dry weight, ergosterol content and hyphal length.

Table 3. Colony diameters of fungi grown on media of various water activities at 25°C

Species	Colony diameter (mm) at 7 days on		
	CYA 0.997 a_w	CY20S 0.98 a_w	MY50G 0.89 a_w
Aspergillus flavus	68	84	16
Byssochlamys fulva	78	58	0
Byssochlamys nivea	43	-[a]	-
Eurotium chevalieri	19	50	36
Fusarium oxysporum	90	78	6
Mucor plumbeus	63	85	4
Penicillium commune	32	44	18
Penicillium roqueforti	52	-	-
Xeromyces bisporus	0	0	12

[a]not tested

Table 4. Growth of four species of fungi on Czapek yeast extract agar as measured by four techniques, and ratios derived from those measurements[a]

Species	Time (d)	Hyphal length (m ×1000)	Colony diam (mm)	Ratio CD/HL	Mycelium dry wt (mg)	Ratio MDW/HL	Ergo-sterol (μg)	Ratio E/HL	Ratio E/MDW
Mucor	3	2.04	51	25	16.5	8.1	11.2	5.5	0.68
plumbeus	6	2.62	73	27.9	30.9	11.8	62.0	23.7	2.00
	16	4.34	83	19.2	36.0	8.3	81.2	18.7	2.26
Fusarium	4	10.45	44	4.2	40.4	3.9	128.6	12.3	3.18
oxysporum	6	32.38	70	2.1	133.8	4.1	444.3	13.7	3.32
	9	83.32	86	1.0	308.6	3.7	598.3	7.1	1.94
Byssochlamys	5	2.51	33	13.1	14.7	5.9	77.5	30.9	5.27
fulva	8	19.83	71	3.6	119.6	6.0	379.6	19.1	3.17
	9	30.75	86	2.8	185.4	6.0	977.7	31.8	5.27
Penicillium	4	3.23	29	9.0	19.6	6.1	43.6	13.5	2.22
roqueforti	7	8.36	57	6.8	103.4	12.3	168.6	20.2	1.63
	14	19.22	86	4.5	238.4	12.4	280.5	14.6	1.18

[a]Ratio CD/HL, ratio of colony diameter (mm) / hyphal length (m ×1000); ratio MDW/HL, ratio of mycelial dry weight (mg) / hyphal length (m ×1000), ratio E/HL, ergosterol (μg) / hyphal length (m ×1000).

3.4. Estimates of Fungal Growth by Mycelial Dry Weight and Ergosterol Compared with Hyphal Length

To provide a common reference point, colonies of diameter 83-86 mm, i.e. virtually full plate growth from a single inoculum point, were selected where possible. To take account of type of medium, two data sets were developed, for colonies on CYA and PDA. Species other than *Penicillium commune, E. chevalieri* and *X. bisporus* produced full plate colonies on both media, although after varied incubation periods. *P. commune* reached 86 mm on PDA after 21 days, but a maximum of only 39 mm on CYA, after 17 days. *E. chevalieri* colonies reached a maximum of only 46 mm diameter on PDA, after 16 days, and were smaller on CYA. For comparisons with other species on CYA, therefore, *E. chevalieri* colonies on CY20S were used. As expected, *X. bisporus* did not grow on either CYA or CY20S, so data from MY50G were used, where this species reached a maximum of 64 mm after 42 days incubation. The overall results from analyses of hyphal length, mycelium dry weight and ergosterol under these conditions are given in Table 5.

3.5. Hyphal Length

Hyphal lengths, estimated for colonies of similar diameters as set out in Table 5, varied widely between species. On CYA, hyphal length varied almost 20 fold between *F. oxysporum* (83,000 m) and *B. nivea* (4200 m). Results were more uniform on PDA, with less than 6 fold variation among the species. *F. oxysporum* produced only 23,000 m of hyphae on PDA. *B. nivea* produced the greatest hyphal length on PDA (28,000 m) but the lowest on CYA (4200 m). Despite profuse growth, *M. plumbeus* produced less than 6000 m of hyphae under any of the varied conditions used. Mycelial dry weights were also lower for *M. plumbeus* than for many of the other species studied.

3.6. Ergosterol Content

Ergosterol content per colony varied widely between genus, between medium and even within genus, reflecting differences in growth density and membrane composition. When grown on CYA, *B. fulva* produced the highest ergosterol content and *M. plumbeus* the

Table 5. Comparison of measurements of mature growth of various fungi on CYA[a]

Species	Colony diameter mm (incubation time, d)	Hyphal length (m ×1000)	Mycelium dry weight (mg)	Ratio MDW/HL	Ergosterol (µg)	Ratio E/HL	Ratio E/MDW
Medium: CYA							
Aspergillus flavus	86 (11)	6.42	297	46.3	298	46.4	1.00
Byssochlamys fulva	86 (9)	30.8	185	6.01	977	31.7	5.28
B. nivea	86 (17)	4.17	23.5	5.63	64.5	15.5	2.74
Fusarium oxysporum	86 (9)	83.3	309	3.71	598	7.17	1.93
Mucor plumbeus	83 (13)	4.34	36	8.29	81.2	18.7	2.25
Penicillium commune	39 (17)	7.67	160	20.9	440	57.4	2.75
P. roqueforti	86 (14)	19.2	238	12.4	281	14.6	1.18
Average				14.7		21.4	2.44
Medium: PDA							
Aspergillus flavus	86 (11)	4.78	186	38.9	84.3	17.6	0.45
Byssochlamys fulva	86 (7)	22.2	156	7.02	1463	65.9	9.37
B. nivea	86 (9)	28.4	180	6.33	183	6.4	1.02
Eurotium chevalieri	46 (16)	8.6	86	10.0	669	7.8	7.78
Fusarium oxysporum	86 (7)	23.3	171	7.33	1884	80.9	11.0
Mucor plumbeus	86 (4)	5.2	109	21.0	259	49.8	2.36
Penicillium commune	86 (2 1)	7.02	108	15.4	818	116.5	7.57
P. roqueforti	86 (11)	7.48	222	29.7	509	68	2.29
Average				17.0		51.6	5.23
Medium: CY20S							
Eurotium chevalieri	86 (14)	6.59	354	53.7	25.1	3.8	0.07
Medium: MY50G							
Xeromyces bisporus	64 (42)	4	28.9	7.22	42.7	10.7	1.47
Overall average				17.6		36.4	4.1

[a]Ratio CD/HL, ratio of colony diameter (mm)/ hyphal length (m ×1000); ratio MDW/HL, ratio of mycelial dry weight (mg)/ hyphal length (m ×1000), ratio E/HL, ergosterol (µg) /hyphal length (m × 1000).

lowest; about a 12 fold difference. On PDA, *F. oxysporum* produced the highest ergosterol content, with *A. flavus* the lowest, about a 22 fold difference.

For some species, medium composition greatly affected ergosterol content. This was most evident for *E. chevalieri,* which produced 25 times as much ergosterol on PDA (670 μg/colony) from colonies less than 50 mm in diameter, than from 86 mm diameter colonies on CY20S (25 μg), despite comparable hyphal lengths on the two media. In contrast, ergosterol production by *A. flavus* was much higher on CYA than on PDA, again with comparable hyphal lengths.

3.7. Mycelium Dry Weights

Mycelium dry weights of 83-86 mm diameter colonies varied between species. On CYA, *A. flavus* and *F. oxysporum* produced colonies with a high mycelium dry weight (approximately 300 mg/plate). On CY20S, *E. chevalieri* colonies were equally heavy. The weights of *M. plumbeus* colonies, however were only 12% of these values (Table 5). For the vigorously growing *Aspergillus, Penicillium* and *Fusarium* species, mycelium dry weights on PDA were lower than values obtained on CYA. However, mycelium weights for *M. plumbeus* and *B. nivea* were much higher on PDA than on CYA. In the case of *B. nivea*, this difference was more than 8 fold.

3.8. Relationship Between Hyphal Length and Mycelium Dry Weight

The relationship between hyphal length and mycelium dry weight over time of growth was reasonably constant within species (Table 4) except for very small colonies of *Penicillium roqueforti*, and varied only four fold between the four species shown in Table 4. When all nine species were compared, on more than one medium, much greater variability was seen. This appears to be due mostly to variation in hyphal length measurements, which is not so precise as the other techniques used here. Increases in hyphal length sometimes occurred with little increase in mycelial dry weight, e.g. for *M. plumbeus* when grown on CYA and *E. chevalieri* grown on PDA. These observations were reproducible (data not shown). As discussed by Schnürer (1993), vacuole formation and autolysis of cell contents may occur in aging cultures which would lead to a reduction in weight per unit length.

On the other hand, some species sporulated heavily in age, e.g. *P. roqueforti* grown on PDA, *A. flavus* on CYA, and *E. chevalieri* on CY20S. Here large increases in mycelial dry weight were accompanied by little or no hyphal growth. The *Penicillium, Aspergillus*, and *Eurotium* species showed ratios above 12 mg/1000 m, while for the *Mucor, Fusarium* and *Byssochlamys* species ratios were 8 mg/1000 m or below. On PDA, ratios ranged from 6.3 mg/1000 m (*B. nivea*) to 38.9 mg/1000 m (*A. flavus*).

3.9. Relationship Between Hyphal Length and Ergosterol Content

The ratios of ergosterol production (μg) to hyphal length (m ×1000) for colonies of various ages were found to be more variable within species than those for mycelial dry weight over hyphal length (Table 4). No pattern with age of cultures was apparent. On PDA, values varied from 6.4 (*B. nivea*) to 116.5 (*P. commune*), an 18 fold difference (Table 5). Low ergosterol production on PDA by *A. flavus* and on CY20S by *E. chevalieri* (see above) was reflected in very low ratios of ergosterol to hyphal length.

3.10. Relationship Between Ergosterol Content and Mycelial Dry Weight

Reasonable agreement was seen between the ratios of ergosterol (μg) to mycelium dry weight (mg) both for cultures of different age in each species and between species (Table 4). If the very low value for small colonies of *M. plumbeus* (0.68) is omitted, ratios varied between 1.18 and 5.27, less than 5 fold. When the effect of medium is considered with the full range of species (Table 5), ratios were again reasonably constant for colonies grown on CYA and PDA. With omission again of a very low figure (0.45 for *A. flavus* on PDA), values varied from 1.0 to 11.0. The average for all species was 4.1 (Table 5). Most species produced higher amounts of ergosterol and mycelium dry weight on PDA than on CYA except for *A. flavus*, which produced higher ergosterol and mycelium dry weight on CYA than on PDA.

However, the very low ratio (0.07) observed from growth of *E. chevalieri* on CY20S is anomalous. No doubt this is due to the very low level of ergosterol produced on this medium by this species. The HPLC spectrum showed several peaks eluting close to ergosterol, so other sterols may have been present.

4. DISCUSSION

This study attempted to validate ergosterol measurements as an index of fungal growth for important food spoilage fungi, chosen because of the great differences in their growth patterns, using hyphal length and mycelium dry weight as standards.

Colony diameter, hyphal length, mycelium dry weight and ergosterol content all gave useful information about the growth of the species examined. However, each technique exhibited advantages and limitations. Colony diameter is a sensitive technique, in that a colony as small as 2 mm is easily measured. It was not possible to determine accurately dry weight, hyphal length or ergosterol concentration on such small amounts of material. However, colony diameter did not show a consistent correlation with the other parameters, especially as colony diameters became larger and more mature. In particular, sporulation caused a reduced correlation between colony diameters and the other parameters. If colony diameters are measured from relatively early growth, e.g. using the Petrislide technique of Pitt and Hocking (1977), then correlations could be higher.

Values for hyphal length obtained from single colonies on standard Petri dishes by Schnürer (1993) ranged from 10,000 m for *Rhizopus stolonifer* to 54,000 m for *Fusarium culmorum*. In this study, comparable fungi produced more hyphae: 43,000 m for *Mucor plumbeus* and 83,000 m for *Fusarium oxysporum*. These differences may be due to the lower nutritional value of the medium used by Schnürer (1993).

In this study, the average ratio of mycelium dry weight over hyphal length was 17.6 mg dry weight/1000 m of mycelium. Schnürer (1993) found values of 4.2 to 6.7 mg/1000 m, calculated from the mycelium and hyphal volume, respectively. Values obtained by us for *Fusarium oxysporum*, *Byssochlamys* spp. grown on CYA and PDA, and *Mucor plumbeus* when grown on CYA were comparable with those of Schnürer (1993). Much larger differences were seen with the rapidly growing and/or highly sporulating *Aspergillus* and *Penicillium* species studied here. Given the differences in fungi studied and media used, the data of Schnürer (1993) and our data are comparable. Working with aquatic fungi, ratios of ergosterol to mycelium dry weight of 2.3 to 11.5 were given by Gessner and Chauvet (1993), similar to those of Schnürer (1993).

Reports have suggested that ergosterol content increases as colonies age (Nout et al., 1987; Torres et al., 1992). However, as discussed by Schnürer (1993), vacuole formation and autolysis of cell contents may occur in aging cultures which would lead to a reduction in weight per unit of length and, consequently, to an increased ergosterol to dry weight ratio. Differences in degree and type of sporulation by the species studied here also clearly play a major part in the variations in the ratios of mycelial dry weight to hyphal length and ergosterol content to hyphal length observed here. For example, *Aspergillus flavus* and the *Penicillium* species produce relatively little vegetative mycelium relative to conidiophores and conidia, increasing mycelial dry weight and probably decreasing ergosterol in relation to hyphal length.

Mycelium dry weight is often considered to be a basic measure of fungal growth but fundamental questions remain unanswered. In this study, mycelium was separated from agar medium using a heat treatment. Cochrane (1958) criticised the separation of agar medium from fungal biomass using hot water because the water may extract soluble fungal components, resulting in a loss of dry weight. However, it is also important to note that dry weights measured without prior extraction are greatly affected by the variation in internal solutes caused by different medium formulations, especially in media of reduced a_w (Hocking and Norton, 1983).

An alternative approach is to scrape or peel the fungal growth from the medium surface. This may lead to incomplete removal and underestimation of dry weight. The technique of Hocking (1986) where fungi were grown on dialysis membrane on the surface of agar media enables ready separation of fungus from medium, and is a notable improvement. However, in the current study, where some species sporulated profusely, a wet extraction technique was considered preferable for safety reasons.

In this study, extracted mycelium dry weight showed a reasonably good correlation with hyphal length (Tables 4, 5), indicating the value of this parameter as a measure of fungal growth in media. However, the measurement of mycelium dry weight is not readily applicable to the estimation of growth of fungi in foods.

Schnürer (1993) noted differences in growth patterns between different fungal species. For a nonsporulating *Fusarium culmorum*, good agreement was found between hyphal length, colony counts and ergosterol content. For *Penicillium rugulosum* and *Rhizopus stolonifer*, changes in ergosterol level were related more closely to changes in hyphal length rather than to production of spores or colony counts. In the present study, hyphal length correlated rather poorly with ergosterol content (Table 5). Schnürer and Jonsson (1992) found reasonable correlation between ergosterol and mould viable counts in Swedish grains under certain conditions, but Saxena et al. (2001) found that viable counts and ergosterol did not correlate well for pure cultures of *Aspergillus ochraceus* and *Penicillium verrucosum*.

The accuracy of the hyphal length technique is affected by factors including variation in hyphal width for different species, degree of sporulation, formation of reproductive structures (e.g. cleistothecia), fragmentation during homogenization, and clumping of hyphae. The intersection technique of Olsen (1950) is probably only statistically sound when large numbers of microscopic fields are counted. Despite the laboriousness of the procedure used here for hyphal length estimations (ten fields from two colonies), this rigour is lacking. All of these factors lead to variation in estimates of hyphae length.

Ergosterol content was a sensitive indication of fungal biomass. As little as 0.01 µg of ergosterol could be detected from mycelium in a colony of 4 mm diameter. However, the amount of ergosterol found in the fungi varied with the growth medium, species and culture incubation time. This variation was reflected in the data shown in Tables 4 and 5. The ratio of ergosterol over mycelium dry weight ranged from 0.07 µg/mg for *E. chevalieri* on CY20S to 11.0 µg/mg for *F. oxysporum* on PDA, a 150 fold variation. Even when these two extreme values are omitted, variation of about 20 fold remained (i.e. 0.45 µg/mg for *A. flavus* to 9.37 µg/mg for *B. fulva*, both on PDA).

Weete (1974) noted that in general sterol levels in fungi varied with medium composition and culture conditions. Four to 10 fold variations have been reported in the ergosterol content of the same fungus under different growth conditions (Newell et al., 1987; Nout et al., 1987). Increased nutritional complexity of the medium, the presence of free fatty acid precursors of the ergosterol biosynthetic pathway and increased availability of oxygen all gave mycelium with increased ergosterol contents.

The measurement of ergosterol alone does not give the absolute amount of fungus present. For this, it is necessary to convert ergosterol values into biomass in terms of mycelium dry weight. After studying 14 aquatic hyphomycetes, Gessner and Chauvet (1993) gave the range of ratios of ergosterol to mycelium dry weight as 2.3 -11.5 µg/g, figures similar to those derived in this work (0.45 -11.0 µg/g). This ratio varies between fungal species and with growth condition, limiting the direct use of ergosterol as a means of calculating mycelium dry weight.

A further explanation for variation in the ergosterol content of fungi is that sterols other than ergosterol can be produced by some species. For example, ergosterol and 22-dihydroergosterol have been reported as the predominant sterol in *A. flavus* (Vacheron and Michel,

1968; Weete, 1973). Other sterols identified as products of deuteromycetous fungi include cerevisterol, ergosterol peroxide, lanosterol, 24-methylenelophenol and 14-dehydroergosterol (Weete, 1973). In this study, *A. flavus* produced a low level of ergosterol despite its high production of biomass, and showed extra peaks in its extract. Similarly, colonies of *E. chevalieri* were found to contain only low amounts of ergosterol, despite high amounts of biomass. Additional peaks seen in the HPLC profile of *E. chevalieri* extracts also indicate that it may produce sterols other than ergosterol. The ergosterol content of *X. bisporus* mycelium was also low, and several additional peaks were observed in the HPLC trace from its extract. Further studies are needed to determine if the additional peaks found in the HPLC profiles are sterols.

The low level of ergosterol produced by *E. chevalieri* has important consequences. *Eurotium* species are very common in stored grains, in which ergosterol has been used to estimate fungal growth. The overall average ratio of ergosterol to mycelium dry weight was 4.1, about 60 times higher than that obtained for *E. chevalieri* on CY20S. Hypothetically, a sample of grain infected by *E. chevalieri*, estimated to contain a particular ergosterol content, could contain 150 times as much fungal biomass as one infected by *Fusarium oxysporum*, which gave an ergosterol to mycelium dry weight ratio of 11 μg/mg in this study.

Estimation of ergosterol content, colony diameter, mycelium dry weight and hyphal length were shown to be good indices for measuring fungal growth, but it is important to keep in mind the limitations of each techniques. The most reliable information about fungal growth will be obtained by using two or more techniques for quantification.

5. ACKNOWLEDGMENTS

The authors wish to thank to Mr N. Tobin and Ms S. L. Leong of Food Science Australia, North Ryde, for helpful advice on chemical analyses and hyphal length measurement, respectively, and to Fundação de Amparo à Pesquisa do Estado de São Paulo (FAPESP) for funding the PhD program for M.H.T.

6. REFERENCES

Bermingham, S., Maltby, L., and Cooke, R. C., 1995, A critical assessment of the validity of ergosterol as an indicator of fungal biomass, *Mycol. Res.* **99**:479-484.

Brancato, F. P., and Golding, N. S., 1953, The diameter of the mold colony as a reliable indicator of growth, *Mycologia* **45**:848-864.

Cochrane, V. W., 1958, *Physiology of Fungi*, John Wiley, New York.

Deploey, J. J., and Fergus, C. L., 1975, Growth and sporulation of thermophilic fungi and actinomycetes in O_2-N_2 atmospheres, *Mycologia* **67**:780-797.

de Ruiter, G. A., Notermans, S. H. W., and Rombouts, F. M., 1993, New methods in food mycology, *Trends Food Sci. Technol.* **4**:91-97.

Gessner, M. O. and Chauvet, E., 1993, Ergosterol-to-biomass conversion factors for aquatic Hyphomycetes, *Appl. Environ. Microbiol.* **59**:502-507.

Gibson, A. M., Baranyi, J., Pitt, J. I., Eyles, M. J., and Roberts, T. A., 1994, Predicting fungal growth: the effect of water activity on *Aspergillus flavus* and related species, *Int. J. Food Microbiol.* **23**:419-431.

Hartog, B. J. and Notermans, S., 1988, The detection and quantification of fungi in food, in: *Introduction to Food-borne Fungi*. 4th edition, R. A. Samson and E. S. van Reenen-Hoestra, eds, Centraalbureau voor Schimmelcultures, Baarn, Netherlands, pp. 222-230.

Hocking, A. D., 1986, *Some Physiological Responses of Fungi Growing at Reduced Water Activities,* PhD thesis, University of New South Wales, Kensington, NSW.

Hocking, A. D., and Norton, R. S., 1983, Natural abundance ^{13}C nuclear magnetic resonance studies on the internal solutes of xerophilic fungi, *J. Gen. Microbiol.* **129**:2915-2925.

Matcham, S. E., Jordan, B. R., and Wood, D. A., 1984, Methods for assessment of fungal growth on solid substrates, in: *Microbiological Methods for Environmental Biotechnology*. J. M. Grainger and J. M. Lynch, eds, Academic Press, London, pp. 5-18.

Miller, J. D., and Young, J. C., 1997, The use of ergosterol to measure exposure to fungal propagules in indoor air, *Am. Ind. Hyg. Assoc. J.* **58**:39-43.

Newell, S. Y., 1992, Estimating fungal biomass and productivity in decomposing litter, in: *The Fungal Community*, G. C. Carroll and D. T. Wicklow, eds, Marcel Dekker, Inc., New York, pp. 521-561.

Newell, S. Y., Miller, J. D., and Fallon, R. D., 1987, Ergosterol content of salt-marsh fungi: effect of growth conditions and mycelial age, *Mycologia* **79**:688-695.

Nout, M. J. R., Bonants-van Laarhoven, T. M. G., de Jongh, P., and Koster, P. G., 1987, Ergosterol content of *Rhizopus oligosporus* NRRL 5905 grown in liquid and solid substrates, *Appl. Microbiol. Biotechnol.* **26**:456-461.

Nylund, J. E., and Wallander, H., 1992, Ergosterol analysis as a means of quantifying mycorrhizal biomass, in: *Methods in Microbiology, Vol 24, Techniques for the Study of Mycorrhiza*. J. R. Norris, D. J. Read, and A. K. Varma, eds, Academic Press, London, pp.77-88.

Olson, F. C. W., 1950, Quantitative estimates of filamentous algae, *Trans. Am. Microsc. Soc.* **69**:272-279.

Paster, N., Lisker, N., and Chet, I., 1983, Ochratoxin A production by *Aspergillus ochraceus* Wilhelm grown under controlled atmospheres, *Appl. Environ. Microbiol.* **45**:1136-1139.

Pitt, J. I., 1984, The significance of potentially toxigenic fungi in foods, *Food Technol. Aust.* **36**:218-219.

Pitt, J. I., and Hocking, A. D., 1977, Influence of solute and hydrogen ion concentration on the water relations of some xerophilic fungi, *J. Gen. Microbiol.* **101**:35-40.

Pitt, J. I., and Hocking, A. D., 1997, *Fungi and Food Spoilage,* 2nd edition, Blackie Academic and Professional, London.

Pitt, J. I., and Miscamble, B. F., 1995, Water relations of *Aspergillus flavus* and closely related species, *J. Food Prot.* **58**:86-90.

Ride, J. P., and Drysdale, R. B., 1972, A rapid method for the chemical estimation of filamentous fungi in plant tissue, *Physiol. Plant Pathol.* **2**:7-15.

Samson, R. A., Hocking, A. D., Pitt J. I., and King, A. D., 1992, *Modern Methods in Food Mycology,* Elsevier Publishers, Amsterdam.

Saxena, J., Munimbazi, C., and Bullerman, L. B., 2001, Relationship of mould count, ergosterol and ochratoxin A production, *Int. J. Food Microbiol.* **71**: 29-34.

Schnürer, J., 1993, Comparison of methods for estimating the biomass of three foodborne fungi with different growth patterns, *Appl. Environ. Microbiol.* **59**:552-555.

Schnürer, J., and Jonsson, A., 1992, Ergosterol levels and mould colony forming units in Swedish grains of food and feed grade, *Acta Agric. Scand. Sect B.* **42**:240-245.

Schwardorf, K., and Muller, H. M., 1989, Determination of ergosterol in cereals, mixed feed components, and mixed feeds by liquid chromatography, *J. Assoc. Off. Anal. Chem.* **72**:457-462.

Seitz, L. M., Mohr, H. E., Burroughs, R., and Sauer D. B., 1977, Ergosterol as an indicator of fungal invasion in grain, *Cereal Chem.* **54**:1207-1217.

Seitz, L. M., Sauer, D. B., Burroughs, R., Mohr, H. E., and Hubbard J. D., 1979, Ergosterol as a measure of fungal growth, *Phytopathology* **69**:1202-1203.

Torres, M., Viladrich, R., Sanchis, V., and Canela, R., 1992, Influence of age on ergosterol content in mycelium of *Aspergillus ochraceus*, *Lett. Appl. Microbiol.* **15**:20-22.

Vacheron, M. J., and Michel, G., 1968, Composition en sterols et en acides gras de deux souches d'*Aspergillus flavus*, *Phytochemistry* **7**:1645-1651.

Weete, J. D., 1973, Sterols of fungi: distribution and biosynthesis, *Phytochemistry* **12**:1843-1864.

Weete, J. D., 1974, Distribution of sterols in the fungi. 1. Fungal spores, *Lipids* **9**: 578-581.

Wells, J. M., and Uota, M., 1970, Germination and growth of five fungi in low-oxygen and high-carbon dioxide atmospheres, *Phytopathologia* **60**:50-53.

Williams, A. P., 1989, Methodological developments in food mycology, *J. Appl. Bacteriol.* **67**: Symp. Suppl. 61S-67S.

Zill, G., Engelhardt, G., and Wallnofer, P. R., 1988. Determination of ergosterol as a measure of fungal growth using Si 60 HPLC, *Z. Lebensm. Unters. Forsch.* **187**: 246-249.

EVALUATION OF MOLECULAR METHODS FOR THE ANALYSIS OF YEASTS IN FOODS AND BEVERAGES

Ai Lin Beh, Graham H. Fleet, C. Prakitchaiwattana and Gillian M. Heard[*]

1. INTRODUCTION

The analysis of yeasts in foods and beverages involves the sequential operations of isolation, enumeration, taxonomic identification to genus and species, and strain differentiation. Although well established cultural methods are available to perform these operations, many molecular methods have now been developed as alternatives. These newer methods offer various advantages, including faster results, increased specificity of analysis, decreased workload, computer processing of data and possibilities for automation. Molecular methods for yeast analysis are now at a stage of development where they can move from the research laboratory into the quality assurance laboratories of the food and beverage industries. However, many practical questions need to be considered for this transition to progress. A diversity of molecular methods with similar analytical objectives are available. Which methods should the food analyst choose and what principles should be used to guide this choice? Food analysts are required to make judgements and decisions about the microbiological quality and safety of consignments of products often worth many millions of dollars in national and international trade. Moreover,

* Food Science and Technology, School of Chemical Engineering and Industrial Chemistry, University of New South Wales, Sydney, New South Wales, Australia, 2052. Correspondence to: g.fleet@unsw.edu.au

these decisions need to conform to the requirements of supplier and customer contracts, and government legislation. Consequently, they have the potential to encounter intense legal scrutiny (Fleet 2001). For these reasons, the food analyst requires basic information about method standardisation, accuracy, reproducibility, precision, specificity and detection sensitivity (Cox and Fleet, 2003). While these criteria have guided the selection and choice of currently accepted cultural methods, they have not been critically applied to the newer molecular techniques.

This Chapter has the following goals: (1) to provide an overview of the diversity of molecular methods that are finding routine application to the analysis of yeasts in foods and beverages, (ii) to outline the variables that affect the performance and reliability of these methods, and (iii) to suggest strategies for the international standardisation and validation of these methods. Molecular methods have found most application to the identification of yeast species and to strain differentiation, but there is increasing use of culture-independent methods to detect and monitor yeasts in food and beverage ecosystems.

2. MOLECULAR METHODS FOR YEAST IDENTIFICATION

The traditional, standard approach to yeast identification has been based on cultural, phenotypic analyses. The yeast isolate is examined for a vast range of morphological, biochemical and physiological properties which are systematically compared with standard descriptions to give a genus and species identity. Generally, it is necessary to conduct approximately 100 individual tests to obtain a reasonably reliable identification (Kurtzman and Fell, 1998; Barnett et al., 2000). Consequently, the entire process is very labour-intensive, lengthy and costly. Although various technical and diagnostic innovations have been developed to facilitate this process, they are not universal in their application and the data generated are not always equivalent (Deak and Beuchat, 1996; Deak 2003; Robert, 2003; Kurtzman et al., 2003).

Molecular methods based on DNA analysis are now being used to quickly identify yeasts to genus and species level. The workload is minimal and, usually, reliable data can be obtained within 1-2 days. Several approaches are being used. The most definitive and universal assay determines the sequence of bases in segments of the ribosomal DNA. Other approaches are based on determination of restriction fragment length polymorphisms (RFLP) of segments of ribosomal DNA, hybridisation with specific nucleic acid probes, and polymerase chain reaction (PCR) assays with species-specific primers. Aspects of these methods and their application to food and beverage yeasts have been reviewed by Loureiro and Querol (1999), Giudici and Pulvirenti (2002), Loureiro and Malfeito-Ferreira (2003), and van der Vossen et al. (2003).

2.1. Sequencing of Ribosomal DNA

The discovery that ribosomal RNA is highly conserved throughout nature but has certain segments which are species variable, has lead to the widespread use of ribosomal DNA sequencing in developing microbial phylogeny and taxonomy. As a consequence, ribosomal DNA sequences are known for most microorganisms, including yeasts, and are now routinely used for diagnostic and identification objectives (Valente et al., 1999). The ribosomal DNA repeat unit found in yeasts is schematically shown in Figure 1. It consists of conserved and variable regions which are arranged in tandem repeats of several hundred copies per genome. The conserved sequences are found in genes encoding for small (18S), 5.8S, 5S, and large (25-28S) subunits of ribosomal RNA. Within each cluster, variable spacer regions occur between the subunits, called internal transcribed spacer (ITS) regions, and between gene clusters, called the intergenic spacer regions (IGS) or the non-transcribed spacer region (NTS). All of these regions have some potential for differentiating yeast genera and species, but most focus has been on the 18S, 26S and ITS regions (Valente et al., 1999; Kurtzman, 2003).

The D1/D2 domain of the large subunit (26S) ribosomal DNA consists of about 600 nucleotides and has been sequenced for virtually all known yeast species. Databases of these sequences can be accessed through GenBank (http://www.ncbi.nlm.nih.gov/), DataBank of Japan (http://www.ddbj.nig.ac.jp/) or the European Molecular Biology Laboratory (*http://www.ebi.ac.uk/embl/*). There is sufficient variation in these sequences to allow differentiation of most ascomycetous (Kurtzman and Robnett, 1998, 2003) and basidiomycetous (Fell et al.

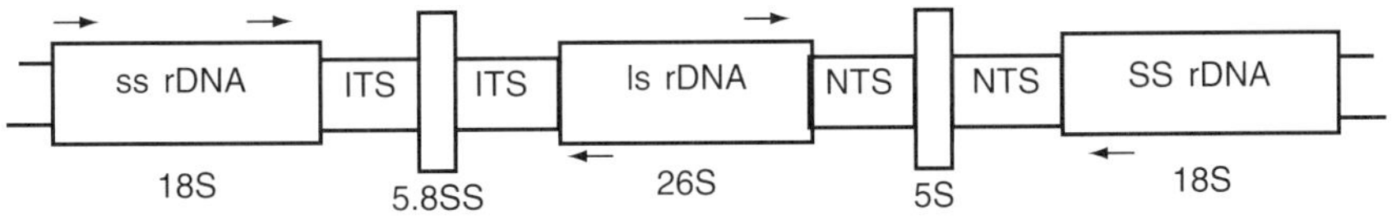

Figure 1. The ribosomal DNA repeat unit

2000; Scorzetti et al., 2002) yeast species. Sequencing of the D1/D2 domain of the 26S ribosomal DNA is now widely used for the routine identification of yeasts and the construction of phylogenetic taxonomy. Sequence comparisons have also been done for the small subunit, 18S ribosomal DNA but, so far, the databases are not extensive and sequence differences may not be sufficient to allow the discrimination of closely related species (James et al., 1997; Naumov et al., 2000; Daniel and Meyer, 2003). The ribosomal spacer regions (ITS) show higher rates of sequence divergence than the D1/D2 domain of the 26S subunit and have proven useful for species differentiation (James et al., 1996; Naumov et al., 2000; Cadez et al., 2003). For example, the *Hanseniaspora uvaurm-guillermondii* cluster is poorly resolved, and species of *Saccharomyces pastorianus/Saccharomyces bayanus*, and *Cryptococcus magnus/Filobasidium floriforme/Filobasidium elegans* are indistinguishable using D1/D2 sequences. Sequencing of the ITS region can provide the required level of differentiation.

Sequencing of mitochondrial and protein encoding genes are also being used to determine phylogenetic relationships among yeasts. These genes include the translation elongation factor 1α, actin-1, RNA polymerase II, pyruvate decarboxylase, beta tubulin gene, small subunit rDNA and cytochrome oxidase II (Daniel et al., 2001; Kurtzman and Robnett 2003; Daniel and Meyer 2003).

The basic protocol for sequencing ribosomal DNA segments is: (i) prepare a pure culture of the yeast isolate, (ii) extract and purify the DNA, (iii) perform PCR amplification of the region to be sequenced, (iv) verify the amplified product by gel eletrophoresis, and (v) sequence the product using internal or external primers. Procedures for conducting these operations are well established but are not standardised, and may vary from one laboratory to another. Primer sequences used to amplify the different segments of the rDNA have been tabulated in White et al. (1990), Valente et al. (1999), Sipiczki (2002) and Kurtzman and Robnett (2003). Table 1 lists some key publications on the identification of yeasts by ribosomal DNA sequencing. Some yeasts are not reliably identified by sequencing single gene segments and it is suggested that sequences be obtained for multiple genes or gene segments for more reliable data (Kurtzman, 2003).

2.2. Restriction Fragment Length Polymorphism (RFLP)

RFLP analysis of the ribosomal DNA segments is emerging as one of the most useful methods for rapidly identifying food and beverage

Table 1. Application of gene sequencing technology to the identification of species of food and beverage yeasts

Region	Application	References
18S	Phylogenetic relationships; *Zygosaccharomyces* and *Torulaspora* species	James et al. (1994)
	Brettanomyces, *Dekkera*, *Debaryomyces*, *Kluyveromyces* species	Cai et al. (1996)
	Candida, *Pichia*, *Citeromyces* species	Suzuki and Nakase (1999)
	Saccharomyces genus; new species *S. kunashirensis*, *S. martiniae*	James et al. (1997)
	Saccharomyces sensu lato group; new species *S. naganishii*, *S. humaticus*, *S. yukushimaensis*	Mikata et al. (2001)
18S, ITS	Phylogenetic relationships of *Saccharomyces sensu stricto* complex; new species *S. cariocanus*, *S. kudriavzevii*, *S. mikatae*	Naumov et al. (2000)
18S; 834-1415	Identification of yeast from dairy products	Cappa and Cocconcelli (2001)
D1/D2 of 26S	Systematics of ascomycetous yeasts	Kurtzman and Robnett (1998)
	Systematics of basdidiomycetous yeasts	Fell et al. (2000)
D1/D2 of 26S, ITS	Systematics of basdidiomycetous yeasts	Scorzetti et al. (2002)
D1/D2 of 26S	*Candida davenportii* sp. nov. from a wasp in a soft-drink production facility	Stratford et al. (2002)
D1/D2 of 26S, ITS	*Tetrapisispora fleetii* sp. nov. from a food processing plant	Kurtzman et al. (2004)
D1/D2 of 26S	Identification of yeast species; in Sicilian sourdough	Pulvirenti et al. (2001)
	in orange juice	Arias et al. (2002)
	in spontaneous wine fermentation	van Keulen et al. (2003)
	from bark of cork oak	Villa-Carvajal et al. (2004)
	from Malbec grape berries	Combina et al. (2005)
	from fermentation of West African cocoa beans	Jespersen et al. (2005)
	contaminant in carbonated orange juice production chain	Pina et al. (2005)
ITS	Phylogenetic relationships of *Kluyveromyces marxianus* group	Belloch et al. (2002)
ITS	Phylogenetic relationships of *Zygosaccharomyces* and *Torulaspora* species	James et al. (1996)

Table 1. Application of gene sequencing technology to the identification of species of food and beverage yeasts—*cont'd*

Region	Application	References
ITS	Phylogenetic analysis of the *Saccharomyces* species	Oda et al. (1997)
	Phylogenetic analysis of the *Saccharomyces sensu stricto* complex	Montrocher et al. (1998)
ITS	Identification of yeast species;	
	in orange fruit and orange juice	Heras-Vazques et al. (2002)
	from Italian sourdough baked products	Foschino et al. 2004
ITS1	Separation of *S. cerevisiae* strains in African sorghum beer	Naumova et al. 2003
IGS	Separation of *Clavispora opuntiae* varieties	Lachance et al. 2000
ITS, IGS	Intraspecies diversity of *Mrakia* and *Phaffia* species	Diaz and Fell 2000
Actin	Phylogenetic relationships of anamorphic *Candida* and related teleomorphic genera	Daniel et al. 2001
mt COX II	Phylogeny of the genus *Kluyveromyces*	Belloch et al. 2000
Multigenes	Ascomycete phylogeny Ascomycete species separation	Kurtzman and Robnett (2003), Kurtzman (2003), Daniel and Meyer (2003)
Multigenes	Taxonomic position of the biotherapeutic agent *Saccharomyces boulardii*	van der Aa Kühle and Jespersen (2003)

yeasts. The preferred region for analysis represents the ITS1-5.8S-ITS2 segment (Figure 1). Using appropriate primers, this segment is specifically amplified by PCR. The PCR product is then cleaved with specific restriction endonucleases, and the resulting fragments are separated by gel electrophoresis. The size (number of base pairs) of the ITS amplicon itself can be useful in discriminating between yeast species. The number (usually 1-4) and size (base pairs) of the fragments as determined by banding patterns on the gel are the main principles used to discriminate between yeast species. Generally, more than one restriction enzyme needs to be used in order to obtain unequivocal discrimination. Some restriction enzymes commonly used are: *Cfo* I, *Hae* III, *Hinf* I, *Hpa* II, *Scr* FI, *Taq* I, *Nde* II, *Dde* I, *Dra* I and *Mbo* II. Several hundred species of food and beverage yeasts have now been examined by this method and databases of fragment profiles for the different restriction enzymes and yeast species have been established

(Esteve-Zarzoso et al., 1999; Granchi et al., 1999; Arias et al., 2002; Heras-Vazquez et al., 2003; Dias et al., 2003; Naumova et al., 2003).

PCR-RFLP analyses have several advantages that are attractive to quality assurance analysis in the food and beverage industries. Once a pure yeast culture has been obtained, identification to species level can be done in several hours. Essentially, DNA is extracted from the yeast biomass, amplified by specific PCR, amplicons are digested with the restriction nucleases and the products separated by gel electrophoresis. The work load and equipment needs are minimal and data are generally reproducible. The expense and time for sequencing are avoided.

Although PCR-RFLP analysis of the ITS1-5.8S-ITS2 region has attracted most study to date, there is increasing interest in the PCR-RFLP analysis of other ribosomal segments. These include the 18S-ITS region, 18S-ITS-5.8S region, the 26S and NTS regions. It is not evident at this stage whether targeting these regions offers any advantage over the ITS1-5.8S-ITS2 region and further studies evaluating the different approaches are required. Table 2 lists some key reports on the application of the PCR-RFLP analysis of ribosomal DNA regions to food and beverage yeasts.

2.3. Nucleic acid probes and species-specific primers

Nucleic acid probes are short, single-stranded nucleotides (usually 20-100 bases) that are designed to complement a specific sequence in the DNA/RNA of the target organism. They are usually labelled with a marker molecule to enable their detection. Probes are used in hybridization reactions, and are applied in a number of formats (Hill and Jinneman, 2000; Cox and Fleet, 2003).

In whole cell hybridization protocols (FISH, CISH), yeasts are directly visualised *in situ*, and identified with fluorescently-labelled, specific probes that bind to rRNA, located in the ribosomes. In ecological studies, this technique is particularly useful for identifying morphological types, for quantifying target species and monitoring microbial community structure and dynamics, for example, in examining the spatial relationships on surfaces of leaves and in biofilms (Table 3).

Probes are also used in dot blot, slot blot and colony blot hybridisation assays of yeast biomass on membranes. Detection is achieved by a labeled DNA probe that hybridises to the DNA/RNA of the immobilised sample (Hill and Jinneman, 2000). Another strategy is to coat the probe onto a solid substrate such as the wells of a microtitre tray, and use a modified ELISA format to detect the target

Table 2. Application of PCR-restriction fragment length polymorphism (RFLP) to the identification of food and beverage yeasts

Method: region; primers; restriction enzymes	Applications	References
ITS2; (primers ITS3 and ITS4); *Ase*I, *Ban*I, *Eco*RI, *Hinc*II, *Sty*I	Medically-important yeast species	Chen et al. (2000)
ITS1-5.8S rRNA-ITS2; (primers ITS1 and ITS4); *Cfo*I, *Hae*III, *Hinf*I	Yeast species from wine fermentations	Guillamón et al. (1998), Ganga and Martinez. (2004), Combina et al. (2005)
	Yeast species from Irish cider fermentations	Morrissey et al. (2004)
	Yeast species associated with orange juice	Arias et al. (2002)
ITS1-5.8S rRNA-ITS2; (primers ITS1 and ITS4); *Cfo*I, *Hae*III, *Hinf*I, *Dde*I	132 species from food and beverages	Esteve-Zarzoso et al. (1999)
	Yeast species from wine fermentations	Granchi et al. (1999), Rodríguez et al. (2004), Clemente-Jimenez et al. (2004)
	Yeast species during fermentation and ageing of sherry wines	Esteve-Zarzoso et al. (2001)
	Yeast species from orange fruit and orange juice	Heraz-Vazquez et al. (2003)
1) ITS1-5.8S rRNA-ITS2; (primers ITS1 and ITS4); *Hinf*I, *Rsa*I, *Nde*II, *Hae*III 2) NTS 2; (primers r-1234 and r-2156); *Alu*I, *Ban*I	Yeast species from Sicilian sourdoughs	Pulvirenti et al. (2001)
ITS1-5.8S rRNA-ITS2; (primers ITS1 and ITS4); *Hae*III, *Hpa*II, *Scr*FI, *Taq*I	*Saccharomyces sensu stricto* strains from African sorghum beer	Naumova et al. (2003)
ITS1-5.8S rRNA-ITS2; (primers ITS1 and ITS4); *Hae*III, *Mae*I	Differentiation of *S. bayanus*, *S. cerevisiae*, *S. paradoxus* isolates from botrytised grape must	Antunovics et al. (2005)
ITS1-5.8S rRNA-ITS2; (primers ITS1 and ITS4); *Hinf*I, *Dde*I, *Mbo*II	Species within the genera of *Hanseniaspora* and *Kloeckera*	Cadez et al. (2003)

1) ITS; (primers ITS1 and NL2); *Mse*I, *Taq*I	Discrimination/diversity of *S. cerevisiae* strains	Baleiras Couto et al. (1996a)
2) NTS; (primers JV51ET and JV52ET); *Mse*I, *Taq*I		
18S rRNA-ITS1; (primers NS1 and ITS2); *Hae*III, *Msp*I, *Alu*I, *Rsa*I	128 species from food, wine, beer and soft drinks	Dlauchy et al. (1999)
18S rRNA-ITS1; (primers NS1 and ITS2); *Hae*III, *Msp*I	Yeast species from Hungarian dairy products	Vasdinyei and Deak (2003)
1) 18S rRNA; (primers p108 and M3989); *Hae*III, *Msp*I	Discrimination of *C. stellata*, *M. pulcherrima*, *K. apiculata* and *S. pombe*	Capece et al. (2003)
2) NTS; (primers NTSF and NTSR); *Hae*III, *Msp*I		
18SrDNA and ITS1; (primers NS1 and ITS2); *Hae*III, *Msp*I	Differentiation of *S. cerevisiae* and *S. paradoxus* isolates from Croatian vineyards	Redzepovic et al. (2002)
18SrDNA and ITS1; (primers NS1 and ITS2); *Cfo*I, *Hae*III, *Hinf*I, *Msp*I	Separation of *Saccharomyces sensu stricto* and *Torulaspora* species	Smole Mozina et al. (1997)
ITS1-5.8S rRNA-ITS2-part 18SrRNA ; (primers NS3 and ITS4); *Scr*FI, *Hae*III, *Msp*I	Differentiation of brewery yeasts; *S. carlsbergensis*, *S. pastorianus, S. bayanus*, *S. cerevisiae*, *S. brasiliensis*, *S. exiguus*	Barszczewski and Robak (2004)
3′ ETS and IGS; (primers 5S2 and ETS2); *Msp*I, *Scr*FI	Discrimination of *S. cerevisiae, S. carlsbergensis* and *S. pastorianus*	Molina et al. (1993)
MET2 ; *Eco*RI, *Pst*I	Differentiation of *S. uvarum and S. cerevisiae* from wine	Demuyter et al. (2004)
	Separation of *S. bayanus*, *S. cerevisiae* and *S. paradoxus* wine strains	Antunovics et al. (2005)
26S rRNA; (primers NL1 and NL4); *Alu*I	Yeast species from wine fermentations	van Keulen et al. (2003)
26S rRNA; (primers NL1 and LRS); *Mse*I, *Apa*I, *Hinf*I	Yeast species from wine fermentations	Baleiras Couto et al. (2005)

Table 3. Application of nucleic acid probes and specific primers for the detection of food and beverage yeasts

Probe/primer Format	Application	References
Whole cell hybridization; Fluorescent/chemiluminescent in situ hybridization (FISH/CISH)		
FISH; PNA probe in D1/D2 26S rRNA	*D. bruxellensis* isolates from wine	Stender et al. (2001), Dias et al. (2003)
CISH; PNA probe in D1/D2 26S rRNA	*D. bruxellensis* isolates from winery air samples	Connell et al. (2002)
CISH; PNA probes in 18S rRNA and 26S rRNA	*S. cerevisiae*, *Z. bailii*, *D. bruxellensis* colonies on filter membranes	Perry-O'Keefe et al. (2001)
FISH; DNA probes in 18S rRNA	*S. cerevisiae*, *P. anomala*, *D. bruxellensis* and *D. hansenii* isolates, detection in yoghurts	Kosse et al. (1997)
FISH; DNA probes in 18S rRNA	Detection and quantification of *A. pullulans* on leaf surfaces	Spear et al. (1999), Andrews et al. (2002)
Dot blot/slot blot hybridization		
Dot blot hybridization	*D. hansenii* isolates from cheese, differentiation of *hansenii* and *fabryii* varieties.	Corredor et al. (2000)
RNA slot blot hybridization	*Candida* sp. EJ1 in wine samples *B. bruxellensis* in wine	Mills et al. (2002) Cocolin et al. (2004)
PCR-ELISA		
Probes in D1/D2 region Probes immobilized on plates, PCR with biotinylated primers Probes in ITS2 region	Detection of marine yeast species; *P. guillermondii, R. diobovatum*, *R. sphaerocarpum*, *K.thermotolerans*-like, *C. parapsilosis*, *C. tropicalis*, *D. hansenii*	Kiesling et al. (2001)
Probes labelled with DIG	Detection of *C. albicans*, *C. tropicalis*, *C. krusei* in blood	Fujita et al. (1995)
Capture PCR amplicons on strepavidin coated plates	Identification of 18 *Candida* species	Elie et al. (1998)

Species-specific primers		
Universal and species-specific primers (V3 region of LSU)	Pathogenic yeasts *C. neoformans*, *T. cutaneum*, *R. mucilaginosa*	Fell (1995)
Universal and species-specific primers (D1/D2 region of LSU)	Detection of several *Candida* species	Mannarelli and Kurtzman (1998)
1) species-specific primer pairs	Identification of *Z. bailii*, *Z. bisporus*, *Z. rouxii* and *T. delbrueckii* isolates from fruit	Sancho et al. (2000)
2) one species-specific, the other universal (ITS region)		
Nested PCR	*D. bruxellensis* strains from isolates and sherry	Ibeas et al. (1996)
Multiplex PCR 5 primers; 1 universal, 4 species-specific (ITS region)	Identification of *Dekkera* isolates; differentiation of *B. bruxellensis*, *B. anomala*, *B. custersianus*, *B. naardenensis*	Egli and Henick-Kling (2001)
PCR and RT-PCR Specific primers (D1/D2 26S LSU)	Detection of *B. bruxellensis*/*B. anomalus* from wine samples	Cocolin et al. (2004)
RT-PCR; Specific primer pairs (cs 1) gene	Detection of viable *C. krusei* in fruit juice	Casey and Dobson (2003)
RT-PCR; Specific primers (ITS and LSU region)	Identification of *S. cerevisiae* and *S. bayanus*/*pastorianus* isolates	Josepa et al. (2000)
PCR, Multiplex PCR; 4 primers; 2 species-specific pairs (YBR033w region)	Detection of *S. cerevisiae*, *S. bayanus* and *S. pastorianus*	Torriani et al. (2004)
Real time PCR		
Primer pairs (D1/D2 LSU)	Detection and enumeration of *D. bruxellensis* in wines	Phister and Mills (2003)
Primer pairs (rad4 gene)	Detection and quantification of *B. bruxellensis* in wines	Delaherche et al. (2004)
Specific primer pairs (cs 1) gene	Quantification of *C. krusei* from fruit juice	Casey and Dobson (2004)
Universal primer pairs ITS3 and ITS4 (5.8S and ITS2)	Differentiation of *Z. bailii*, *Z. rouxii*, *C. krusei*, *R. glutinis* and *S. cerevisiae* by difference in Tm	Casey and Dobson (2004)

DNA in PCR amplicons (Kiesling et al., 2002) (Table 3). Species-specific primers are used in PCR assays to generate amplicons. Production of the amplicon means that the particular target species is present in the sample. Qualitative detection of the target amplicon is done by its visualisation in gel electrophoresis. Real time PCR systems are now being applied to yeasts, and allow the simultaneous detection and quantification of the target species, omitting the electrophoresis step (Table 3).

Nucleic acid probes and specific primers have gained widespread use in the detection of bacterial species. Their application to the detection of yeast species has not been that extensive and further development is needed. Most probes reported to date have been developed around specific sequences in ribosomal DNA, and it would be worthwhile to identify other species-specific genes that could be targeted for probe development.

2.4. Differentiation of Strains Within a Species

The distinctive character and appeal of many foods and beverages (eg. bread, beer, wine) produced by fermentation with yeasts are frequently attributable to the contribution and properties of particular strains. Strain typing is also useful to trace the source of yeast contamination in outbreaks of food spoilage. The ability to differentiate strains within a species is, therefore, an important requirement in quality assurance programs. Over the past 20 years, a diversity of molecular methods has been developed and applied to the differentiation of yeast strains, and some of these are sufficiently robust and convenient for routine use (Table 4).

Electrophoretic karyotyping of genomic DNA using pulse field gel electrophoresis (PFGE) and RFLP analysis of genomic DNA have been widely applied to "fingerprint" yeast strains with very good success and confidence, but they require significant attention to DNA preparation and extraction, as well as to subsequent electrophoretic analyses (Cardinali and Martini, 1994; Deak, 1995; van der Aa Kühle et al., 2001). Analysis of mitochondrial DNA by RFLP produces fragment profiles that give excellent strain discrimination. Simplified methods for extraction and processing of the mitochondrial DNA have greatly improved the convenience and reliability of this assay, and consequently it has found significant application to the analysis of food and beverage yeasts (Querol et al., 1992; López et al., 2001; see review of Loureiro and Malfieto-Ferreira, 2003).

Table 4. Application of PCR-based methods for strain and species differentiation of yeasts associated with foods and beverages

Primers	Application	References
AFLP		
EcoRI-C/Mse-AC	Species and strain differentiation of *Saccharomyces* and non-*Saccharomyces* wine yeasts	de Barros Lopes et al. (1999, 2002)
MseI-C/PstI-AA, -AC, -AT		
EcoRI/MseI, nine primer pairs	Differentiation of wine, brewing, bakery and sake strains of *Saccharomyces* species	Azumi and Goto-Yamamoto (2000)
MseI-EcoRI four primer pairs	Genetic analysis of *S. cerevisiae* wine strains	Gallego et al. (2005)
EcoRI-C/MseI-AC	Identification of pathogenic *Candida* species, subspecies of *C. albicans* and *C. dublinensis*	Borst et al. (2003), Ball et al. (2004)
EcoRI-MseI four primer pairs	Intraspecific variability among *A. pullulans*	De Curtis et al. (2004)
RAPD and micro/minisatellites		
$(GTG)_5$ $(CAG)_5$ and M13	Differentiation of *S. cerevisiae*, *S. pastorianus*, *S. bayanus*, *S. willianus*	Lieckfeld et al. (1993)
$(GTG)_5$, $(GACA)_4$ and phage M13 core sequence	Differentiation of *C. neoformans*, *C. albidus*, *C. laurentii* and *R. rubra* species, and strains of *C. neoformans*	Meyer et al. (1993)
Primers 15, 18, 20, 21	Differentiating *S. cerevisiae* and *Z. bailii*	Baleiras Couto et al. (1994)
Preliminary screening of primers	Characterisation of wine yeasts; *R. mucilaginosa*, *S. cerevisiae*, *S. exiguus*, *P. membranifaciens*, *P. anomala*, *T. delbrueckii*, *C. vini* strains	Quesada and Cenis (1995)
Decamer 1; ACG GTG TTG G	Distinguish species within genus *Saccharomyces*	Molnár et al. (1995)
Decamer 2; TGC CGA GCT G		
Decamer 3; GGG TAA CGC C	Distinguish species within genus *Metschnikowia*	Lopandic et al. (1996)
	Yeast species isolated from floral nectaries	Herzberg et al. (2002)
M13	Identify yeast species from cheese; *D. hansenii*, *S. cerevisiae*, *I. orientalis*, *K. marxianus*, *K. lactis*, *Y. lipolytica*, *C. catenulata*, *G. candidum*	Prillinger et al. (1999)

Continued

Table 4. Application of PCR-based methods for strain and species differentiation of yeasts associated with foods and beverages—*cont'd*

Primers	Application	References
Decamer 1; ACG GTG TTG G Decamer 2; TGC CGA GCT G Decamer 3; TGC AGC GTG G Decamer 4; GGG TAA CGC C		
M13	Yeast species from dairy products; *S. cerevisiae*, *K. marxianus*, *K. lactis*, *D. hansenii*, *Y. lipolytica*, *T. delbrueckii*	Andrighetto et al. (2000)
RF2		
	Yeast species from sourdough products	Forshino et al. (2004)
M13V universal	Yeast species from Greek sourdough; *P. membranifaciens*, *S. cerevisiae*, *Y. lipolytica*	Paramithiotis et al. (2000)
M13	Yeast species from artisanal Fiore Sado cheese; *C. zeylanoides*, *D. hansenii*, *K. lactis*, *C. lambica*, *G. candidum*	Fadda et al. (2004)
	Strain typing of *D. hansenii* and *G. candidum*	Vasdinyei and Deak (2003)
Primers 24, 28, OPA11	Discrimination of *S. cerevisiae* strains	Baleiras Couto et al. (1996a)
$(GAC)_5$, $(GTG)_5$		
$(GTG)_5$ and $(CAG)_5$	Differentiation of strains of *Z. bailii* and *Z. bisporus*	Baleiras Couto et al. (1996b)
	Separation of *S. cerevisiae* from *K. apiculata*	Caruso et al. (2002)
	Separation of *C. stellata*, *M. pulcherrima*, *K. apiculata* and *Schiz. pombe*	Capece et al. (2003)
$(GTG)_5$, $(CAG)_5$ and M13	Genotyping the *R. glutinis* complex	Gadanho and Sampaio (2002)
Primer P24, $(GTG)_5$ and $(GAC)_5$	Typing of *P. galeiformis* strains in orange juice production	Pina et al. (2005)
$(GTG)_5$ $(ATG)_5$ and M13	Differentiation of *Hanseniaspora* species	Cadez et al. (2002)
OPA03, OPA 18		

ERIC1R, ERIC2	*C. boidinii, C. mesenterica, C. sake, C. stellata, D. anomala, D. bruxellensis, H. uvarum, I. terricola, S. ludwigii, Schiz. pombe, T. debrueckii, Z. bailii, S. bayanus, S. cerevisiae* from wine	Hierro et al. (2004)
REPIR1, REP2I		
Intron splice site primer EI1		
Intron splice primers	Differentiation of commercial wine *S. cerevisiae* strains. Differentiation of non-*Saccharomyces* wine species and strains; *S. cerevisiae, S. bayanus, T. delbrueckeii, I. orientalis, H. uvarum, H. guillermondii, M. pulcherrima, P. fermentans, P. membranaefaciens*	de Barros Lopes et al. (1996, 1998)
EI1, EI2, LA1, LA2		
S. cerevisiae and sensu stricto strains		
Primers δ1 and δ2	Differentiation of *S. cerevisiae* strains, *S. douglassi, S. chevalierii, S. bayanus*	Ness et al. (1993)
Primers δ12 and δ2		Lavallée et al. (1994), Fernández-Espinar et al. (2001), Schuller et al. (2004)
	Identification/authentication of commercial wine *S. cerevisiae* strains	
	Characterisation of wild *S. cerevisiae* strains from grapes and wine fermentations	Versavaud et al. (1995), Cappello et al. (2004), Demuyter et al. (2004)
	Differentiating *S. cerevisiae* strains from sourdough	Pulvirenti et al. (2001)
	Genetic relatedness between clinical and food *S. cerevisiae* strains	de Llanos et al. (2004)
Introns in COX1 mitochondrial gene	Monitor wine starter *S. cerevisiae* strains during fermentation	López et al. (2003)

Continued

Table 4. Application of PCR-based methods for strain and species differentiation of yeasts associated with foods and beverages—*cont'd*

Primers	Application	References
Microsatellite markers/Sequence-Tagged Site markers		
(ScTAT1) chromosome XIII	Differentiation of *S. cerevisiae* strains	Gallego et al. (1998)
Locus SCYOR267, Locus SC8132X, Locus SCPTSY7	Differentiation of industrial *S. cerevisiae* wine strains	González Techera et al. (2001)
Locus SC8132X	Monitoring the populations of *S. cerevisiae* strains during grape juice fermentation	Howell et al. (2004)
Multiplex 1; Loci ScAAT2, ScAAT3, ScAAT5	Differentiation of *S. cerevisiae* isolates from spontaneous wine fermentations	Pérez et al. (2001), Gallego et al. (2005)
Multiplex 2; Loci ScAAT1, ScAAT4, ScAAT6	Differentiation of commercial *S. cerevisiae* wine strains	Schuller et al. (2004)

Minisatellite core sequence (wild type phage) GAG GGT GGC GGT TCT; M13V universal; GTT TCC CCA GTC ACG AC; Phage M13 core sequence; GAG GGT GGX GGX TCT

2.5. PCR-based Fingerprinting

PCR technology has provided new opportunities for developing faster, more convenient methods for typing yeasts and fungi. Two of the first methods developed were Random Amplified Polymorphic DNA (RAPD) analysis and Amplified Fragment Length Polymorphism (AFLP) analysis (Baleiras Couto et al., 1994, 1995, 1996a; Vos et al., 1995; van der Vossen et al., 2003). These methods have the capability of analyzing an extensive portion of the genome, and reveal polymorphisms that differentiate at both the species and strain levels (Table 4).

In the case of RAPD, DNA template is subject to PCR amplification with single, short primers (10-15bp of random/arbitary sequences) that hybridize to a set of arbitary loci in the genome. Some primers used for this purpose are listed in van der Vossen et al. (2003). PCR cycles are performed under conditions of low stringency. The amplicons produced are separated on gel electrophoresis and give profiles that fingerprint the strain or species. AFLP is a variation of RAPD. The approach firstly involves digestion of genomic DNA with two restriction nucleases (usually *Eco*RI and *Mse*I). The fragments are then ligated with end-specific adapters, followed by two successive rounds of PCR. Pre-selective PCR amplifies fragments using primers complimentary to the adapter sequences. A second, selective PCR is performed with primers containing additional nucleotide bases at the 3′ end (selective bases are user defined). The resulting products are resolved by gel electrophoresis or by capillary electrophoresis. There are extra steps involved in AFLP, but this method can potentially generate more extensive information from a single restriction/ligation reaction than other PCR strategies (de Barros Lopes et al., 1999; Lopandic et al., 2005). Both RAPD and AFLP analyses gave excellent strain and species differentiation. However, the main hurdles to routine application are the need for stringent standardisation of conditions in order to obtain reproducible data, and the workload involved.

Micro- and minisatellites are short repeat motifs of about 15-30 and 2-10bp, respectively. Primers targeting these sequences are used in PCR assays to generate an array of amplicons, the profile or "fingerprint" of which reflects the polymorphism of these regions and the distance between them. Some commonly used primers for food yeasts are $(GTG)_5$, $(GAC)_5$ and M13 phage core sequences. This method has had good success in differentiating strains of several food spoilage and wine yeasts (Baleiras Couto, et al., 1996b), and for differentiating species of yeasts from dairy sources (Prillinger et al., 1999) (Table 4).

Intron splice sequences are also known for their polymorphism and have been examined as sites that could give strain differentiation. De Barros Lopes et al. (1996, 1998) reported a PCR method for the analysis of these sites. The method was simple, quick (several hours), robust, reproducible and gave profiles enabling the differentiation of winemaking strains of *Saccharomyces*. Building on this concept, López et al. (2002) developed a multiplex PCR assay based on introns of the COX1 mitochondrial gene. The assay, which was simple and fast (8 hours), gave good differentiation of wine strains of *Saccharomyces*. Other repetitive elements that are targeted in PCR-mediated fingerprinting techniques include the delta repeat, the repetitive extragenic palindromic (REP) and enterobacterial repetitive intergenic consensus (ERIC) sequences (Hierro et al., 2004).

The application of these PCR methods to food and beverage yeasts (Table 4) generally gives good species and strain characterisation. However, the level of resolution achieved is greatly influenced by the choice of primers and the taxa under study.

3. MOLECULAR STRATEGIES FOR MONITORING YEAST COMMUNITIES IN FOODS AND BEVERAGES

The diversity of yeast species associated with foods and beverages is usually determined by culturing homogenates of the product on plates of agar media (Fleet, 1992; Deak, 2003). Yeast colonies are then isolated and identified. As mentioned in previous sections, molecular methods have now found widespread application in identifying these isolates. New, culture-independent methods based on PCR-denaturing gel gradient electrophoresis (DGGE) and PCR-temperature gradient gel electrophoresis (TGGE) are now being used to determine the ecological profile of yeasts in foods and beverages (Muyzer and Smalla, 1998). The basic strategy is outlined in Figure 2. Total DNA is extracted from samples of the product. Using universal fungal primers (or genus-specific primers), yeast ribosomal DNA within the extract is specifically amplified by PCR. Generally, the D1/D2 domain of the 26S subunit is targeted, but other regions such as the 18S subunit may be used. The amplicons produced by PCR are next separated by either DGGE or TGGE, which resolve the different DNA amplicons on the basis of their sequence/melting domains. DGGE uses a polyacrylamide gel containing a linear gradient of denaturant (mix-

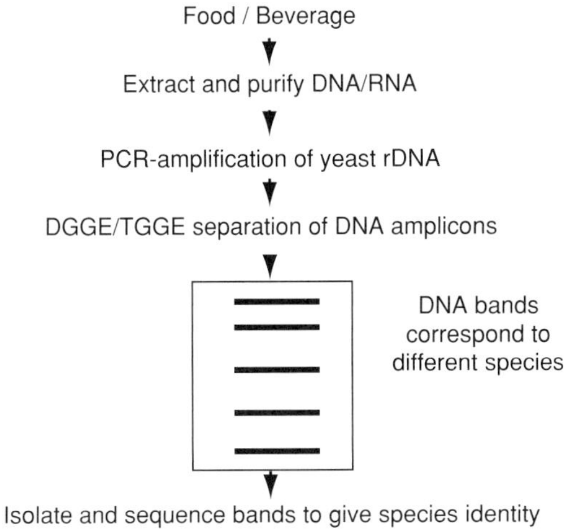

Figure 2. A culture-independent approach for determining the yeast ecology of foods and beverages using PCR-DGGE/TGGE analyses

ture of urea and formamide), while TGGE uses a gradient of temperature to denature the strands of DNA amplicons. Usually, each DNA band found in the gel corresponds to a yeast species. The band is excised from the gel and sequenced to give the species identity. Thus, a profile of the species associated with the ecosystem is obtained, without the need for agar culture. It is believed that this molecular approach overcomes the bias of culture methods, and reveals species that might fail to produce colonies on agar media. Consequently, it is considered that a more accurate representation of the diversity of yeast species in the food product is obtained (Giraffa, 2004). Table 5 lists a range of studies where PCR-DGGE/TGGE have been applied to the analysis of food and beverage yeasts.

Generally, there has been good agreement between yeast species detected in foods and beverages by PCR-DGGE/TGGE and culture on agar media, but some discrepancies have been noted. In some cases, yeasts were recovered by DGGE/TGGE analyses but not by culture. These observations have led to suggestions that viable but non-culturable yeasts may be present (Mills et al., 2002; Meroth et al., 2003; Masoud et al., 2004; Prakitchaiwattana et al., 2004; Nielsen et al., 2005). However, DNA from non-viable yeast cells or DNA released from autolysed yeast cells could account for these findings, and caution is needed when interpreting the DGGE/TGGE data. To address this limitation, Mills et al. (2002) and Cocolin and Mills

Table 5. Application of PCR-denaturation gradient gel electrophoresis and PCR-temperature gradient gel electrophoresis to ecological studies of yeasts in foods and beverages.

Method: region/primers	Application	References
PCR-DGGE	Succession of yeast species in wine fermentations;	
Nested PCR; NL1-NL4,		
NL1gc-LS2; 26S rDNA		
	model wine fermentation;	Cocolin et al. (2000)
	commercial Dolce wine fermentations;	Cocolin et al. (2001)
	continuous wine fermentation;	Cocolin et al. (2002a)
PCR-DGGE and RT PCR-DGGE	Yeast species in *Botrytis*-affected wine fermentations	Mills et al. (2002)
26S rDNA/rRNA; nested PCR;	Inhibition of wine yeast species by sulphur dioxide	Cocolin and Mills (2003)
NL1-NL4, NL1gc-LS2		
PCR-TGGE	Discrimination of wine yeast species	Hernán-Gómez et al. (2000)
18S rDNA;		
YUNIV1gc-YUNIV3	Diversity of yeast species in wine fermentations	Fernández-Gonzáles et al. (2001)
PCR-DGGE	Yeast species in sourdough starters mixtures, rye flour and sourdough samples	Meroth et al. (2003)
26S rDNA;		
U1gc-U2		
PCR-DGGE	Yeast species on wine grapes	Prakitchaiwattana et al. (2004)
26S rDNA; nested PCR;		
NL1-NL4, NL1gc-LS2		

PCR-DGGE 18S rDNA; nested PCR; NS1-NS8, NS1gc-NS2+10	Phyllospheric yeast species and fungicide treatments on russetting of Elstar apples	Gildemacher et al. (2004)
PCR-DGGE 26S rDNA; NL1gc-LS2	Yeast species in raw milk	Cocolin et al. (2002b)
PCR-DGGE 26S rDNA; nested PCR; NL1-NL4, NL1gc-LS2; or NL1-LS2, NL1gc-LS2	Yeast species involved in coffee fermentation	Masoud et al. (2004)
PCR-DGGE 26S rDNA; NL1gc-LS2	Yeast populations associated with Ghanaian cocoa fermentations	Nielsen et al. (2005)
PCR-DGGE ITS2; Schafgc-Schar (*Saccharomyces sensu-stricto* specific)	Differentiation of *Saccharomyces sensu strictu* strains; *S. cerevisiae*, *S. paradoxus* and *S. bayanus*/*S. pastorianus*	Manzano et al. (2004)
PCR-TGGE ITS2; Schafgc-Schar	*S. cerevisiae* and *S. paradoxus* in wine fermentation	Manzano et al. (2005)

(2003) have proposed the use of RT-PCR protocols that target RNA rather than DNA templates from the food or beverages.

In several studies, culture methods have revealed the presence of yeast species not found by PCR-DGGE (Mills et al., 2002; Masoud et al., 2004; Prakitchaiwattana et al., 2004). Such yeasts were present at populations at less than 10^2-10^3 CFU/ml or g of product and this is considered to be the lower limit of detection by PCR-DGGE. Nevertheless, there are reports where yeast species present at 10^4-10^5 CFU/ml or g product have not been detected by PCR-DGGE. This can occur when there is a mixture of different yeast species in the sample, with one species being numerically present at populations 100-1000 times more than other species. The DNA of the dominant species may bind primers more favorably, resulting in weaker amplification of the minority species. Also, when universal primers are used, there may be stronger hybridization of the primers with the DNA of some species over others (Mills et al., 2002; Prakitchaiwattana et al., 2004; Nielsen et al., 2005).

While the relative mobility of DNA bands on DGGE/TGGE gels can discriminate between closely related species (e.g. those in *Saccharomyces sensu stricto*, Manzano et al., 2004), exceptions can occur. In some cases, different species have produced bands with similar mobilities (Hernán-Gómez et al., 2000; Gildemacher et al., 2004). There are several reports where multiple bands have been found for the one species (eg. two bands for *Candida* sp. (Mills et al., 2002), three bands for *Pichia kluyveri* (Masoud et al., 2004), and several bands for *Metschnikowia pulcherrima* (Prakitichaiwattana et al., 2004). The reasons for multiple banding within the one species are not well understood, but may reflect artefacts of PCR using primers with GC clamps and DNA denaturation kinetics during electrophoresis. Multiple bands could arise from nucleotide variations among multiple rDNA copies within a single strain, or could also indicate the presence of different strains within a species. To address these anomalies, it is good practice therefore, to isolate individual bands from DGGE/TGGE gels and confirm their identity by sequencing.

Finally, food samples often contain large amounts of DNA from plants, animals and other microbial groups (eg. bacteria and filamentous fungi) that have the potential to interfere with the specific PCR-amplification of yeast DNA and compromise the reliability and quality of the data obtained by DGGE/TGGE. For example, we have experienced particular difficulty in detecting yeasts in mould ripened cheeses with PCR-DGGE and PCR-TGGE because of the large amounts of fungal (*Penicillium*) DNA that occur in the cheese extracts (Nurlinawati, Cox and Fleet, unpublished data).

4. FACTORS AFFECTING PERFORMANCE OF MOLECULAR METHODS FOR THE ANALYSIS OF YEASTS

As mentioned previously, molecular methods need to meet certain performance and practical criteria before they will gain acceptance for routine use in the quality assurance laboratories of the food and beverage industries. First, they will need to give accurate, reliable and reproducible data at the appropriate levels of sensitivity and selectivity, and meet the appropriate tolerances for false positive and false negative results. Second, they will need to be simple, convenient and inexpensive to use, and give results relatively quickly (Cox and Fleet, 2003).

PCR assays form the basis of most molecular methods used to analyse yeasts. Consequently, it is important to identify and understand the various factors which affect the performance of this technology. The principles of PCR can be found in many text books (eg. McPherson and Møller, 2000). More specific discussions of its application in microbiological analysis are given by Bridge et al. (1998), Hill and Jinneman (2000), Sachse and Frey (2003) and Cox and Fleet (2003). The following sections highlight some of the conceptual and practical variables that affect its performance as applied to the analysis of food and beverage yeasts. More general discussions of these factors are given by Edel (1998), Hill and Jinneman (2000), Sachse (2003), Radström et al. (2003), Bretagne (2003) and Lübeck and Hoofar (2003).

PCR assay involves the following operations; (i) sample preparation (ii) extraction and preparation of DNA (iii) amplification of DNA by PCR (iv) detection of PCR products and (v) processing and interpretation of the data.

4.1. Sample Preparation and DNA Extraction

For many applications, the sample is a pure culture of a yeast isolate. Consequently, sampling is not an issue provided that the culture has proven purity. Nevertheless, the culture needs to be grown to provide biomass for DNA extraction. Variables here include the culture medium and time of incubation, and these have not been given proper consideration in previous literature. Carry over of media ingredients could inhibit the PCR assay (Rossen et al., 1992). The physiological age of the yeast cells (e.g. exponential, stationary phase, autolysing, dead) at the time of assay can affect the efficiency of DNA extraction and the quality of DNA template for PCR. Depending on culture age, various cell proteins, for example, may interact with the genomic DNA, thereby affecting primer annealing to the template, or they can affect the activity of the DNA polymerase (de Barros Lopes et al., 1996). Some basidiomycetous yeasts may have tougher cell walls than ascomycetous yeasts and require more vigorous procedures for equivalent DNA extraction (Prakitchaiwattana et al., 2004).

The purity and concentration of template DNA become more critical when analysing yeast cells associated with food or beverage matrices, as for example, in studies using DGGE or TGGE (Table 5). It is well known that plant polysaccharides, humic components and other polyphenolic materials can co-purify with DNA and inhibit the PCR reaction (Wilson, 1997; Marshall et al., 2003). The relative ratio of yeast DNA to other DNA species (e.g. that from filamentous fungi, bacteria, plants) is another factor that is not properly understood in the performance of PCR assays. The design of primers would be very important here to minimise or prevent their binding to non-target DNA.

Many "in-house" methods have been described to extract and purify DNA from yeast cells, whether they be biomass originating from a pure culture or biomass extracted directly from the food matrix. The methods include freezing and boiling of cells, mechanical disruption by shaking with glass beads or zirconium, digestion with lytic enzymes and extraction with chemical solvents (Hill and Jinneman, 2000; Haugland et al., 2002). Ideally, this "front-end" part of the analytical process needs to be simple and convenient, but yield template DNA that will perform satisfactorily in PCR assays and give the required detection sensitivity. Critical evaluation and some degree of standardization of these methods are needed.

4.2. PCR Amplification of Template DNA

PCR assays are enzymatic reactions. Accordingly, their performance and progress follow the basic principles of enzymology. The substrates are the template DNA, oligonucleotide primers and equimolar amounts of each nucleotide base; ATP, GTP, TTP and CTP, the enzyme to catalyze the reaction is DNA polymerase, and the product is newly synthesized DNA. Like all enzymatic processes, the reaction is highly specific, and its kinetics are determined by factors such as pH, temperature, concentration of reactants, requirements for co-factors and the presence of any inhibitors. Magnesium ions are critical co-factors, the concentration of which determines the specificity, efficiency and fidelity of the reaction.

Two factors make PCR more complex than other enzyme reactions. First, temperature control requires cyclic variation according to the following program; starting at 91-97'C for the denaturation of double stranded DNA; decreasing to 40-65'C for primer annealing to single strands of template DNA; and increasing to 68-74'C for DNA strand extension by DNA polymerase. About 30-40 cycles are conducted. Second, the DNA product also becomes an enzyme substrate as the reaction progresses. The great diversity of PCR applications also introduces specific variables that require understanding and optimization. In particular, these are the length and sequence of primers, the amount of template DNA, the amount of non-target/background DNA, and carry-over material from the sample matrix that could inhibit primer annealing and DNA polymerase activity. Table 6 summarizes some of the key variables that affect the performance of PCR assays. Because of the broad range of PCR applications, it is difficult to prescribe one set of optimum conditions. Consequently, optimisation must be done on the basis of each application. The key aims of optimisation are to increase diagnostic specificity and diagnostic sensitivity (detection limit). Some good discussions of these variables can be found in Edel (1998), McPherson and Møller (2000), Sachse (2003), Radström et al. (2003), Bretagne (2003) and Lübeck and Hoofar (2003). Wilson (1997) has reviewed various factors that inhibit and facilitate PCR.

The conditions of electrophoresis used to separate and detect the DNA amplicons represent another suite of variables that need

Table 6. Factors affecting DNA amplification by PCR

1. Primer to template ratio
2. Efficiency of primer annealing
3. Enzyme to template ratio
4. Length and sequence of primers
5. Concentration of non-target DNA
6. Inhibitors from sample matrix
7. Temperature cycling protocol
8. Source of DNA polymerase
9. Reaction facilitators

to be optimised and managed. Variables, here include the gelling agent (agarose or polyacrylamide) and its concentration, running time, temperature and voltage, and composition of buffer (Andrews, 1986; Hames, 1998). The extent of cross-linking and pore size in polyacrylamide gels affected the resolution of DNA bands and detection of yeast species on grapes by PCR-DGGE (Prakitchaiwattana et al., 2004).

5. STANDARDIZATION OF MOLECULAR METHODS FOR ANALYSIS FOR YEASTS IN FOODS AND BEVERAGES

The commercial significance of yeast in foods and beverages has major implications in national and international trade (Fleet, 2001). Companies trading in foods and beverages will usually have contractual arrangements that specify criteria for the presence of yeasts. In this context, molecular methods used for yeast analyses will need to meet the rigours of legal or forensic scrutiny. They will need some form of standardization and international acceptance. While there have been major advances in the harmonization and standardization of cultural methods for the analysis of microorganisms in food and beverages (Food Control, 1996; Scotter et al., 2001; Langton et al., 2002), this need remains a challenge for molecular methods.

The lack of standardization and validation for molecular analyses of microorganisms in foods and beverages is well recognized and strategies are being developed to address this need, especially for bacteria of public health significance (Schafer et al., 2001; Lübeck and Hoofar, 2003; Lübeck et al., 2003). Hoofar and Cook (2003) and Malorny et al. (2003) have outlined the principles and protocols for achieving this goal, based upon the FOOD-PCR project of the European Commission (http://www.PCR.dk). These initiatives equally apply to the molecular analyses of yeasts in foods and provide a framework upon which to develop similar projects for yeasts and other fungi. The first stage is to define the specific application to be evaluated. This is followed by evaluating and defining the conditions of the assay (e.g. sample treatment and DNA extraction, primer selection, PCR conditions, detection limit), developing positive and negative control assays, selecting procedures for data analysis, and, finally, developing protocols for "in-house" and inter-laboratory validation trials. From these evaluations, consensus and standardization should emerge. Leuschner et al. (2004) reported the results of an interlaboratory evaluation of a PCR method for the detection and identification of probiotic strains of *S. cerevisiae* in animal feed. Good agreement (but not 100%) was obtained between laboratories and an "official" method for analysing animal feed for these yeasts was proposed.

6. REFERENCES

Andrews, A. T., 1986, *Electrophoresis Theory, Techniques and Biochemical and Clinical Applications*, Oxford University Press, UK.

Andrews, J. H., Spear, R. N., and Nordheim, E. V., 2002, Population biology of *Aureobasidium pullulans* on apple leaf surfaces, *Can. J. Microbiol.* **48**:500-513.

Andrighetto, G., Psomas, E., Tzanetakis, N., Suzzi, G., and Lombardi, A., 2000, Randomly amplified polymorphic DNA (RAPD) PCR for the identification of yeasts isolated from dairy products, *Letters in Appl. Microbiol.* **30**:5-9.

Antunovics, Z., Irinyi, L., and Sipiczki, M., 2005, Combined application of methods to taxonomic identification of *Saccharomyces* strains in fermenting botrytized grape must, *J. Appl. Microbiol.* **98**:971-979.

Arias, C. R., Burns, J. K., Friedrich, L. M., Goodrich, R. M., and Parish, M. E., 2002, Yeast species associated with orange juice: Evaluation of different identification methods, *Appl. Environ. Microbiol.* **68**:1955-1961.

Azumi, M., and Goto-Yamamoto, N., 2001, AFLP analysis of type strains and laboratory and industrial strains of *Saccharomyces sensu stricto* and its application to phenetic clustering, *Yeast*. **18**:1145-1154.

Baleiras Couto, M. M., van der Vossen, J. M. B. M., Hofstra, H., and Huis in't Veld, J. H. J., 1994, RAPD analysis: a rapid technique for differentiation of spoilage yeasts, *Int. J. Food Microbiol.* **24**:249-260.

Baleiras Couto, M. M., Vogels, J. T., Hofstra, H., Huis in't Veld, J. H. J., and van der Vossen, J. M. B. M.,1995, Random amplified polymorphic DNA and restriction enzyme analysis of PCR amplified rDNA in taxonomy: two identification techniques for food-borne yeasts, *J. Appl. Bacteriol.* **79**:525-535.

Baleiras Couto, M. M., Eijsma, B., Hofstra, H., Huis in't Veld, J. H. J., and van der Vossen, J. M. B. M., 1996a, Evaluation of molecular typing techniques to assign genetic diversity among *Saccharomces cerevisiae* strains, *Appl. Environ. Microbiol.* **62**:41-46.

Baleiras Couto, M. M., Hartog, B. J., Huis in't Veld, J. H. J., Hofstra, H., and van der Vossen, J. M. B. M., 1996b, Identification of spoilage yeasts in food-production chain by microsatellite polymerase chain reaction fingerprinting, *Food Microbiol.* **13**:59-67.

Balerias Couto, M. M., Reizinho, R. G., and Duarte, F. L., 2005, Partial 26S rDNA restriction analysis as a tool to characterise non-*Saccharomyces* yeasts present during red wine fermentations, *Int. J. Food Microbiol.* **102**:49-56.

Ball, L. M., Bes, M. A., Theelen, B., Boekhout, T., Egeler, R. M., and Kuijper, E. J., 2004, Significance of amplified fragment length polymorphism in identification and epidemiological examination of *Candida* species colonization in children undergoing allogeneic stem cell transplantation, *J. Clin. Microbiol.* **42**:1673-1679.

Barnett, J. A., Payne, R. W., Yarrow, D., 2000, *Yeasts: Characteristics and Identification*, 3rd edition, Cambridge University Press.

Barszczewski, W., and Robak, M., 2004, Differentiation of contaminating yeasts in brewery by PCR-based techniques, *Food Microbiol.* **21**:227-231.

Belloch, C., Querol, A., Garcia, M. D., and Barrio, E., 2000, Phylogeny of the genus *Kluyveromyces* inferred from the mitochondrial cytochrome-c oxidase II gene, *Int. J. Syst. Evol. Microbiol.* **50**:405-416.

Belloch, C., Fernández-Espinar, T., Querol, A., Dolores Garcia, M., and Barrio, E., 2002, An analysis of inter- and intraspecific genetic variabilities in the *Kluyveromyces marxianus* group of yeast species for the reconsideration of the *K. lactis* taxon, *Yeast.* **19**:257-268.

Borst, A., Theelen, B., Reinders, E., Boekhout, T., Fluit, A.C., and Savelkoul, P.H., 2003, Use of amplified fragment length polymorphism analysis to identify medically important *Candida* spp., including *C. dubliniensis*, *J. Clin. Microbiol.* **41**:1357-1362.

Bretagne, S., 2003, Molecular diagnostics in clinical parasitology and mycology: limits of the current polymerase chain reaction (PCR) assays and interest of the real-time PCR assays, *Clin. Microbiol. Infect.* **9**:505-511.

Bridge, P. D., Arora, D. K., Reddy, C. A., and Elander, R. P., 1998, *Applications of PCR in Mycology*, CAB International, Wallingford, UK.

Cadez, N., Poot, G. A., Raspor, P., and Smith, M. T., 2003, *Hanseniaspora meyeri* sp. nov., *Hanseniaspora clermontiae* sp. nov., *Hanseniaspora lachancei* sp. nov. and *Hanseniaspora opuntiae* sp. nov., novel apiculate yeast species, *Int. J. Syst. Evol. Microbiol.* **53**:1671-1680.

Cadez, N., Raspor, P., de Cock, A. W. A. M., Boekhout, T., and Smith, M. T., 2002, Molecular identification and genetic diversity within species of the genera *Hanseniaspora* and *Kloeckera*, *FEMS Yeast Res.* **1**:279-289.

Cai, J., Roberts, I. N., and Collins, M. D., 1996, Phylogenetic relationships among members of the ascomycetous yeast genera *Brettanomyces*, *Debaromyces*, *Dekkera*, and *Kluyberomyces* deduced by small-subunit rRNA gene sequences, *Int. J. Syst. Bacteriol.* **46**:542-549.

Capece A., Salzano G., and Romano P., 2003, Molecular typing techniques as a tool to differentiate non-Saccharomyces wine species, *Int. J. Food Microbiol.* **84**:33-39.

Cappa, F., and Cocconcelli, P. S., 2001, Identification of fungi from dairy products by means of 18S rRNA analysis, *Int. J. Food Microbiol.* **69**:157-160.

Cappello, M. S., Bleve, G., Grieco, F., Dellaglio, F., and Zacheo, G., 2004, Characterization of *Saccharomyces cerevisiae* strains isolated from must of grape grown in experimental vineyard, *J. Appl. Microbiol.* **97**:1274-1280.

Cardinali, G., and Martini, A., 1994, Electrophoretic karyotypes of authentic strains of the *sensu stricto* group of the genus *Saccharomyces*, *Int. J. Syst. Bacteriol.* **44**:791-797.

Caruso, M., Capece, A., Salzano, G., and Romano, P., 2002, Typing of *Saccharomyces cerevisiae* and *Kloeckera apiculata* strains from Aglianico wine, *Letters Appl. Microbiol.* **34**:323-328.

Casey, G. D., and Dobson, A. D. W., 2003, Molecular detection of *Candida krusei* contamination in fruit juice using the citrate synthase gene *cs1* and a potential role or this gene in the adaptive response to acetic acid, *J. Appl. Microbiol.* **95**:13-22.

Casey, G. D., and Dobson, A. D. W., 2004, Potential of using real-time PCR-based detection of spoilage yeast in fruit juice: a preliminary study, *Int. J. Food Microbiol.* **91**:327-335.

Chen, Y. C., Eisner, J. D., Kattar, M. M., Rassoulian-Barrett, S. L., LaFe, K., Yarfitz, S. L., Limaye, A. P., and Cookson, B. T., 2000, Identification of medically important yeasts using PCR-based detection of DNA sequence polymorphisms in the internal transcribed spacer 2 region of the rRNA genes, *J. Clin. Microbiol.* **38**:2302-2310.

Clemente-Jimenez, J. M., Mingorance-Cazorla, L., Martínez-Rodríguez, S., Heras-Vàzquez, F. J., and Rodríguez-Vico, F., 2004, Molecular characterization and oenological properties of wine yeasts isolated during spontaneous fermentation of six varieties of grape must, *Food Microbiol.* **21**:149-155.

Cocolin, L., Bisson, L. F., and Mills, D. A., 2000, Direct profiling of yeast dynamics in wine fermentations, *FEMS Microbiol. Letters.* **189**:81-87.

Cocolin, L., Heisey, A., and Mills, D. A., 2001, Direct identification of the indigenous yeasts in commercial wine fermentations, *Am. J. Enol. Vitic.* **52**:49-53.

Cocolin, L., Manzano, M., Rebecca, S., and Comi, G., 2002a, Monitoring of yeast population changes during a continuous wine fermentation by molecular methods, *Am. J. Enol. Vitic.* **53**:24-27.

Cocolin, L., Aggio, D., Manzano, M., Cantoni, C., and Comi, G., 2002b, An application of PCR-DGGE analysis to profile the yeast populations in raw milk, *Int. Dairy J.* **12**:407-411.

Cocolin, L., and Mills, D. A., 2003, Wine yeast inhibition by sulfur dioxide: A comparison of culture-dependent and independent methods, *Am. J. Enol. Vitic.* **54**:125-130.

Cocolin, L., Rantsiou, K., Iacumin, L, Zironi, R., and Comi, G., 2004, Molecular detection and identification of *Brettanomyces/Dekkera bruxellensis* and *Brettanomyces/Dekkera anomalus* in spoiled wines, *Appl. Environ. Microbiol.* **70**:1347-1355.

Combina, M., Mercado, L., Borgo, P., Elia, A., Jofré, V., Ganga, A., Martinez, C., and Catania, C., 2005, Yeasts associated to Malbec grape berries from Mendoza, Argentina, *J. Appl. Microbiol.* **95**:1055-1061.

Connell, L., Stender, H., and Edwards, C. G., 2002, Rapid detection and identification of *Brettanomyces* from winery air samples based on peptide nucleic acids, *Am. J. Enol Vitic.* **53**:322-323.

Corredor, M., Davila, A. M., Gaillardin, C., and Casaregola, S., 2000, DNA probes specific for the yeast species *Debaryomyces hansenii*: useful tools for rapid identification, *FEMS Microbiol Letters.* **193**:171-177.

Cox, J. M., and Fleet, G. H., 2003, New directions in the microbiological analysis of foods, in: *Foodborne Microorganisms of Public Health Significance*, 6th edition, A. D. Hocking, ed, Food Microbiology Group, Australian Institute of Food Science and Technology, Sydney, pp. 103-162.

Daniel, H. M., Sorrell, T. C., and Meyer, W., 2001, Partial sequence analysis of the actin gene and its potential for studying the phylogeny of *Candida* species and their teleomorphs, *Int. J. Syst. Evol. Microbiol.* **51**:1593-1606.

Daniel, H. M., and Meyer, W., 2003, Evaluation of ribosomal RNA and actin gene sequences for the identification of ascomycetous yeasts, *Int. J. Food Microbiol.* **86**:61-78.

de Barros Lopes M., Soden A., Henschke P. A., and Langridge P., 1996, PCR differentiation of commercial yeast strains using intron splice site primers, *Appl. Environ. Microbiol.* **62**:4514-4520.

de Barros Lopes M., Soden A., Martens A. L., Henschke P. A., and Langridge P., 1998, Differentiation and species identification of yeasts using PCR, *Int. J. Syst. Bacteriol.* **48**:279-286.

de Barros Lopes M., Rainieri S., Henschke P. A., and Langridge P., 1999, AFLP fingerprinting for analysis of yeast genetic variation, *Int. J. Syst. Bacteriol.* **49**: 915-924.

de Barros Lopes M., Bellon J. R., Shirley N. J., and Ganter P. F., 2002, Evidence for multiple interspecific hybridization in *Saccharomyces sensu stricto* species, *FEMS Yeast Res.* **1**:323-331.

de Llanos, R., Querol, A., Planes, A. M., and Fernández-Espinar, M. T., 2004, Molecular characterization of clinical *Saccharomyces cerevisiae* isolates and their association with non-clinical strains, *System. Appl. Microbiol.* **27**:427-435.

De Curtis, F., Caputo, L., Castoria, R., Lima, G., Stea, G., and De Cicco, V., 2004, Use of fluorescent amplified length polymorphism (fAFLP) to identify specific molecular markers for the biocontrol agent *Aureobasidium pullulans* strain LS30, *Postharvest Biol. Technol.* **34**:179-186.

Deak, T. and Beuchat, L. R., 1996, *Handbook of Food Spoilage Yeasts*, CRC Press, Boca Raton.

Deak, T., 1995, Methods for the rapid detection and identification of yeasts in foods, *Trend Food Sc. Technol.*, 6: 287-292.

Deak, T., 2003, Detection, enumeration and isolation of yeasts, in: *Yeasts in Food, Beneficial and Detrimental Aspects*, T. Boekhout and V. Robert, eds, Behr's-Verlag, Hamburg, pp. 39-68.

Delaherche A., Claisse O., and Lonvaud-Funel, A., 2004, Detection and quantification of *Brettanomyces bruxellensis* and 'ropy' *Pediococcus damnosus* strains in wine by real-time polymerase chain reaction, *J. Appl. Microbiol.* **97**:910-915.

Demuyter, C., Lollier, M., Legras, J.-L., and Le Jeune, C., 2004, Predominance of *Saccharomyces uvarum* during spontaneous fermentation, for three consecutive years, in an Alsatian winery, *J. Appl. Microbiol.* **97**:1140-1148.

Dias, L., Dias, S., Sancho, T., Stender, H., Querol, A., Malfeito-Ferreira M., and Loureiro, V., 2003, Identification of yeasts isolated from wine-related environments and capable of producing 4-ethylphenol, *Food Microbiol.* **20**:567-574.

Diaz, M. R., and Fell, J. W., 2000, Molecular analyses of the IGS and ITS regions of rDNA of the psychrophilic yeasts in the genus *Mrakia*, *Antonie van Leeuwenhoek*, **77**:7-12.

Dlauchy, D., Tornai-Lehoczki, J., and Péter, G., 1999, Restriction enzyme analysis of PCR amplified rDNA as a taxonomic tool in yeast identification, *System. Appl. Microbiol.* **22**:445-453.

Edel, V., 1998, Polymerase chain reaction in mycology: an overview, in: *Applications of PCR in Mycology*. P. D. Bridge, D. K. Arora, C. A. Reddy, R. P. Elander, eds, CAB International, Wallingford, UK, pp.1-20.

Egli, C. M., and Henick-Kling, T., 2001, Identification of *Brettanomyces/Dekkera* species based on polymorphisms in the rRNA internal transcribed spacer region, *Am. J. Enol. Vitic.* **52**:241-247.

Elie, C. M., Lott, T. J., Reiss, E., and Morrison, C. J., 1998, Rapid identification of *Candida* species with species-specific DNA probes, *J. Clin. Microbiol.* **36**: 3260-3265.

Esteve-Zarzoso B., Belloch, C., Uruburu, F., and Querol, A., 1999, Identification of yeasts by RFLP analysis of the 5.8S rRNA gene and the two ribosomal internal transcribed spacers, *Int. J. Syst. Bacteriol.* **49**:329-337.

Esteve-Zarzoso B., Peris-Torán, M. J., García-Maiquez, E., Uruburu, F., and Querol, A., 2001, Yeast population dynamics during the fermentation and biological aging of sherry wines, *Appl. Environ. Microbiol.* **67**:2056-2061.

Fadda, M. E., Mossa, V., Pisano, M. B., Deplano, M., and Cosentino, S., 2004, Occurrence and characterization of yeasts isolated from artisanal Fiore Sardo cheese, *Int. J. Food Microbiol.* **95**:51-59.

Fell, J. W., 1995, rDNA targeted oligonucleotide primers for the identification of pathogenic yeasts in a polymerase chain reaction, *J. Ind. Microbiol.* **14**:475-477.

Fell, J. W., Boekhout, T., Fonseca, A., Scorzetti, G., and Statzell-Tallman, A., 2000, Biodiversity and systematics of basidiomycetous yeasts as determined by large-subunit rDNA D1/D2 domain sequence analysis, *Int. J. Syst. Evol. Microbiol.* **50**:1351-1371.

Fernández-Espinar, M. T., López, V., Ramón, D., Bartra, E., and Querol, A., 2001, Study of the authenticity of commercial wine yeast strains by molecular techniques, *Int. J. Food Microbiol.* **70**:1-10.

Fernández-Gonzáles, M., Espinosa, J. C., Úbeda, J. F., and Briones, A. I., 2001, Yeast present during wine fermentation: comparative analysis of conventional plating and PCR-TTGE, *System. Appl. Microbiol.* **24**:634-638.

Fleet, G. H., 1992, Spoilage yeasts, *Crit. Rev. Biotechnol.* **12**:1-44.

Fleet, G. H., 2001, Food microbiology, in: *Expert Evidence*, I. Freckelton and H. Selby, eds, Lawbook Co. Thomson Legal and Regulatory, Sydney, Chapter 116, pp. 9-2171-9-2511.

Food Control, 1996, Special issue on validation of rapid methods in food microbiology, G. Campbell-Platt and D. C. Archer, eds, Elsevier, UK, volume 7, pp. 3-58.

Foschino, R., Gallina, S., Andrighetto, C., Rossetti, L., and Galli, A., 2004, Comparison of cultural methods for the identification and molecular investigation of yeasts from sourdoughs for Italian sweet baked products, *FEMS Yeast Res.* **4**:609-618.

Fujita S., Lasker B. A., Lott T. J., Reiss, E., and Morrison, C. J., 1995, Microtitration plate enzyme immunoassay to detect PCR-amplified DNA from *Candida* species in blood. *J. Clin. Microbiol.* **33**:962-967.

Gadanho, M., and Sampaio, J. P., 2002, Polyphasic taxonomy of the basidiomycetous yeast genus *Rhodotorula*: *Rh. glutinis sensu stricto* and *Rh. dairenensis* comb. nov., *FEMS Yeast Res.* **2**:47-58.

Gallego, F. J., Perez, M. A., Martinez, I., and Hidalgo, P., 1998, Microsatellites obtained from database sequences are useful to characterize *Saccharomyces cerevisiae* strains, *Am. J. Enol. Vitic.* **49**:350-351.

Gallego, F. J., Pérez, M. A., Núñez, Y., and Hidalgo, P., 2005, Comparison of RAPDs, AFLPs, and SSR markers for the genetic analysis of yeast strains of *Saccharomyces cerevisiae*, *Food Microbiol.* **22**:561-568.

Ganga, M. A., and Martínez, C., 2004, Effect of wine yeast monoculture practice on the biodiversity of non-*Saccharomyces* yeasts, *J. Appl. Microbiol.* **96**:76-83.

Gildemacher, P. R., Heijne, B., Houbraken, J., Vromans, T., Hoekstra, E. S., and Boekhout, T., 2004, Can phyllosphere yeasts explain the effect of scab fungicides on russetting of Elstar apples?, *Eur. J. Plant Pathology*. **110**:929-937.

Giraffa, G., 2004, Studying the dynamics of microbial populations during food fermentation – a review, *FEMS Microbiol.* 28: 251-260.

Giudici, P., and Pulvirentii, A., 2002, Molecular methods for identification of wine yeasts, in: *Biodiversity and Biotechnology of Wine Yeasts*, M. Ciani, ed., Research Signpost, Kerala, pp. 35-52.

Gonzáles Techera, A., Jubany, S., Carrau, F. M., and Gaggero, C., 2001, Differentiation of industrial wine yeast strains using microsatellite markers, *Lett. Appl. Microbiol.* **33**:71-75.

Granchi, L., Bosco, N. Messini, A., and Vincenzini M., 1999, Rapid detection and quantification of yeast species during spontaneous wine fermentation by PCR-RFLP analysis of the rDNA ITS region, *J. Appl. Microbiol.* **87**:949-956.

Guillamón, J. M., Sabaté, J., Barrio, E., Cano, J., and Querol, A., 1998, Rapid identification of wine yeast species based on RFLP analysis of the ribosomal internal transcribed spacer (ITS) region, *Arch. Microbiol.* **169**:387-392.

Hames, B. D., 1998, *Gel Electrophoresis of Proteins: A Practical Approach*, Oxford University Press, NY.

Haugland, R. A., Brinkman, N., and Vesper, S. J., 2002, Evaluation of rapid DNA extraction methods for the quantitative detection of fungi using real-time PCR analysis, *J. Microbiol. Methods*. **50**:319-323.

Heras-Vazquez, F. J., Mingorance-Cazorla, L., Clemente-Jimenez, J. M., and Rodriguez-Vico, F., 2003, Identification of yeast species from orange fruit and juice by RFLP and sequence analysis of the 5.8S rRNA gene and the two internal transcribed spacers, *FEMS Yeast Res*. **3**:3-9.

Hernán-Gómez, S., Espinosa, J. C., and Ubeda, J. F., 2000, Characterization of wine yeasts by temperature gradient gel electrophoresis (TGGE), *FEMS Microbiol. Letters*. **193**:45-50.

Herzberg, M., Fischer, R., and Titze, A., 2002, Conflicting results obtained by RAPD-PCR and large-subunit rDNA sequences in determining and comparing yeast strains isolated from flowers: a comparison of two methods, *Int. J. Syst. Microbiol.* **52**:1423-1433.

Hierro, N., González, Á., Mas, A., and Guillamón, J. M., 2004, New PCR-based methods for yeast identification, *J. Appl. Microbiol.* **97**:792-801.

Hill, W. E., and Jinneman, K. C., 2000, Principles and applications of genetic techniques for detection, identification and subtyping of food-associated pathogenic microorganisms, in: *The Microbiological Safety and Quality of Food*, B. M. Lund, T. C. Baird-Parker and G. W. Gould, eds, Aspen Publishers Inc., Maryland, pp. 1813-1851.

Hoofar, J., and Cook, N., 2003, Critical aspects of standardization of PCR, in: *Methods in Molecular Biology, Vol. 216, PCR Detection of Microbial Pathogens: Methods and Protocols*. K. Sachse and J. Frey, eds, Humana Press, Totowa, NJ, pp. 51-64.

Howell, K. S., Bartowsky, E. J., Fleet, G. H., and Henschke, P. A., 2004, Microsatellite PCR profiling of *Saccharomyces cerevisiae* strains during wine fermentation, *Lett. Appl. Microbiol.* **38**:315-320.

Ibeas, J. Lozano, J., Perdigones, F., and Jimenez, J., 1996, Detection of *Dekkeral Brettanomyces* strains in sherry by a nested PCR method, *Appl. Environ. Microbiol.* **62**:998–1003.

James, S. A., Collins, M. D., and Roberts, I. N., 1994, Genetic interrelationship among species of the genus *Zygosaccharomyces* as revealed by small-subunit rRNA gene sequences, *Yeast*. **10**:887-881.

James, S. A., Cai, J., Roberts, I. M., and Collins, M. D., 1996, Use of an rRNA internal transcribed spacer region to distinguish phylogenetically closely related species of the genera *Zygosaccharomyces* and *Torulaspora*, *Int. J. Syst. Bacteriol.* **46**:189-194.

James, S. A., Cai, J., Roberts, I. N., and Collins, M. D., 1997, A phylogenetic analysis of the genus *Saccharomyces* based on 18S rRNA genes sequences: description of *Saccharomyces kunashirensis* sp. nov. and *Saccharomyces martiniae* sp. nov., *Int. J. Syst. Bacteriol.* **47**:453-460.

Josepa S., Guillamon J. M., and Cano J., 2000, PCR differentiation of *Saccharomyces cerevisiae* from *Saccharomyces bayanus/Saccharomyces pastorianus* using specific primers, *FEMS Microbiol. Lett* **193**:255-259.

Jespersen, L., Nielsen, D. S., Hønholt, S., and Jakobsen, M., 2005, Occurrence and diversity of yeasts involved in fermentation of West African cocoa beans, *FEMS Yeast Res.* **5**:441-453.

Kiesling, T., Diaz, M. R., Statzell-Tallman, A., and Fell J. W., 2002, Field identification of marine yeasts using DNA hybridization macroarrays, in: *Fungi in Marine Environments.* K. D. Hyde, S. T. Moss, and L. L. P. Vrijmoed, eds, Fungal Diversity Press, Hong Kong, pp. 69-80.

Kosse, D., Seiler, H., Amann, R., Ludwig, W., and Scherer, S., 1997, Identification of yoghurt-spoiling yeasts with 18S rRNA-targeted oligonucleotide probes, *System. Appl. Microbiol.* **20**:468–480.

Kurtzman, C. P., 2003, Phylogenetic circumscription of *Saccharomyces*, *Kluyveromyces* and other members of *Saccharomycetaceae* and the proposal of the new genera *Lachancea*, *Nakaseomyces*, *Vanderwaltozyma* and *Zygotorulaspora*, *FEMS Yeast Res.*, 4: 232-245.

Kurtzman, C. and Fell, J. W., 1998, *The Yeasts – a Taxonomic Study*, 4th edition, Elsevier, Amsterdam.

Kurtzman, C. P., and Robnett, C. J., 1998, Identification and phylogeny of ascomycetous yeasts from analysis of nuclear large subunit (26S) ribosomal DNA partial sequences, *Antonie van Leeuwenhoek.* **73**:331-371.

Kurtzman, C. P., and Robnett, C. J., 2003, Phylogenetic relationships among yeasts of the '*Saccharomyces* complex' determined from multigene sequence analyses, *FEMS Yeast Res.* **3**:417-432.

Kurtzman, C. P., Boekhout, T., Robert, V., Fell, J. W., and Deak, T., 2003, Methods to identify yeasts, in: *Yeasts in Food, Beneficial and Detrimental Aspects.* T. Boekhout, and V. Robert, eds, Behr's-Verlag, Hamburg, pp. 69-121.

Kurtzman, C. P., Statzell-Tallman, A., and Fell, J. W., 2004, *Tetrapisispora fleetii* sp. nov., a new member of the *Saccharomycetaceae*, *Stud. Mycol.* **50**:397-400.

Lachance, M. A., Starmer, W. T., Bowles, J. M., Phaff, H. J., and Rosa, C. A., 2000, Ribosomal DNA, species structure, and biogeography of the cactophilic yeast *Clavispora opuntiae*, *Can. J. Microbiol.* **46**:195-210.

Langton, S. D., Chevennement, R., Nagelkerke, N., and Lombard, B., 2002, Analysing collaborative trials for qualitative microbiological methods: accordance and concordance, *Int. J. Food Microbiol.* **79**:175-181.

Lavallée, F., Salvas, Y., Lamy, S., Thomas, D. Y. Degré, R., and Dulau, L., 1994, PCR and DNA fingerprinting used as quality control in the production of wine yeast strains, *Am. J. Enol. Vitic.* **45**:86-91.

Leuschner, R. G. K., Bew, J., Fourcassier, P., and Bertin, G., 2004, Validation of the official control method based on polymerase chain reaction (PCR) for identification of authorised probiotic yeast in animal feed, *System. Appl. Microbiol.* **27**: 492-500.

Lieckfeldt, E., Meyer, W., and Börner, T., 1993, Rapid identification and differentiation of yeasts by DNA and PCR fingerprinting, *J. Basic Microbiol.* **33**:413-426.

Lopandic, K., Prillinger. H., Molnár, O., and Giménez-Jurado, G., 1996, Molecular characterization and genotypic identification of *Metschnikowia* species, *System. Appl. Microbiol.* **19**:393-402.
Lopandic, K., Molnár, O., and Prillinger, H., 2005, Application of ITS sequence analysis, RAPD and AFLP fingerprinting in characterising the yeast genus *Fellomyces*, *Microbiol. Res.* **160**:13-26.
López V., Fernández-Espinar M. T., Barrio E., Ramon D., and Querol A., 2003, A new PCR-based method for monitoring inoculated wine fermentations, *Int. J. Food Microbiol.* **81**:63-71.
López, V., Querol, A., Ramon, D., and Fernández-Espinar, M. T., 2001, A simplified procedure to analyse mitochondrial DNA from industrial yeasts, *Int. J. Food Microbiol.* 68: 75-81.
Loureiro, V., and Malfeito-Ferreira, M., 2003, Spoilage yeasts in the wine industry, *Int. J. Food Microbiol.* 86: 23-50.
Loureiro, V., and Querol, A., 1999, The prevalence and control of spoilage yeasts in foods and beverages, *Trends Food Sci. Technol.* **10**: 356-365.
Lübeck, P. S., and Hoofar, J., 2003, PCR technology and applications to zoonotic food-borne bacterial pathogens, in: *Methods in Molecular Biology, Vol. 216, PCR Detection of Microbial Pathogens: Methods and Protocols*. K. Sachse and J. Frey, eds, Humana Press, Totowa, NJ, pp. 65-84.
Lübeck, P. S., Cook, N., Wagner, M., Fach, P., and Hoofar, J., 2003, Toward an international standard for PCR-based detection of food-borne thermotolerant Campylobacters: validation in a multicenter collaborative trial, *Appl. Environ. Microbiol.* **69**:5670-5672.
Malorny, B., Tassios, P. T., Rådström, P., Cook, N., Wagner, M., and Hoofar, J., 2003, Standardization of diagnostic PCR for the detection of foodborne pathogens, *Int. J. Food Microbiol.* **83**:39-48.
Mannarelli, B. M., and Kurtzman, C. P., 1998, Rapid identification of *Candida albicans* and other human pathogenic yeasts by using short oligonucleotides in a PCR, *J. Clin. Microbiol.* **36**:1634-1641.
Manzano, M., Cocolin, L., Iacumin, L., Cantoni, C., and Comi, G., 2005, A PCR-TGGE (Temperature Gradient Gel Electrophoresis) technique to assess differentiation among enological *Saccharomyces cerevisiae* strains, *Int. J. Food Microbiol.* **101**:333-339.
Manzano, M., Cocolin, L., Longo, B., and Comi, G., 2004, PCR-DGGE differentiation of strains of *Saccharomyces sensu stricto*, *Antonie van Leeuwenhoek*. **85**:23-27.
Marshall, M. N., Cocolin, L., Mills, D. A., and VanderGheynst, J. S., 2003, Evaluation of PCR primers for denaturing gradient gel electrophoresis analysis of fungal communities in compost, *J. Appl. Microbiol.* **95**:934-948.
Masoud, W., Cesar, L. B., Jespersen, L., and Jakobsen, M., 2004, Yeast involved in fermentation of *Coffea arabica* in East Africa determined by genotyping and by direct denaturing gradient gel electrophoresis, *Yeast* **21**:549-556.
McPherson, M. J., and Møller, S. G., 2000, *PCR*, BIOS Scientific Publishers, Oxford, UK.
Meroth, C. B., Hammers, W. P., and Hertel, C., 2003, Identification and population dynamics of yeasts in sourdough fermentation processes by PCR-denaturing gradient gel electrophoresis, *Appl. Environ. Microbiol.* **69**:7453-7461.
Meyer, W., Mitchell, T. G., Freedman, E., and Vilgalys, R., 1993, Hybridization probes for conventional DNA fingerprinting used as single primers in the poly-

merase chain reaction to distinguish strains of *Cryptococcus neoformans*, *J. Clin. Microbiol.* **31**:2274-2280.

Mikata, K., Ueda-Nishimura, K., and Hisatomi, T., 2001, Three new species of Saccharomyces *sensu lato* van der Walt from Yaku Island in Japan: *Saccharomyces nagaishii* sp. nov., *Saccharomyces humaticus* sp. nov. and *Saccharomyces yakushimaensis* sp. nov., *Int. J. Syst. Evol. Microbiol.* **51**:2189-2198.

Mills, D. A., Johannsen, E. A., and Cocolin, L., 2002, Yeast diversity and persistence in botrytis-affected wine fermentations, *Appl. Environ. Microbiol.* **68**:4884-4893.

Molina, F., Jong, S.-C., and Huffman, J. L., 1993, PCR amplification of the 3' external transcribed and intergenic spacers of the ribosomal DNA repeat unit in three species of *Saccharomyces*, *FEMS Microbiol. Lett.* **108**:259-264.

Molnár, O., Messner, R., Prillinger, H., Stahl, U., and Slavikova, E., 1995, Genotypic identification of *Saccharomyces* species using random amplified polymorphic DNA analysis, *System. Appl. Microbiol.* **18**:136-145.

Montrocher, R., Verner, M-C., Briolay, J., Gautier, C., and Marmeisse, R., 1998, Phylogenetic analysis of the *Saccharomyces cerevisiae* group based on polymorphisms of rDNA spacer sequences, *Int. J. Syst. Bacteriol.* **48**:295-303.

Morrissey, W. F., Davenport, B., Querol, A., and Dobson, A. D. W., 2004, The role of indigenous yeasts in traditional Irish cider fermentations, *J. Appl. Microbiol.* **97**:647-655.

Muyzer, G., and Smalla, K., 1998, Aplication of denaturing gradient gel electrophoresis (DGGE) and temperature graident gel electrophoresis (TGGE) in microbial ecology, *Antonie van Leeuwenhoek.* **73**:127-141.

Naumov, G. I., James, S. A., Naumova, E. S., Louis, E. J., and Roberts, I. W., 2000, Three new species of *Saccharomyces sensu stricto* complex: *Saccharomyces cariocanus*, *Saccharomyces kudriavzevii* and *Saccharomyces mikatae*, *Int. J. Syst. Evol. Microbiol.* **50**:1931-1942.

Naumova, E. S., Korshunova, I. V., Jespersen, L., and Naumov, G. I., 2003, Molecular genetic identification of *Saccharomyces sensu stricto* strains from African sorghum beer, *FEMS Yeast Res.* **3**:177-184.

Ness, F., Lavallée, F., Dubourdieu, D., Aigle, M., and Dalau, L., 1993, Identification of yeast strains using the polymerase chain reaction, *J. Sci. Food and Agriculture.* **62**:89-94.

Nielsen, D.S., Hønholt, S., Tano-Debrah, K., and Jespersen, L., 2005, Yeast populations associated with Ghanaian cocoa fermentations analysed using denaturing gradient gel electrophoresis (DGGE), *Yeast.* **22**:271-284.

Oda, Y., Yabuki, M., Tonomura, K., and Fukunaga, M., 1997, A phylogenetic analysis of *Saccharomyces* species by the sequence of 18S-28S rRNA spacer regions, *Yeast.* **13**:1243-1250.

Paramithiotis, S., Müller, M. R. A., Ehrmann, M. A., Tsakalidou, E., Seiler, H., Vogel, R., and Kalantzopoulos, G., 2000, Polyphasic identification of wild yeast strains isolated from Greek sourdoughs, *System. Appl. Microbiol.* **23**:156-164.

Pérez, M. A., Gallego, F. J., Martínez, I., and Hidalgo, P., 2001, Detection, distribution and selection of microsatellites (SSRs) in the genome of the yeast *Saccharomyces cerevisiae* as molecular markers, *Letters in Appl. Microbiol.* **33**: 461-466.

Perry-O'Keefe, H., Stender, H., Broomer, A., Oliveira, K., Coull, J., and Hyldig-Nielsen, J. J., 2001, Filter-based DNA *in situ* hybridization for rapid detection, identification and enumeration of specific micro-organisms, *J. Appl. Microbiol.* **90**:180-189.

Phister, T. G., and Mills, D. A., 2003, Real-time PCR assay for detection and enumeration of *Dekkera bruxellensis* in wine, *Appl. Environ. Microbiol.* **69**:7430-7434.

Pina, C., Teixeiró, P., Leite, P., Villa, M., Belloch, C., and Brito, L., 2005, PCR-fingerprinting and RAPD approaches for tracing the source of yeast contamination in a carbonated orange juice production chain, *J. Appl. Microbiol.* **98**:1107-1114.

Prakitchaiwattana, C. J., Fleet, G. H., and Heard, G. M., 2004, Application and evaluation of denaturing gradient gel electrophoresis to analyse the yeast ecology of wine grapes, *FEMS Yeast Res.* **4**:865-877.

Prillinger, H., Molnár, O., Eliskases-Lechner, F., and Lopandic, K., 1999, Phenotypic and genotypic identification of yeasts from cheese, *Antonie van Leeuwenhoek* **75**:267-283.

Pulvirenti, A., Caggia, C., Restuccia, C., Gullo, M., and Giudici, P., 2001, DNA fingerprinting methods used for identification of yeasts isolated from Sicilian sourdoughs, *Annals Microbiol.* **51**:107-120.

Querol, A., Barrio, E., and Ramon, D., 1992, A comparative study of different methods of yeast strains characterisation, *Syst. Appl. Microbiol.*, **58**:2948-2953.

Quesada, M. P., and Cenis, J. L., 1995, Use of random amplified polymorphic DNA (RAPD-PCR) in the characterization of wine yeasts, *Am. J. Enol. Vitic.* **46**: 204-208.

Rådström, P., Knutsson, R., Wolffs, P., Dahlenborg, M., and Löfström, C., 2003, Pre-PCR processing of samples, in: *Methods in Molecular Biology, Vol. 216, PCR Detection of Microbial Pathogens: Methods and Protocols*. K. Sachse and J. Frey, eds, Humana Press, Totowa, NJ, pp. 31-50.

Redzepovic, S., Orlic, S., Sikora, S., Majdak, A., and Pretorius, I. S., 2002, Identification and characterization of *Saccharomyces cerevisiae* and *Saccharomyces paradoxus* strains isolated from Croatian vineyards, *Lett. Appl. Microbiol.* **35**:305-310.

Robert, V., 2003, Data processing, in: *Yeasts in Foods, Beneficial and Detrimental Aspects*. T. Boekhout and V. Robert, eds, Behr's-Verlag, Hamburg, pp. 139-170.

Rodríguez, M. E., Lopes, C. A., van Broock, M., Valles, S., Ramón, D., and Caballero, A. C., 2004, Screening and typing of Patagonian wine yeasts for glycosidase activities. *J. Appl. Microbiol.* **96**:84-95.

Rossen, L., Nørskov, P., Holmstrøm, K., and Rasmussen, O.F., 1992, Inhibition of PCR by components of food samples, microbial diagnostic assays and DNA-extraction solutions, *Int. J. Food Microbiol.* **17**:37-45.

Sachse, K., 2003, Specificity and performance of diagnostic PCR assays, in: *Methods in Molecular Biology, Vol. 216, PCR Detection of Microbial Pathogens: Methods and Protocols*. K. Sachse and J. Frey, eds, Humana Press, Totowa, NJ, pp. 3-29.

Sachse, K., and Frey, J., 2003, *Methods in Molecular Biology, Vol. 216, PCR Detection of Microbial Pathogens: Methods and Protocols*, Humana Press, Totowa, NJ.

Sancho, T., Giménez-Jurado, G., Malfeito-Ferreira M., and Loureiro, V., 2000, Zymological indicators: a new concept applied to the detection of potential spoilage yeast species associated with fruit pulps and concentrates, *Food Microbiol.* **17**:613–624.

Schafer, R. W., Hertogs, K., Zolopa, A., Warford, A., Bloor, S., Betts, B. J., Merigan, T. C., Harrigan, R., and Larder, B. A., 2001, High degree of interlaboratory reproducibility of human immunodeficiency virus type 1 protease and reverse transcriptase sequencing of plasma samples from heavily treated patients. *J. Clin. Microbiol.* **39**:1522-1529.

Schuller, D., Valero, E., Dequin, S., and Caral, M., 2004, Survey of molecular methods for the typing of wine yeast strains, *FEMS Microbiol. Lett.* **231**:19-26.

Scorzetti, G., Fell, J. W., Fonseca, A., and Statzell-Tallman, A., 2002, Systematics of basidiomycetous yeasts: a comparison of large subunit D1/D2 and internal transcribed spacer rDNA regions, *FEMS Yeast Res.* **2**:495-517.

Scotter, S. L., Langton, S., Lombard, B., Lahellec, C., Schultern, S., Nagelkerke, N., in't Veld, P. H., and Rollier, P., 2001, Validation of ISO method 11290 Part 2. Enumeration of *Listeria monocytogenes* in foods, *Int. J. Food Microbiol.* **70**: 121-129.

Sipiczki, M., 2002, Taxonomic and physiological diversity of *Saccharomyces bayanus* in: *Biodiversity and Biotechnology of Wine Yeasts*. M. Ciani, ed., Research Signpost, Kerala, India, pp. 53-69.

Smole Mozina, S., Dlauchy, D., Deak, T., and Raspor, P., 1997, Identification of *Saccharomyces sensu stricto* and *Torulaspora* yeasts by PCR ribotyping, *Lett. Appl. Microbiol.* **24**:311-315.

Spear, R. N., Li, S., Nordheim, E. V., and Andrews, J. H., 1999, Quantitative imaging and statistical analysis of fluorescence *in situ* hybridization (FISH) of *Aureobasidium pullulans*, *J. Microbiol. Methods* **35**:101-110.

Stender, H., HHKurtzman, C., Hyldig-Nielsen, J. J., Sorensen, D., Broomer, A., Oliveira, K., Perry-O'Keefe, H., Sage, A., Young, B., and Coull, J., 2001, Identification of *Dekkera bruxellensis* (*Brettanomyces*) from wine by fluorescence in situ hybridization using peptide nucleic acid probes, *Appl. Environ. Microbiol.* **67**:938–941.

Stratford, M., Bond, C. J., James, S. A., Roberts, I. N., and Steels, H., 2002, *Candida davenportii* sp. nov., a potential soft-drinks spoilage yeast isolated from a wasp, *Int. J. Syst. Evol. Microbiol.* **52**:1369-1375.

Suzuki, M., and Nakase, T., 1999, A phylogenetic study of ubiquinone Q-8 species of the genera *Candida, Pichia,* and *Citeromyces* based on 18S ribosomal DNA sequence divergence, *J. Gen. Appl. Microbiol.* **45**:239-246.

Torriani S., Zapparoli G., Malacrino P., Suzzi G., and Dellaglio, F., 2004, Rapid identification and differentiation of *Saccharomyces cerevisiae*, *Saccharomyces bayanus* and their hybrids by multiplex PCR, *Lett. Appl. Microbiol.* **38**:239-244.

Valente, P., Ramos, J. P., and Leoncini, O., 1999, Sequencing as a tool in yeast molecular taxonomy, *Can. J. Microbiol.* **45**:949-958.

van Keulen, H., Lindmark, D. G., Zeman, K. E., and Gerlosky, W., 2003, Yeasts present during spontaneous fermentation of Lake Erie Chardonnay, Pinot Gris and Riesling, *Antonie van Leeuwenhoek* **83**:149-154.

van der Aa Kühle, A., Jespersen, L., Glover, R. L. F., Diawara, B., and Jakobsen, M., 2001, Identification and characterization of *Saccharomyces* strains from West African sorghum beer, *Yeast* **18**:1069-1079.

van der Aa Kühle, A., and Jespersen, L., 2003, The taxonomic position of *Saccharomyces boulardii* as evaluated by sequence analysis of the D1/D2 domain of 26S rDNA, the ITS1-5.8S rDNA-ITS2 region and the mitochondrial cytochrome-c oxidase II gene, *System. Appl. Micobiol.* **26**:564-571.

van der Vossen, J. M. B. M., Rahaoui, H., de Nijs, M. W. C., and Hartog, B. J., 2003, PCR methods for tracing and detection of yeasts in the food chain, in: *Yeasts in Food, Beneficial and Detrimental Aspects*, T. Boekhout and V. Robert, eds, Behr's-Verlag, Hamburg, pp. 123-138.

Vasdinyei R., and Deák, T., 2003, Characterization of yeast isolates originating from Hungarian dairy products using traditional and molecular identification techniques, *Int. J. Food Microbiol.* **86**:123-130.

Versavaud A., Courcoux P., Roulland C., Dulau L., and Hallet J. N., 1995, Genetic diversity and geographical distribution of wild *Saccharomyces cerevisiae* strains from the wine-producing area of Charentes, France. *Appl. Environ. Microbiol.* **61**:3521-3529.

Villa-Carvajal, M., Coque, J. J. R., Álvarez-Rodríguez, M. L., Uruburu, F., and Belloch, C., 2004, Polyphasic identification of yeasts isolated from bark of cork oak during the manufacturing process of cork stoppers, *FEMS Yeast Res.* **4**:745-750.

Vos, P., Hogers, R., Bleeker, M., Reijans, M., van de Lee, T., Hornes, M., Frijters, A., Pot, J., Peleman, J., Kuiper, M., and Zabeau, M., 1995, AFLP: a new technique for DNA fingerprinting, *Nucleic Acids Res.* **23**:4407-4414.

White, T. J., Bruns, T., Lee, S., and Taylor, J., 1990, Amplification and direct sequencing of fungal ribosomal RNA genes for phylogenentics, in*: PCR Protocols: A Guide for Methods and Applications*. M. A. Innis, D. H. Gelfand, J. J. Sninsky and T. J. White, eds, Academic Press, San Diego, pp. 315-352.

Wilson, I. G., 1997, Inhibition and facilitation of nucleic acid amplification. *Appl. Environ. Microbiol.* **63**:3741-3751.

STANDARDIZATION OF METHODS FOR DETECTING HEAT RESISTANT FUNGI

*Jos Houbraken and Robert A. Samson**

1. INTRODUCTION

Heat resistant fungi can be defined as those capable of surviving temperatures at or above 75°C for 30 or more minutes. The fungal structures which can survive these temperatures are ascospores, and sometimes chlamydospores, thick walled hyphae or sclerotia (Scholte et al., 2000). During the last three years, spoilage incidents involving heat resistant fungi occurred increasingly in various products examined in our laboratory. *Paecilomyces variotii*, *Fusarium oxysporum*, *Byssochlamys fulva*, *B. nivea*, *Talaromyces trachyspermus* and *Neosartorya* species were often encountered in pasteurized fruit, dairy products and soft drinks. A questionnaire sent to many laboratories showed that inappropriate methods were used for the detection of heat resistant fungi, or that sometimes there was no special protocol at all. The use of inappropriate media, such as Sabouraud agar, wrong incubation conditions and the analysis of inadequately sized samples were often encountered. In addition, accurate identification of the isolated colonies to species level often was not performed.

In the literature many methods are described for the detection of heat resistant fungi (Murdock and Hatcher, 1978; Beuchat and Rice, 1979; Beuchat and Pitt, 1992). Beuchat and Pitt (1992) described two methods: the Petri dish method or plating method and the direct incubation method. In the first method, test tubes are used for the heating of the sample. Subsequently, the sample is poured into large Petri

* Centraalbureau voor Schimmelcultures, PO Box 85167, 3508 AD, Utrecht, The Netherlands. Correspondence to: houbraken@cbs.knaw.nl

dishes (diameter 14 cm) and agar media is added. In the second method, flat-sided bottles are used for heating and these bottles are incubated directly at 30°C. The direct incubation method has the advantage that contamination from aerial spores is minimized; the disadvantage of this method is that the colonies growing in the bottle have to be transferred onto agar media for identification.

During recent years a great variety of products have been investigated in our laboratory for the presence of heat resistant fungi. Some products were liquids (juices, colourants), some with a high viscosity (fruit concentrates), or solid products such as soil, pectin, liquorice, strawberries and cardboard. For several of these products the recommended methods are not suitable. A modified detection method using Stomacher bags is therefore proposed here.

2. METHOD FOR EXAMINATION FOR HEAT RESISTANT MOULDS

2.1. Main Modifications

The modified method proposed here is based onto the protocol of Pitt and Hocking (1997). The major modification is the use of Stomacher bags for the heating step which is the important step in the isolation of heat resistant fungi. The product can be easily homogenized in Stomacher bags, with a low risk of aerial contamination. The ascospores of many species of heat resistant fungi require heat activation before they will germinate (Katan, 1985; Lingappa and Sussman, 1959). The heat treatment also inactivates vegetative cells of fungi and bacteria, as well as less heat resistant spores (Beuchat and Pitt, 1992). Some protocols require that bottles be used for the heating of the sample. If bottles that are circular in cross section rather than flat-sided are used, the heat penetration into the sample is particularly slow. When a Stomacher bag containing 250 ml of sample is sealed half way along its length, the thickness of the bag with the sample will be little more than 1 cm, providing much faster heat penetration than in a 250 ml bottle.

2.2. Description of the Modified Method

Because the concentration of the heat resistant fungi in a sample is normally very low, the analysis of a large amount of sample is recommended. At least 100 g of sample should be examined (Samson et al., 2000).

Homogenize the sample before beginning the analysis. Transfer 100 g of sample into a sterile Stomacher bag. Add sterile water (150 g) to the sample and homogenize using a Stomacher for 2 to 4 min. If the sample is likely to contain a higher concentration of heat resistant fungi (e.g. a soil sample), or the sample cannot be homogenized in 150 grams of water (e.g. solid ingredients such as pectin), then a smaller amount of sample can be used. After the homogenization step, the Stomacher bag should be sealed about half way along its length using a heat-sealer, ensuring that no air bubbles are present. After checking the bag for leakage, heat treat the Stomacher bag for 30 min at 75°C in a water bath (preferably one with the capability of shaking or stirring). The water-bath should be at 75°C before the sample is introduced. The sample should be placed in a horizontal position, totally submerged in the water.

After the heat treatment, cool the samples to approximately 55°C. Aseptically transfer the contents of the Stomacher bag to a Schott bottle, or similar, (500 ml) with 250 ml melted double strength MEA containing chloramphenicol (200 mg/l, Oxoid) tempered to 55°C. Mix thoroughly and distribute the agar and sample mixture into seven large plastic Petri dishes (diameter 14.5 cm). Place the Petri dishes into a polyethylene bag to prevent drying and incubate in an upright position at 30°C in darkness.

The general procedure is illustrated in Figure 1.

2.3. Incubation

Many protocols require plates to be incubated for at least 30 days. This period is too long for quality control in the food and beverage industry. In our experience, heat-resistant fungi will usually form colonies after 5 days and mature within 14 days incubation at 30°C, so check the Petri dishes for the presence of colonies after 7 and 14 days. Subculture if necessary, and identify all colonies using standard methods (Pitt and Hocking, 1997; Samson et al., 2000). In some cases a prolonged incubation period is necessary for the identification of the fungus. An overview of the incubation time needed for some particular species is given in Figure 2.

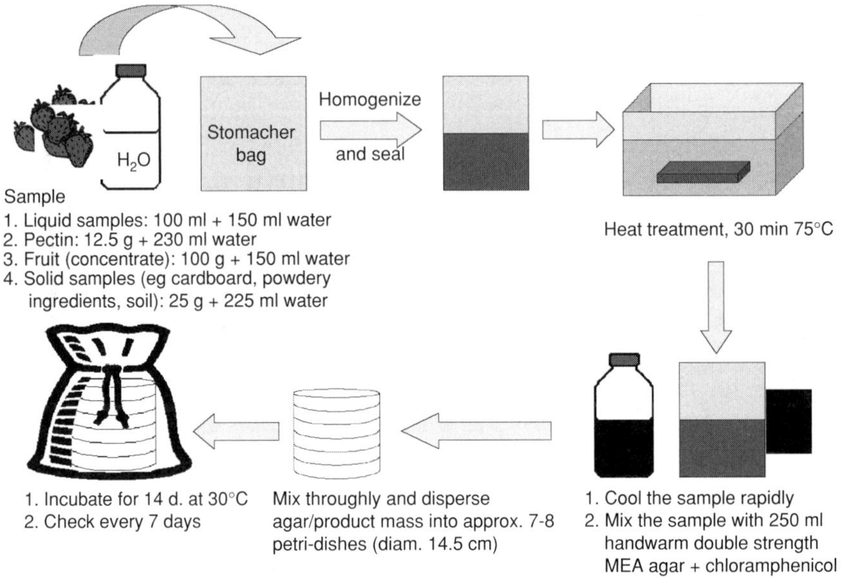

Figure 1. Modified method for the detection of heat resistant fungi

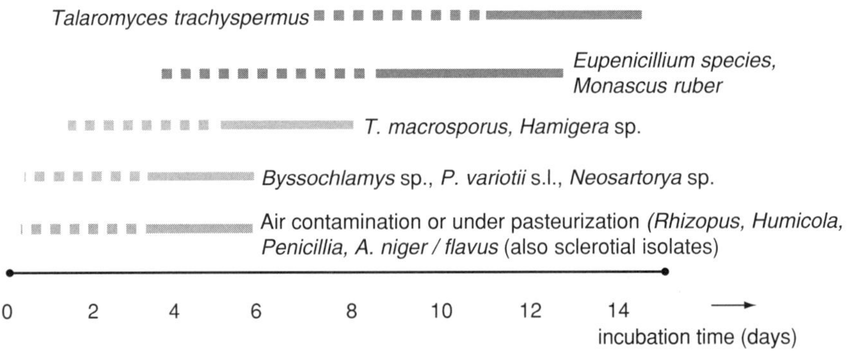

Figure 2. Overview of the incubation time needed to allow growth and development for particular species of heat resistant fungi

3. REFERENCES

Beuchat, L. R., and Rice, S. L., 1979, *Byssochlamys* spp. and their importance in processed fruit syrups, *Trans. Br. Mycol. Soc.* **68**:65-71.

Beuchat, L. R., and Pitt, J. I., 1992, Detection and enumeration of heat-resistant molds, in: *Compendium for the Microbiological Examination of Foods*, C. Vanderzant and D. F. Splittstoesser, eds, American Public Health Association, Washington D. C., pp. 251-263.

Katan, T., 1985, Heat activation of dormant ascospores of *Talaromyces flavus, Trans. Br. Mycol. Soc.* **84**:748-750.

Lingappa, Y., and Sussman, A. S., 1959, Changes in the heat resistance of ascospores of *Neurospora* upon germination, *Am. J. Botany* **49**:671-678.

Murdock, D. I., and Hatcher, W. S., 1978, A simple method to screen fruit juices and concentrates for heat-resistant mold, *J. Food. Prot.* **41**:254-256.

Pitt, J. I., and Hocking, A. D., 1997, *Fungi and Food Spoilage*, 2nd edition, Blackie Academic and Professional, London.

Samson, R. A., Hoekstra, E. S., Lund, F., Filtenborg, O., and Frisvad, J. C., 2000, Methods for the detection, isolation and characterization of food-borne fungi, in: *Introduction to Food- and Airborne Fungi*, 6th edition, R. A. Samson, E. S. Hoekstra, J. C. Frisvad and O. Filtenborg, eds, Centraalbureau voor Schimmelcultures, Utrecht, pp. 283-297.

Scholte, R. P. M., Samson, R. A. and Dijksterhuis, J., 2000, Spoilage fungi in the industrial processing of food, in: *Introduction to Food- and Airborne Fungi*, 6th edition, R. A. Samson, E. S. Hoekstra, J. C. Frisvad, and O. Filtenborg, eds, Centraalbureau voor Schimmelcultures, Utrecht, pp. 339-356.

Section 3. Physiology and ecology of mycotoxigenic fungi

ECOPHYSIOLOGY OF FUMONISIN PRODUCERS IN *FUSARIUM* SECTION *LISEOLA*

Vicente Sanchis, Sonia Marín, Naresh Magan and Antonio J. Ramos[*]

1. INTRODUCTION

Fumonisins were first described as being produced by *Fusarium* Section *Liseola* species (Gelderblom et al., 1988; Marasas et al., 1988), and subsequently a number of toxicological studies have demonstrated their role in animal health problems caused by consumption of contaminated feeds. A relationship has been postulated between frequent ingestion of fumonisin containing maize and incidence of esophageal cancer by humans in certain areas of the world. Thus fumonisins have been classified as possible human carcinogens (Group 2B), by IARC (1993). High levels of fumonisins have been reported in maize in Africa, Asia and South America (Chu and Li, 1994; Doko et al., 1996; Kedera et al., 1999; Ono et al., 1999; Medina-Martinez and Martinez, 2000), sometimes co-occurring with other mycotoxins. Surveys have shown that much of the maize intended for human consumption is contaminated with fumonisins to some extent (Pittet et al., 1992; Sanchis et al., 1994), and these mycotoxins may contaminate a wide range of corn-based foods in our diet (Weidenboerner, 2001).

* V. Sanchis (*vsanchis@tecal.udl.es*), S. Marín and A. J. Ramos, Food Technology Dept, Lleida University, 25198 Lleida, Spain; N. Magan, Cranfield Biotechnology Centre, Cranfield University, Silsoe, England.

2. IMPORTANCE OF THE ECOPHYSIOLOGICAL STUDIES

The importance of the widespread contamination of foods and feeds by fumonisins has led to a proliferation of studies aimed at understanding the ecophysiology of the *Fusarium* spp. involved, particularly *F. verticillioides* and *F. proliferatum*, and the delineation of environmental conditions that allow production of fumonisins. An understanding of the influence of biotic and abiotic factors on germination, growth and fumonisin production by these species is important in managing the problem of fumonisin contamination in the food supply. This study focuses on the impact of different abiotic factors including substrate, water activity (a_w), temperature and preservatives, and biotic factors such as the natural mycoflora present in the grain.

3. INFLUENCE OF ABIOTIC FACTORS ON FUNGAL DEVELOPMENT

3.1. Substrate

Fusarium Section *Liseola* species are much more commonly found in maize than other grains, such as wheat and barley. Studies carried out by Marín et al. (1999a) have shown that even though *Fusarium verticillioides* and *F. proliferatum* are able to grow on a wide variety of substrates, including wheat, barley and maize, high fumonisin biosynthesis only occurs in maize. The assayed isolates were able to grow in the different cereal grains under similar conditions of temperature and a_w, but negligible amounts of fumonisin B_1 (FB_1) were detected.

Although fumonisins are found mainly in corn and corn-based foods and feeds, there are a few reports of fumonisins from other substrates such as 'black oats' animal feed from Brazil (Sydenham et al., 1992), New Zealand forage grass (Scott, 1993), Indian sorghum (Shetty and Bhat, 1997), rice (Abbas et al., 1998), asparagus (Logrieco et al., 1998), beer (Torres et al., 1998), wheat/barley/soybeans (Castella et al., 1999), and tea (Martins et al., 2001).

3.2. Water Activity and Temperature

Our results show that germination of *F. verticillioides* and *F. proliferatum* is possible between 5-37°C at a_w values above 0.88, but the range for growth is slightly narrower (7-37°C above 0.90 a_w) (Figure 1). The lag phases were shorter at 25-30°C and 0.94-0.98 a_w, and they increased to 10-500 h at marginal temperatures (5-10°C). There were some differences between strains. FB_1 production in grain was observed between 10-37°C but only at a_w of 0.93 and above. The optimum conditions for fumonisin production by *F. verticillioides* and *F. proliferatum* were 15-30°C at 0.97 a_w (Marín et al., 1996; Marín et al., 1999b). These two environmental conditions (temperature and moisture availability) are the main factors which control fumonisin production in grain.

The authors have developed detailed two-dimensional profiles of conditions that allow the production of FB_1 (Marín et al., 1999b) (Figure 2).

3.3. Preservatives

Grain preservatives based on propionates have shown some activity in controlling the growth of the *Fusarium* spp., and FB_1 production. Growth rates decreased as preservative concentration increased, regardless of a_w, while fumonisin production decreased only when a_w was 0.93 or lower. In general, only a concentration ≥ 0.07% propionate was effective. In the presence of low propionate concentrations (0.03%), FB_1 production was sometimes stimulated, possibly due to assimilation of these compounds by the moulds. The inhibitory effect of the preservatives is significantly affected by the water activity and temperature of the grain (Marín et al., 1999c).

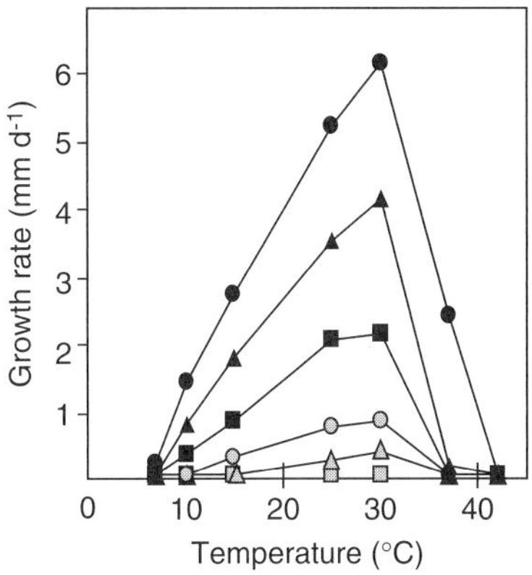

Figure 1. Effect of temperature and water activity on growth rate of *F. verticillioides.* a_w : 0.98 (●), 0.96 (▲), 0.94 (■), 0.92 (●), 0.90 (▲), 0.88 (■)

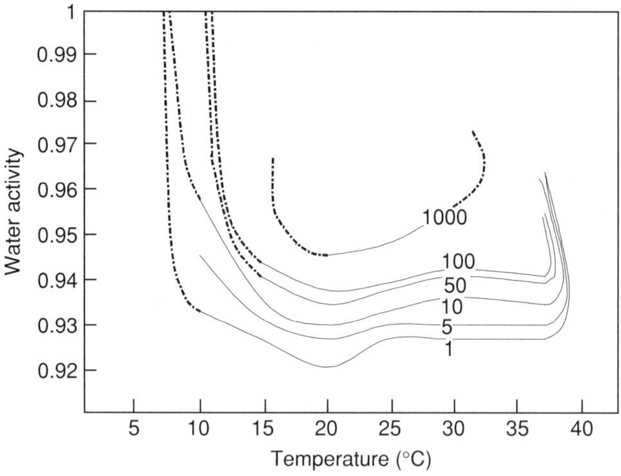

Figure 2. Water activity/temperature profiles for FB_1 accumulation by *F. verticillioides* after 28 days of incubation on irradiated maize (from Marín et al., 1999b).

Essential oils from plants have been also tested for their antimycotoxigenic activity and found to inhibit FB_1 accumulation in maize under moist conditions. Resveratrol, a compound known for its antioxidant properties, decreased FB_1 accumulation in maize at a concentration as low as 23 ppm (Fanelli et al., 2003).

4. INFLUENCE OF BIOTIC FACTORS ON FUNGAL DEVELOPMENT

In general, the presence of other fungal species in mixed cultures inhibited the growth of *F. verticillioides* and *F. proliferatum*. *Aspergillus niger* and *A. flavus* were particularly effective in decreasing the competitiveness of the Fusaria (Marín et al., 1998). However, their effectiveness depended on abiotic factors. In general, the *Fusarium* spp. competed better at higher a_w levels (0.98), and temperatures close to 15°C. Interestingly, FB_1 production was stimulated by certain species at high water availabilities (0.98 a_w), mainly when competing with *A. niger* for occupation of the same niche (Table 1).

Fumonisins are secondary metabolites: if *Fusarium* is an endophyte, fumonisin production may be more important for retaining its niche than for occupying it. However, if contamination occurs from the air

Table 1. Influence of water activity and temperature on production of fumonisin B_1 (ppm) by *Fusarium* spp. on maize grain in the presence of competing mycoflora after a 4-weeks incubation period

Temperature (°C)	15°C			25°C		
Water activity	0.93	0.95	0.98	0.93	0.95	0.98
F. verticillioides	0.8±0.1	7.1±1.3	54.1±10.7	29.3±0.4	5.5±5.9	4.8±5.2
+ *A. niger*	0.2±0.1	0.4±0.1	137.9±1.4	0.5±0.0	5.1±0.3	360.9±9.2
+ *A. ochraceus*	0.6±0.0	1.1±0.1	84.3±1.8	0.3±0.2	47.3±2.9	40.2±0.6
+ *A. flavus*	12.2±6.5	15.8±18.6	12.8±6.0	0.2±0.1	0.2±0.0	10.4±7.1
+ *P. implicatum*	0.1±1.0	0.2±0.1	3.8±1.6	11.5±15.6	0.7±0.1	48.0±11.9
F. proliferatum	17.2±4.8	34.0±8.8	22.8±6.7	5.6±1.0	3.6±4.7	0.7±0.5
+ *A. niger*	19.1±26.4	33.5±6.3	1084.4±33.5	0.2±0.1	3.0±1.4	0.8±0.1
+ *A. ochraceus*	0.1±0.0	1.7±0.1	48.8±0.5	0.9±0.0	2.9±0.1	0.2±0.0
+ *A. flavus*	6.8±2.7	105.9±1.6	602.7±8.1	6.8±8.3	9.8±3.6	64.1±32.9
+ *P. implicatum*	3.6±4.5	284.3±401.4	234.4±308.6	3.3±3.0	21.3±28.5	11.6±6.6

or leaves, establishment may be assisted by fumonisin production. However, field data do not support the hypothesis that *F. verticillioides* gains a competitive advantage via FB_1 (Reid et al., 1999).

5. CONCLUSION

Most fumonisin is produced in maize pre-harvest, but fumonisin control in maize post-harvest can be achieved by effective control of the moisture content (a_w). Temperature control and periodic aeration, along with the natural microflora may act as additional controls.

6. ACKNOWLEDGMENTS

The authors are grateful to the Spanish Government (CICYT, Comision Interministerial de Ciencia y Tecnologia, grant ALI98 0509-C04-01) and the European Union (QLRT-1999-996) for the financial support.

7. REFERENCES

Abbas, H. K, Cartwright, R. D., Shier, W. T., Abouzied, M. M., Bird, C. B., Tice, L. G., Ross, P. F., Sciumbato, G. L., and Meredith, F. I., 1998, Natural occurrence of fumonisins in rice with sheath rot disease, *Plant Dis.* **82**:22-25.

Castella, G., Bragulat, M. R., and Cabanes, F. J., 1999, Surveillance of fumonisins in maize-based feeds and cereals from Spain, *J. Agric. Food Chem.* **47**:4707-4710.

Chu, F. S., and Li, G. Y., 1994, Simultaneous occurrence of fumonisin B_1 and other mycotoxins in moldy corn collected from the People's Republic of China in regions with high incidences of esophageal cancer, *Appl. Environ. Microbiol.* **60**:847-852.

Doko, M. B., Canet, C., Brown, N., Sydenham, E. W., Mpuchane, S., and Siame, B. A., 1996, Natural co-occurrence of fumonisins and zearalenone in cereals and cereal-based foods from eastern and southern Africa, *J. Agric. Food Chem.* **44**:3240-3243.

Fanelli, C, Taddei, F., Trionfetti, P., Jestoi, M., Ricelli, A., Visconti, A., and Fabbri, A., 2003, Use of resveratrol and BHA to control fungal growth and mycotoxin production in wheat and maize seeds, *Aspects Appl. Biol.* **68**:63-71.

Gelderblom, W. C. A., Jaskiewicz, K.; Marasas, W. F. O., Thiel, P. G., Hora, R. M., Vleggaar, R., and Kriek, N. P. J., 1988, Fumonisins -novel mycotoxins with cancer-promoting activity produced by *Fusarium moniliforme*, *Appl. Environ. Microbiol.* **54**:1806-1811.

Kedera, C. J., Plattner, R. D., and Desjardins, A. E., 1999, Incidence of *Fusarium* spp. and levels of fumonisin B_1 in maize in western Kenya, *Appl. Environ. Microbiol.* **65**:41-44.

IARC, 1993, IARC monographs on the evaluation of carcinogenic risks of chemicals to humans; some naturally occurring substances: food items and constituents, heterocyclic aromatic amines and mycotoxins. Ochratoxin A, International Agency for Research on Cancer, Lyon, pp. 26-32.

Logrieco, A., Doko, M. B., Moretti, A., Frisullo, S., and Visconti, A., 1998, Occurrence of fumonisin B_1 and B_2 in *Fusarium proliferatum* infected asparagus plants, *J. Agric. Food Chem.* **46**:5201-5204.

Marasas, W. F. O., Jaskiewicz, K., Venter, F. S., and van Schalkwyk, D. J., 1988, *Fusarium moniliforme* contamination of maize in oesophageal cancer areas in Transkei, *Sth Afr. Med. J.*, **74**:110-114.

Marín, S., Sanchis, V., Teixido, A., Saenz, R., Ramos, A. J., Vinas, I., and Magan, N., 1996, Water and temperature relations and microconidial germination of *Fusarium moniliforme* and *F. proliferatum* from maize, *Can. J. Microbiol.* **42**:1045-1050.

Marín, S., Sanchis, V., Rull, F., Ramos, A. J., and Magan, N., 1998, Colonization of maize grain by *Fusarium moniliforme* and *Fusarium proliferatum* in the presence of competing fungi and their impact on fumonisin production, *J. Food Prot.* **61**: 1489-1496.

Marín, S., Magan, N., Serra, J., Ramos, A. J., Canela, R., and Sanchis V., 1999a, Fumonisin B_1 production and growth of *Fusarium moniliforme* and *Fusarium proliferatum* on maize, wheat, and barley grain, *J. Food Sci.* **64**:921-924.

Marín, S., Magan, N., Belli, N., Ramos, A. J., Canela, R., and Sanchis V., 1999b, Two dimensional profiles of fumonisin B_1 production by *Fusarium moniliforme* and *Fusarium proliferatum* in relation to environmental factors and potential for modelling toxin formation in maize grain, *Int. J. Food Microbiol.* **51**:159-167.

Marín, S., Sanchis, V., Sanz, D., Castel, I., Ramos, A. J., Canela, R., and Magan, N., 1999c, Control of growth and fumonisin B_1 production by *Fusarium verticillioides* and *Fusarium proliferatum* isolates in moist maize with propionate preservatives, *Food Addit. Contam.* **16**:555-563.

Martins, M. L., Martins, H. M., and Bernardo, F., 2001, Fumonisins B_1 and B_2 in black tea and medicinal plants, *J. Food Prot.* **64**:1268-1270.

Medina-Martinez, M. S., and Martinez, A. J., 2000, Mold occurrence and aflatoxin B_1 and fumonisin B_1 determination in corn samples in Venezuela, *J. Agric. Food Chem.* **48**:2833-2836.

Ono, E. Y. S., Sugiura, Y., Homechin, M., Kamogae, M., Vizzoni, E., Ueno, Y., and Hirooka, E.Y., 1999, Effect of climatic conditions on natural mycoflora and fumonisins in freshly harvested corn of the state of Parana, Brazil, *Mycopathologia*, **147**:139-148.

Pittet, A., Parisod, V., and Schellenberg, M., 1992, Occurrence of fumonisins B_1 and B_2 in corn-based products from the Swiss market, *J. Agric. Food Chem.* **40**:1445-1453.

Reid, L. M., Nicol, R. W., Ouellet, T., Savard, M., Miller, J. D., Young, J. C., Stewart, D. W., and Schaafsma, A. W., 1999, Interaction of *Fusarium graminearum* and *F. moniliforme* in maize ears: disease progress, fungal biomass, and mycotoxin accumulation, *Phytopathology,* **89**:1028-1037.

Sanchis, V., Abadias, M., Oncins, L., Sala, N., Vinas, I., and Canela, R., 1994, Occurrence of fumonisins B_1 and B_2 in corn-based products from the Spanish market, *Appl. Environ. Microbiol.* **60**:2147-2148.

Scott, P. M., 1993, Fumonisins, *Int. J. Food Microbiol.* **18**:257-270.

Shetty, P. H., and Bhat, R. V., 1997, Natural occurrence of fumonisin B_1 and its co-occurrence with aflatoxin B_1 in Indian sorghum, maize and poultry feeds, *J. Agric. Food Chem.* **45**:2170-2173.

Sydenham, E. W., Marasas, W. F. O., Shephard, G. S., Thiel, P. G., and Hirooka, E. Y., 1992, Fumonisin concentrations in Brazilian feeds associated with field outbreaks of confirmed and suspected animal mycotoxicoses, *J. Agric. Food Chem.* **40**: 994-997.

Torres, M. R., Sanchis, V., and Ramos, A. J., 1998, Occurrence of fumonisins in Spanish beers analyzed by an enzyme-linked immunosorbent assay method, *Int. J. Food Microbiol.* **39**:139-143.

Weidenboerner, M., 2001, Foods and fumonisins, *Eur. Food Res. Technol.* **212**: 262-273.

ECOPHYSIOLOGY OF *FUSARIUM CULMORUM* AND MYCOTOXIN PRODUCTION

*Naresh Magan, Russell Hope and David Aldred**

1. INTRODUCTION

Fusarium ear blight is a cereal disease responsible for significant reduction in yield and quality of wheat grain throughout the world. In addition to degradation in grain quality, *Fusarium* species produce an array of mycotoxins which may contaminate the grain. This mycotoxin production occurs preharvest and during the early stages of drying (Botallico and Perrone, 2002; Magan et al., 2002). *F. culmorum* is the most common cause of *Fusarium* ear blight in the United Kingdom and some other countries and can produce trichothecenes including deoxynivalenol (DON) and nivalenol (NIV). DON and NIV are harmful to both animals and humans, causing a wide range of symptoms of varying severity, including immunosuppression.

Germination of macroconidia of *F. culmorum* can occur over a wide range of temperatures (5-35°C) with a minimum a_w near 0.86 at 20-25°C based on an incubation period of about 40 days (Magan and Lacey, 1984a). Longer term incubations on other media have suggested limits for germination of about 0.85 a_w (Snow, 1949).

The ecological strategies used by *F. culmorum* to occupy and dominate in the grain niche are not understood. Fungi can have combative (C-selected), stress (S-selected) or ruderal (R-selected) strategies or

* Applied Mycology Group, Biotechnology Centre, Cranfield University, Silsoe, Bedford MK45 4DT, U.K. Correspondence to *n.magan@cranfield.ac.uk*

merged secondary strategies (C-R, S-R, C-S, C-S-R; Cooke and Whipps, 1993). Thus primary resource capture of nutritionally rich food matrices such as grain by *F. culmorum* may depend on germination and growth rate, enzyme production, sporulation and the capacity for producing secondary metabolites to enable effective competition with other fungi.

Attempts to control *F. culmorum* and other *Fusarium* species have relied on the application of fungicides preharvest coupled with effective storage regimens. However, the timing and application of these control measures are critical. Some fungicides are ineffective against *Fusarium* ear blight and may in some cases result in a stimulation of mycotoxin production, particularly under suboptimal fungal growth conditions and low fungicide doses (D'Mello et al., 1999; Jennings et al., 2000; Magan et al., 2002). It has been shown that moisture conditions at anthesis are crucial in determining infection and mycotoxin production by *F. culmorum* on wheat during ripening. Few studies have been carried out to determine the effect that key environmental factors such as water activity (a_w), temperature and time have on fungal growth and mycotoxin production. Some studies have identified the a_w range for germination and growth of *F. culmorum* and other *Fusarium* species (Sung and Cook, 1981; Magan and Lacey 1984a; Magan and Lacey 1984b). The combined effect of a_w and temperature has been found to be significant for growth and fumonisin production by *Fusarium verticillioides* and *F. proliferatum* which infect maize (Marin *et al.* 1999).

Knowledge of the threshold limits for growth and mycotoxin production are very important in controlling the entry of mycotoxins into the food chain. The objective of this study was to examine in detail the impact of water availability, temperature and time on growth and DON and NIV production by an isolate of *F. culmorum* on a medium based on wheat grains. Production of enzymes that may assist *F. culmorum* to dominate in the grain niche was also examined.

2. MATERIALS AND METHODS

2.1. Fungal Isolates and Media

A representative strain of *Fusarium culmorum* (98WW4.5FC, Rothamsted Research Culture Collection, Harpenden, Herts, UK) was chosen from a range examined previously and isolated from UK wheat grain, with a known history of mycotoxin production (Lacey et al., 1999; Magan et al., 2002). The strain produced quantities of mycotoxins comparable with other strains from Europe.

A 2% milled wheat grain agar (Agar No. 3, Oxoid, Basingstoke, UK) was used as the basic medium. The a_w was adjusted with glycerol in the range 0.995-0.850 as described by Dallyn (1978). The equivalent moisture contents of the a_w treatments of 0.995, 0.98, 0.95, 0.90 and 0.85 were 30%, 26%, 22%, 19% and 17.5% for wheat grain. Glycerol was used because of its inherent a_w stability over the temperature range 10-40°C. Media were sterilised by autoclaving for 15 min at 120°C. Media were cooled to 50°C before pouring into 90 mm Petri plates. The a_w of media was confirmed using an Aqualab instrument (Decagon Inc., Washington State, USA). In all cases the a_w levels were checked at both incubation temperatures (15° and 25°C) and were within 0.003 of the desired levels.

For studies in whole grain, winter wheat was gamma irradiated at 12kGy to remove contaminant microorganisms, but conserve germinative capacity. No mycotoxins were found in the grain lot used. Varying amounts of water were added to the grain and an adsorption curve prepared to facilitate accurate modifications of the a_w of the grain comparable with the media-based studies. Grain was placed in sterile flasks and inoculated with appropriate volumes of sterile water to obtain the necessary treatments. The flasks were sealed and left for 24-36 h to equilibrate at 4°C. The grain was then decanted carefully into 90 mm Petri dishes to obtain a monolayer of wheat grain. In all cases the a_w levels were checked at both temperatures as described above and were within 0.003 of the desired levels.

2.2. Inoculation and Growth Measurements

For both agar and grain based studies, replicates of each a_w treatment were inoculated centrally with a 5μl drop of a 10^5 cfu/ml *F. culmorum* macroconidial suspension obtained from a 7 d colony grown on 2% milled wheat agar. Conidia were obtained by flooding the culture with 5 ml sterile distilled water containing 0.5% Tween 80 and agitating the colony surface with a sterile glass rod. Replicates of the same treatment were enclosed in polyethylene chambers together with 500 ml of a glycerol/water solution of the same a_w, closed and incubated at 15° or 25°C for up to 40 days. Growth measurements were taken throughout the incubation period, by taking two diametric measurements of the colonies at right angles. Colonisation rates were determined subsequently by linear regression of the radial extension rates. Three replicates per treatment were removed after 10, 20, 30 or 40 d and analysed for DON and NIV (agar media) and for DON (wheat grain). The experiments were repeated once.

2.3. Mycotoxin Extraction and Analyses

Mycotoxin extraction was adapted from Cooney et al. (2001). The entire agar and mycelial culture or grain sample from each replicate sample was placed in acetonitrile/methanol (14:1; 40 ml) and shaken for 12 h. Aliquots (2 ml) were taken for DON/NIV analysis and passed through a cleanup cartridge comprising a 2 ml syringe (Fisher Ltd.) packed with a disc of filter paper (No. 1 Whatman International Ltd.), a 5 ml luger of glass wool and 300 mg of alumina/activated carbon (20:1). The sample was allowed to gravity feed through the cartridge and residues in the cartridge washed out with acetonitrile/methanol/water (80:5:15; 500 μl). The combined eluate was evaporated (compressed air, 50°C) and then resuspended in methanol/water (5:95; 500 μl).

Quantification of DON/NIV was accomplished by HPLC, using a Luna™ C18 reverse phase column (100 mm × 4.6 mm i.d.) (Phenomenex, Macclesfield, U.K.). Separation was achieved using an isocratic mobile phase of methanol/water (12:88) at an elution rate of 1.5 ml/min. Eluates were detected using a UV detector set at 220 nm with an attenuation of 0.01 AUFS. The retention times for NIV and DON were 3.4 and 7.5 min respectively. External standards were used for quantification (Sigma-Aldrich, Poole, Dorset, U.K.). The limit for quantification was 5 ng/g for DON and 2.5 ng/g for NIV.

2.4. Hydrolytic Enzyme Profiles in Grain

For enzyme extraction subsamples of grain (2 g) were placed in 4 ml potassium phosphate extraction buffer (10 mM; pH 7.2). The bottles were shaken on a wrist action shaker for 1 h at 4°C. Washings were decanted into plastic Eppendorf tubes (1.5 ml) and centrifuged in a bench microcentrifuge for 15 min. The supernatant was decanted and stored in aliquots at –20°C for total and specific enzyme activity determinations. The total activity of seven hydrolytic enzyme activities was assayed using ρ-nitrophenyl substrates (Sigma Chemical Co., UK). Enzyme extract (40 μl), substrate solution (40 μl) and the appropriate buffer (20 μl) were pipetted into the wells of the microtitre plate and incubated at 37°C for 1 h along with the appropriate controls. The reaction was stopped by the addition of 5 μl 1M sodium carbonate solution and left for 3 min. The enzyme activity was measured, using a MRX multiscan plate reader (Dynex Technologies Ltd., Billinghurst, UK), by the increase in optical density at 405 nm caused by the liberation of ρ-nitrophenol by enzymatic hydrolysis of the substrate. Enzyme activity was calculated from a calibration curve of absorbance at 405 nm *vs* ρ-nitrophenol concentration and expressed as μmol ρ-nitrophenol released/min.

For specific activity determinations the protein concentration was obtained using a Bicinchoninic acid protein assay kit (Sigma-Aldrich Ltd, Poole, Dorset, UK). This kit consisted of bicinchoninic acid solution, copper (II) sulphate pentahydrate 4% solution and albumin standard (containing bovine serum albumin (BSA) at a concentration of 1.0 mg/ml). Protein reduces alkaline Cu (II) to Cu (I), which forms a purple complex with bicinchoninic acid (a highly specific chromogenic reagent). The resultant absorbance at 550 nm is directly proportional to the protein concentration. The working reagent was obtained by the addition of 1 part copper (II) sulphate solution to 50 parts bicinchoninic acid solution. The reagent is stable for one day provided it is stored in a closed container at room temperature. Aliquots (10 μl) of each standard or enzyme extracts were placed in the appropriate microtitre plate wells. Potassium phosphate extraction buffer 10 mM pH 7.2 (10 μl) was pipetted into the blank wells. The working reagent (200 μl) was added to each well, shaken and plates incubated at 37°C for 30 min. The plates were allowed to cool to room temperature before measuring the absorbance at 550 nm using a MRX multiscan plate reader. The protein concentrations in the enzyme extracts were obtained from the calibration curve of absorbance at 550 nm against BSA concentration. These values were used to calculate the specific activity of the enzymes in nmol ρ-nitrophenol released per min per μg protein.

2.5. Statistical Analyses

The data were analysed using ANOVA (SigmaStat, SPSS Inc.), with significance values of <0.05 used. Excel 97 (Microsoft) was used to determine growth rates by linear regression.

3. RESULTS

3.1. Impact of Environment on Germination and Growth

The growth rate of *F. culmorum* on wheat-based media compared with that on monolayers of wheat grain is shown in Figure 1. Growth was very similar over the whole range at 15° and 25°C, with a limit of about 0.90 a_w. Mycelial colonisation of grain was much faster at 25° than 15°C, over the range 0.995-0.96 a_w. Previous studies on wheat-based media have identified minimum limits for germination and growth as being at 0.86 and 0.88 a_w at 20-25°C, respectively (Magan and Lacey, 1984a,b). However, germination and growth occurred over a wide temperature range (5-35°C). Studies on *F. graminearum,* also an important wheat pathogen, suggest a similar range of conditions for isolates from both Europe and South America (Hope, 2003; Ramirez et al., 2004).

3.2. Impact of Environment on Production of Deoxynivalenol and Nivalenol

Studies on wheat-based media show that the time course of production of DON varies with temperature and a_w (Figure 2). DON was produced over a narrower range of conditions (0.97-0.995 a_w) at 25° than 15°C (0.95-0.995 a_w). Concentrations produced were similar over a range of a_w levels at 15°C, while at 25°C amounts 100 times greater were produced but over a narrower a_w range. The pattern of production of NIV was different from DON, but again less NIV was produced at 15° that 25°C (Figure 3). Optimum NIV was produced at intermediate a_w levels at 15°C (0.95-0.98 a_w). In contrast, at 25°C a significantly higher concentration was produced but only at 0.98 a_w after 40 d incubation.

On wheat grain, maximum production occurred after 40 days at both 15° and 25°C over a similar a_w range to that on milled wheat media (Figure 4). Again, higher concentrations of DON were produced at 25°C, with about a 5-10 fold reduction at 15°C. However, the a_w range for rapid DON production was wider at 15° than 25°C. Comparisons can be made with production of fumonisins by *Fusarium* Section *Liseola*. Studies by Marin et al. (1999) showed that the limits for fumonisin production on maize grain were about 0.91-0.92 a_w for both *F. verticillioides* and *F. proliferatum* with temperature ranges of 15-30°C.

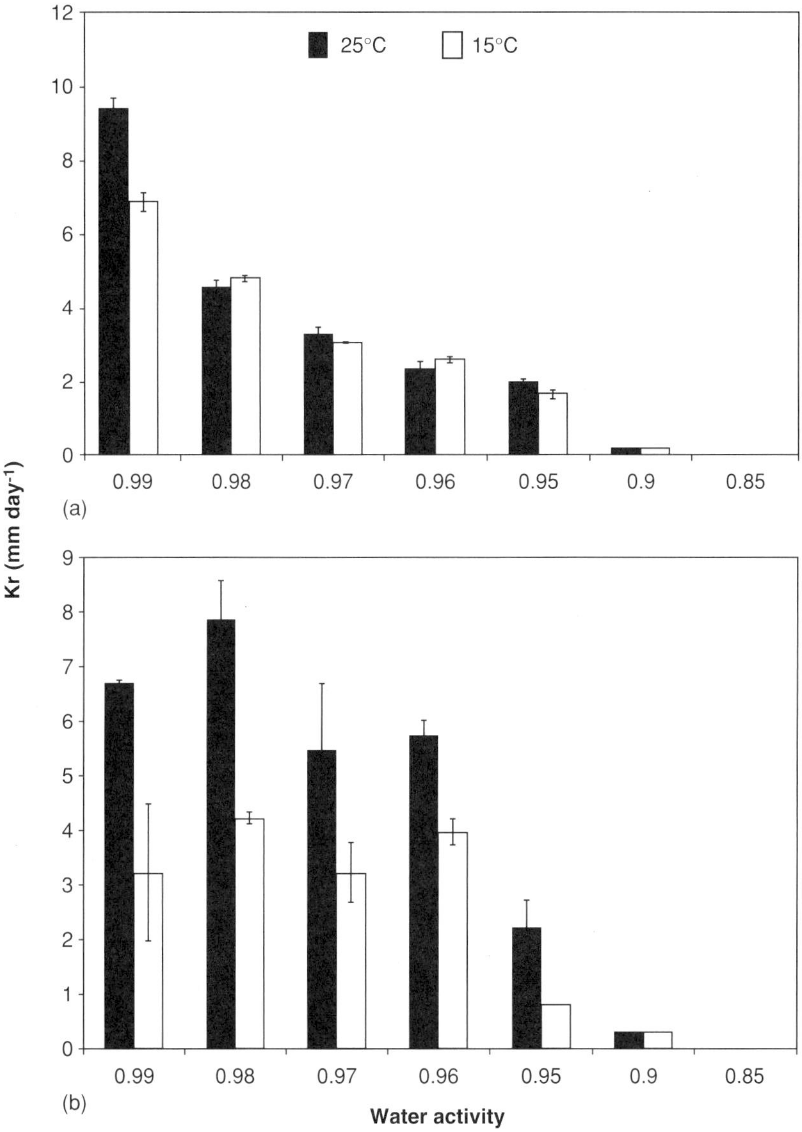

Figure 1. Comparison of growth rates of *F. culmorum* on milled wheat agar (a) and on monolayers of wheat grain (b) at 15° and 25°C (adapted from Hope and Magan, 2003 and Hope, 2003).

For *F. culmorum* the available data from the literature and present work have been combined to provide a two dimensional profile of the combined impact that a_w and temperature have on DON production on wheat. Data on germination (Snow, 1949; Magan and Lacey,

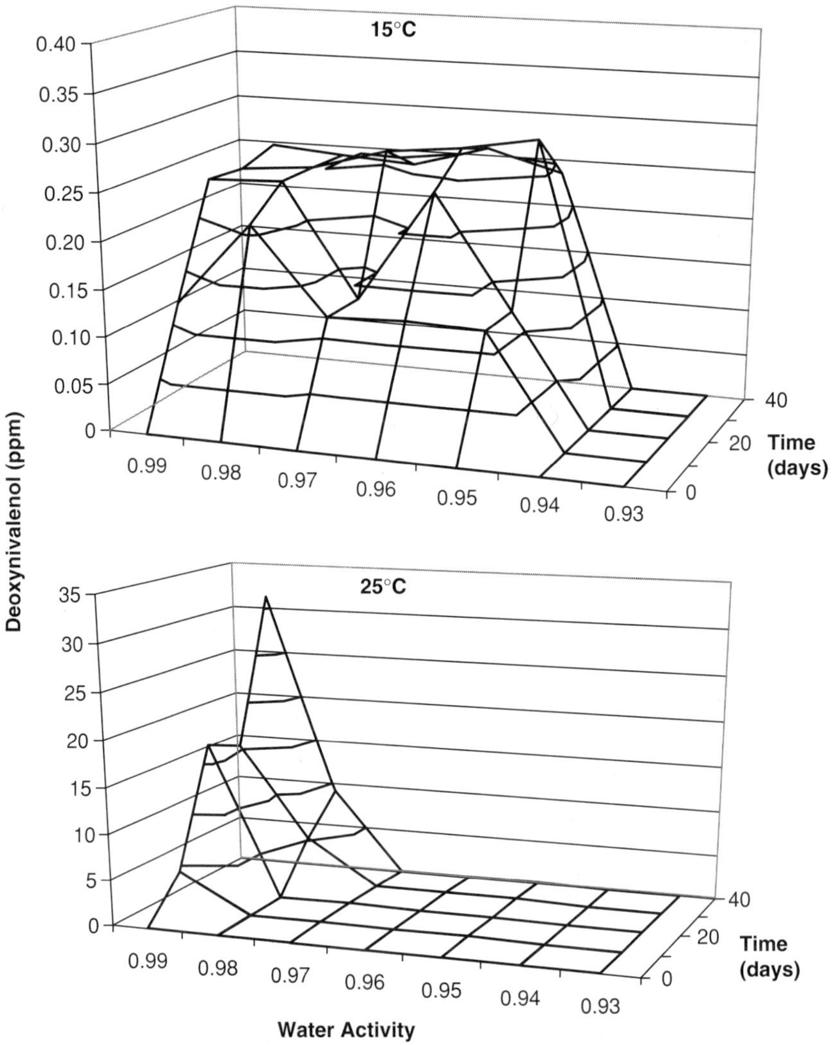

Figure 2. Effect of a_w and time on deoxynivalenol production by *F. culmorum* on milled wheat agar at 15° and 25°C (from Hope and Magan, 2003).

1984a), growth (Magan and Lacey, 1984b) and on DON production (Hope, 2003; Hope and Magan, 2003) have been combined to produce these profiles (Figure 5). This shows that the range of conditions is broader for germination and growth than for DON production. This trend is similar to that observed for other mycotoxigenic fungi (Northolt et al., 1976; 1979) for aflatoxins and *A. flavus* group, and for *Penicillium verrucosum* and ochratoxin.

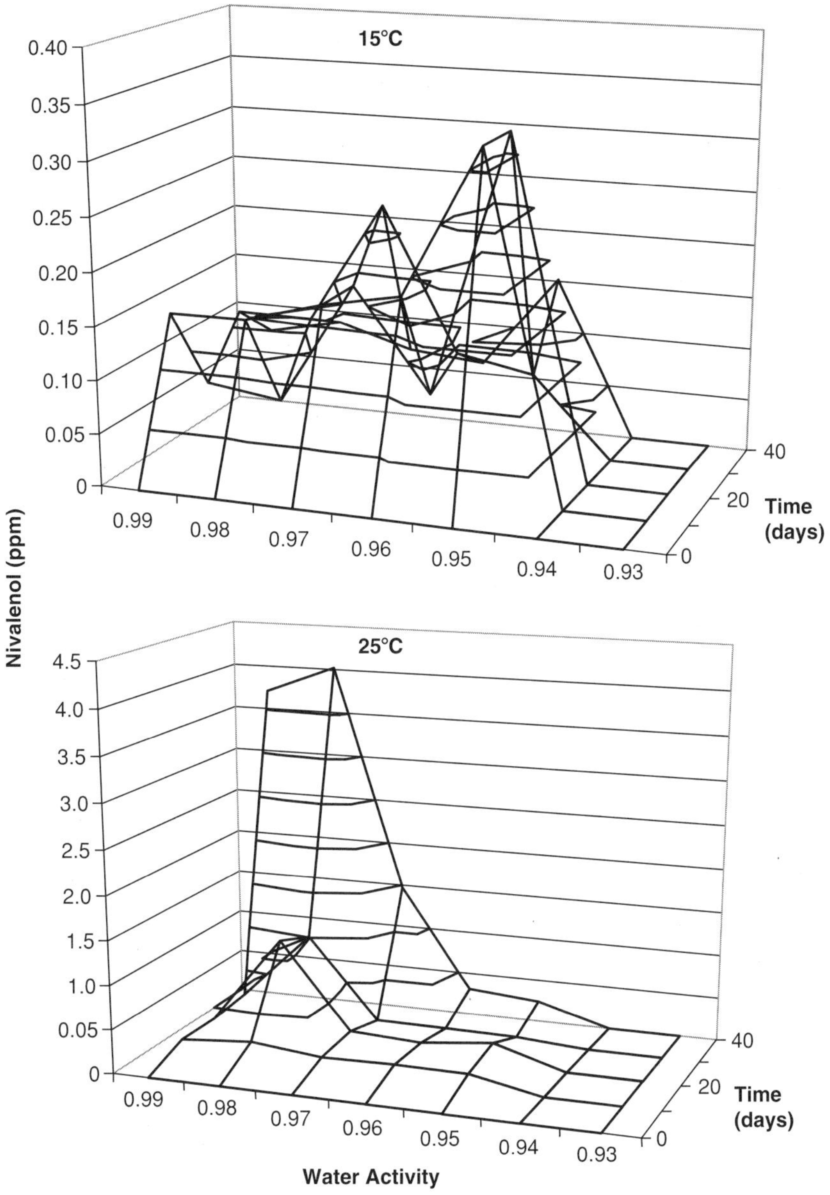

Figure 3. Effect of a_w and time on nivalenol production by *F. culmorum* on milled wheat agar at 15° and 25°C (from Hope and Magan, 2003).

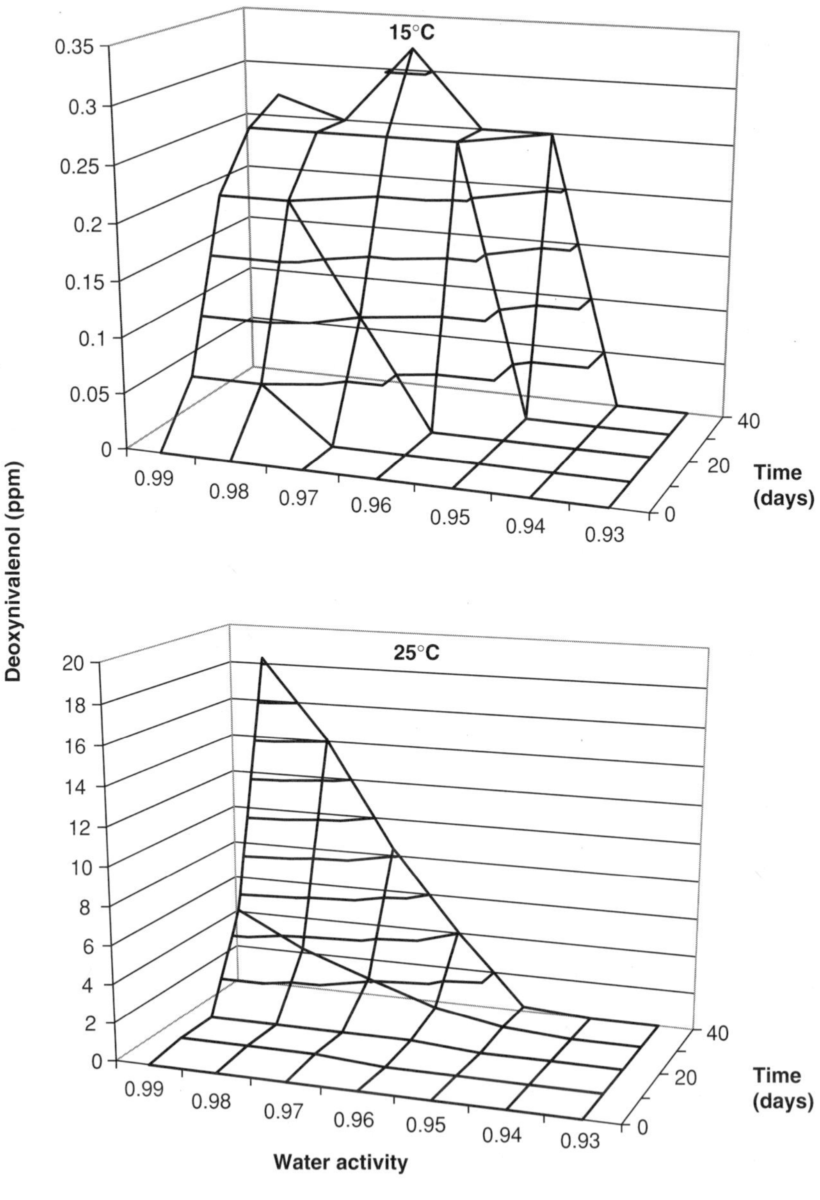

Figure 4. Effect of a_w and time on deoxynivalenol production by *F. culmorum* on monolayers of wheat grain at 15° and 25°C (Hope, 2003).

3.3. Competition and Colonisation by *F. culmorum*

Figure 6 shows that many hydrolytic enzymes are produced by *F. culmorum* which may enable rapid utilization of nutritional

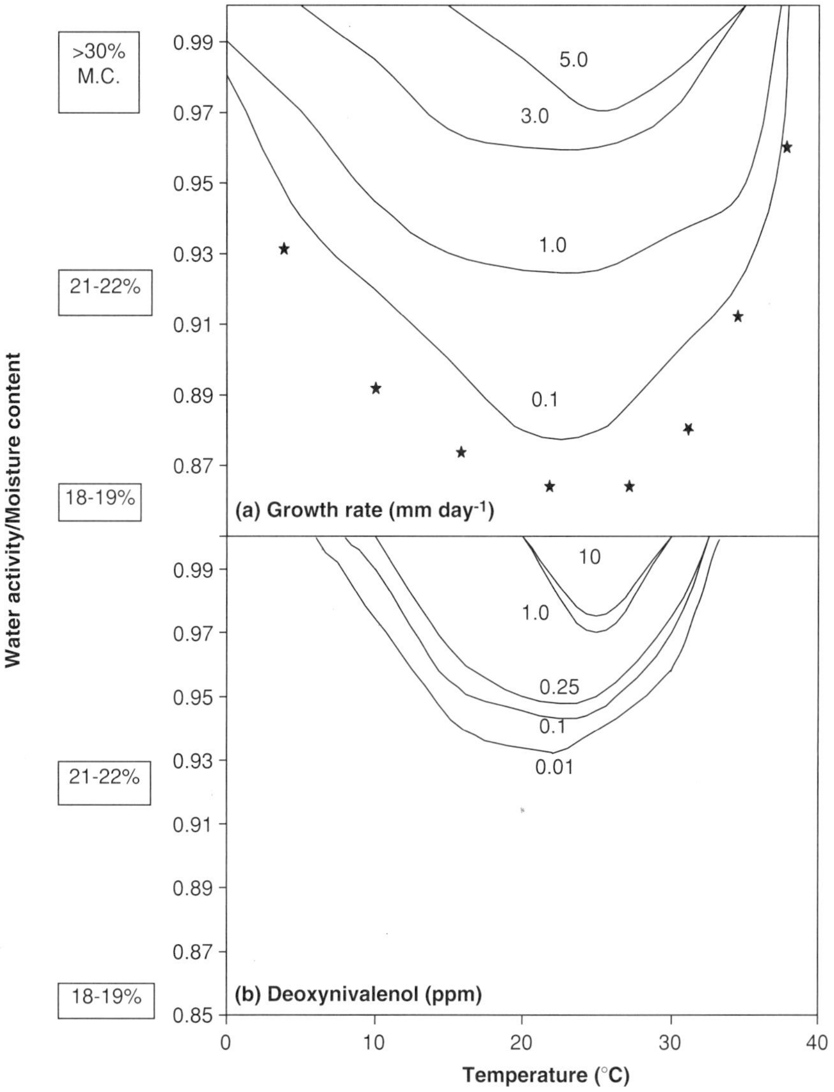

Figure 5. Comparison of profiles and limits for (a) germination (★) and growth (mm/day) and (b) DON (mg/kg) production by *Fusarium culmorum* on wheat grain (compiled from Magan and Lacey, 1984a,b; Magan, 1988; Hope, 2003; Hope and Magan, 2003).

resources over a range of a_w and temperature conditions. Previous studies have demonstrated that *F. culmorum* produces a significant amount of cellulases over a range of a_w levels (Magan and Lynch, 1986). These hydrolytic enzymes may, when combined with secondary

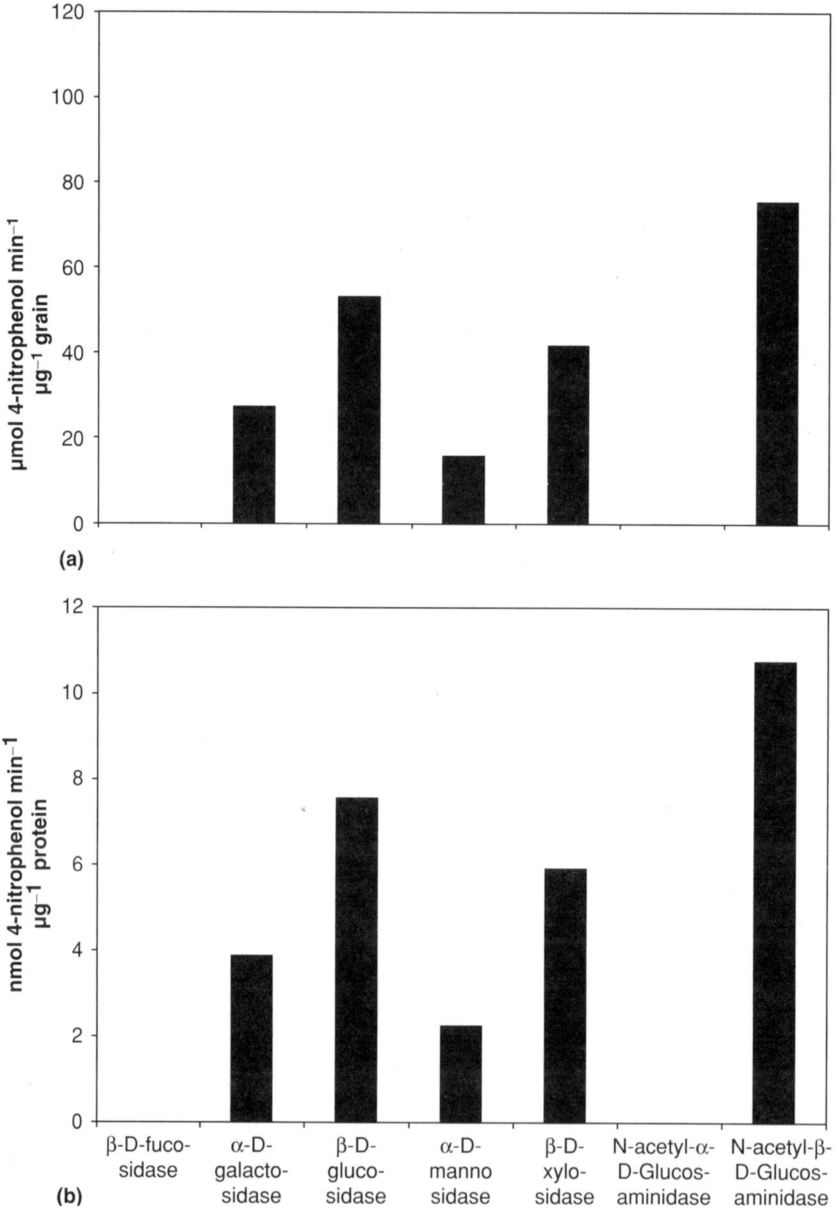

Figure 6. Production of total (a) and specific (b) activity of seven different enzymes by *F. culmorum* on irradiated wheat grain incubated for 14 days at 0.99 a_w and 25°C (Hope, 2003).

metabolite production and tolerance to intermediate moisture conditions facilitate competitiveness in food raw materials. Studies of *F. verticillioides* and *F. proliferatum* have also shown that production of some hydrolytic enzymes by these species during colonisation of maize may be used as an indicator and diagnosis of early infection by these species (Marin et al., 1998).

It is of interest that *F. culmorum* appears less competitive than *F. graminearum* when the two species interact *in vitro* on agar media or on wheat grain (Magan et al., 2003), as both occupy similar ecological niches.

4. CONCLUSIONS

Growth, competitiveness, dominance and mycotoxin production in food matrices are influenced by complex interactions between the environment, the prevailing fungal community, and external factors. The role of mycotoxins is still unclear, but they may enable fungi to occupy a particular niche, or assist in excluding other competitors from the same niche.

6. REFERENCES

Bottalico, A., and Giancarlo, P., 2002, Toxigenic *Fusarium* species and mycotoxins associated with head blight in small-grain cereals in Europe, *Eur. J. Plant Pathol.* **108**:611-624.

Cooke, R. C., and Whipps, J. M., 1993. *Ecophysiology of Fungi*, Blackwell Scientific Publications, Oxford UK. 337 pp.

Cooney, J. M., Lauren, D. R., and di Menna, M. E., 2001, Impact of competitive fungi on trichothecene production by *Fusarium graminearum, J. Agric. Food Chem.* **49**:522-526.

Dallyn, H., 1978, Effect of substrate, water activity on the growth of certain xerophilic fungi, PhD thesis, Polytechnic of the South Bank London, Council for National Academic Awards.

D'Mello, J. P. F., Placinta, C. M., and Macdonald, A. M. C., 1999, *Fusarium* mycotoxins: a review of global implications for animal health, welfare and productivity, *Animal Food Sci. Technol.* **80**:183-205.

Jennings, P., Turner J. A., and Nicholson, P., 2000, Overview of *Fusarium* ear blight in the UK -effect of fungicide treatment on disease control and mycotoxin production, *The British Crop Protection Council. Pests and Diseases 2000* **2**:707-712.

Hope, R., 2003, *Ecology and control of Fusarium species and mycotoxins in wheat grain,* PhD thesis, Institute of BioScience and Technology, Cranfield University, Silsoe, Bedford, U. K.

Hope, R. and Magan, N., 2003, Two-dimensional environmental profiles of growth, deoxynivalenol and nivalenol production by *Fusarium culmorum* on a wheat-based substrate, *Lett. Appl. Microbiol.* **37**:70-74.

Lacey J., Bateman G. L., and Mirocha C. J., 1999, Effects of infection time and moisture on development of ear blight and deoxynivalenol production by *Fusarium* spp. in wheat, *Annals Appl. Biol.* **134**:277-283.

Magan, N., 1988, Effect of water potential and temperature on spore germination and germ tube growth *in vitro* and on straw leaf sheaths, *Trans. Br. Mycol. Soc.* **90**: 97-107.

Magan, N., and Lacey, J., 1984a, The effect of temperature and pH on the water relations of field and storage fungi, *Trans. Br. Mycol. Soc.* **82**:71-81.

Magan, N., and Lacey, J., 1984b, Water relations of some *Fusarium* species from infected wheat ears and grain, *Trans. Br. Mycol. Soc.* **83**:281-285.

Magan, N., and Lynch, J.M., 1986, Water potential, growth and cellulolysis of soil fungi involved in decomposition of crop residues, *J. Gen. Microbiol.* **132**:1181-1187.

Magan, N., Hope, R., Colleate, A., and Baxter, E. S., 2002, Relationship between growth and mycotoxin production by *Fusarium* species, biocides and environment, *Eur. J. Plant Pathol.* **108**:685-690.

Magan, N., Hope, R., Cairns, V., and Aldred, D., 2003, Post-harvest fungal ecology: impact of fungal growth and mycotoxin accumulation in stored grain, *Eur. J. Plant Pathol.* **109**:723-730.

Marin, S., Sanchis, V., and Magan, N., 1998, Effect of water activity on hydrolytic enzyme production by *F. moniliforme* and *F. proliferatum* during early stages of growth on maize, *Int. J. Food Microbiol.* **42**:1-10

Marin, S., Sanchis, V., Ramos, A. J., and Magan, N., 1999, Two-dimensional profiles of fumonisin B_1 production by *Fusarium moniliforme* and *F. proliferatum* in relation to environmental factors and potential for modelling toxin formation in maize grain, *Int. J. Food Microbiol.* **51**:159-167.

Northolt, M. D., van Egmond, H. P., and Paulsch, W. E., 1979, Ochratoxin A production by some fungal species in relation to water activity and temperature, *J. Food Prot.* **42**:485-490.

Northolt, M. D., Verhulsdonk, C. A. H., Soentoro, P. S. S. and Paulsch, W. E., 1976, Effect of water activity and temperature on aflatoxin production by *Aspergillus parasiticus, J. Milk Food Technol.* **39**:170-174.

Ramirez, M. L., Chulze, S. N., and Magan, N., 2004, Impact of environmental factors and fungicides on growth and deoxinivalenol production by *Fusarium graminearum* isolates from Argentinian wheat, *Crop Prot.* **23**:117-125.

Snow, D., 1949, Germination of mould spores at controlled humidities, *Annals Appl. Biol.* **36**:1-13.

Sung, J. M., and Cook, R. J., 1981, Effect of water potential on reproduction and spore germination of *Fusarium roseum* "graminearum", "culmorum" and "avenaceum," *Phytopathology* **71**:499-504.

FOOD-BORNE FUNGI IN FRUIT AND CEREALS AND THEIR PRODUCTION OF MYCOTOXINS

*Birgitte Andersen and Ulf Thrane**

1. INTRODUCTION

The growth of filamentous fungi in foods and food products results in waste and is costly as well as sometimes hazardous. Many different fungal species can spoil food products or produce mycotoxins or both. As each fungal species produces its own specific, limited number of metabolites and is associated with particular types of food products, the number of mycotoxins potentially present in a particular product is limited (Filtenborg et al., 1996). If physical changes occur in a product, changes in the association of fungal species found in the product will also occur. With current understanding it is possible to predict which fungi and mycotoxins a given product may contain, when the type of food product and the history of production and storage are known.

In Europe, fruit has received minor attention in relation to fungal spoilage, whereas fungal spoilage of cereals has been studied extensively, but often with the focus on only one or two fungal genera. Apple juice is one of the few commodities that has caught the attention of the European authorities and regulation of patulin will be in force by the end of 2003 in Denmark (EC, 2004).

* Center for Microbial Biotechnology, BioCentrum-DTU, Technical University of Denmark, DK-2800 Kgs. Lyngby, Denmark. Correspondence to: ba@biocentrum.dtu.dk

Knowledge of the composition and succession of the mycobiota in cereal grains and fruit during maturation, harvest and storage is an important step towards the prediction of possible mycotoxin contamination. Some major spoilage genera on stored apples and cherries, e.g. *Botrytis*, *Cladosporium* and *Rhizopus* are not known to produce significant mycotoxins, while others including *Alternaria*, *Aspergillus*, *Fusarium* and *Penicillium* include species capable of producing a wide range of mycotoxins (Pitt and Hocking, 1997). The production of mycotoxins is often species specific (Frisvad et al., 1998), so accurate identification of fungi to species is of major importance. Historically, identifications have not always been accurate, and incorrect identifications have resulted in confusion and misinterpretations (Marasas et al., 1984; Thrane, 2001; Andersen et al., 2004). In the case of *Alternaria*, where many taxa are still undescribed (Simmons and Roberts, 1993; Andersen et al., 2002), identification is only possible to a species-group level for many isolates. Before a contaminated sample is analysed for mycotoxins, it is important to know which mycotoxins are likely to be present. Metabolite profiles from known species grown in pure culture can provide valuable information about the mycotoxins that may be found in cereals and fruit and their products, once the fungi normally associated with those products are known.

During the last 15 years our group has analysed numerous cereal and fruit samples and recorded the fungal species found. Analysis of that large amount of data has shown that similar fungal species occur on the same product types year after year. The purpose of this paper is to present a list of the fungal species found on apples, cherries, barley and wheat from the Northern temperate zone together with a list of mycotoxins known to be produced by these fungi.

2. MATERIALS AND METHODS

2.1. Media

Dichloran Rose Bengal Yeast Sucrose agar (DRYES; Frisvad, 1983) and V8 juice agar (V8; Simmons, 1992) were used for fungal analyses of fruit, while Czapek Dox Iprodione Dichloran agar (CZID; Abildgren et al., 1987), Dichloran 18% Glycerol agar (DG18; Hocking and Pitt, 1980), DRYES, and V8 were used for analysis of cereals. CZID plates were incubated in alternating light/dark cycle consisting of 12 hours of black fluorescent and cool white daylight and 12 hours darkness at 20-23°C, while DG18 and DRYES plates were incubated at 25°C in darkness and V8 plates in alternating cool white daylight (8 hours light/16 hours darkness) at 20-23°C. *Alternaria* and other dematiaceous hyphomycetes were enumerated on DRYES and/or V8. *Fusarium* species were enumerated on CZID and/or V8, while *Eurotium, Aspergillus* and *Penicillium* species and other hyaline fungi were enumerated on DG18 and/or DRYES.

A wide range of media were used for fungal identification. For *Alternaria* species and other black fungi, DRYES, Potato Carrot Agar (PCA; Simmons, 1992) and V8 were used; for *Eurotium* species, Malt Extract Agar (MEA, Pitt and Hocking, 1997) and Czapek Dox agar (CZ; Samson et al., 2004) were used; for *Fusarium* species, Potato Dextrose agar (PDA; Samson et al., 2004) Yeast Extract Sucrose agar (YES; Samson et al., 2004) and Synthetischer nährstoffarmer agar (SNA: Nirenberg, 1976) were used. For *Penicillium*, Czapek Yeast extract Agar (CYA; Pitt and Hocking, 1997), MEA, YES and Creatine Sucrose agar (CREA; Samson et al., 2004) were used.

2.2. Fruit

Apple flowers with petals removed, apple peel and apples with fungal lesions were plated directly onto DRYES and V8. Sound apples were surfaced disinfected by shaking in freshly prepared 0.4 % NaOCl for 2 minutes and then rinsing with sterile water. The cores of the surface disinfected apples were cut out with a sterile cork borer, cut into 5 pieces and plated on V8 and DRYES.

Cherry flowers with petals and cherries with fungal lesions were plated directly onto DRYES and V8. Some cherries were surfaced disinfected as described above. The stem and calyx ends of the surface disinfected cherries were excised with a sterile scalpel and plated on V8 and DRYES. The plates were incubated as described above and after a week of incubation the fungal colonies were enumerated. Representative colonies were then isolated and identified to species level.

The development in the mycobiota at genus level was followed on apples (variety Jonagored) and two varieties of cherries (Vicky and Van) during one growth season (2001 for cherries and 2002 for apples). The trees were sprayed for fungal diseases (grey mould, *Botrytis* spp.) several time before harvest. Flowers of both apples and cherries were examined for fungal growth before the first spray application. Ten trees were selected from each orchard and sampled (10 units) from each tree. The ten samples from each orchard were pooled (100 units per sample) and screened for the presence of *Alternaria*, *Botrytis*, *Cladosporium*, *Fusarium* and *Penicillium* species.

2.3. Cereals

Barley and wheat samples were plated directly with and without surface disinfection. Kernels were surfaced disinfected by shaking in freshly prepared 0.4 % NaOCl for 2 minutes and then rinsing with sterile water. Each sample consisted of 500 grains, that were not surface disinfected. All samples were plates on DG18, DRYES, V8 and CZID. The plates were incubated as described above and after a week of incubation the fungal colonies were enumerated. Representative colonies were then isolated and identified to species level. They were screened for the presence of *Alternaria*, *Bipolaris*, *Cladosporium*, *Eurotium/Aspergillus*, *Fusarium* and *Penicillium* species.

The development in the mycobiota at genus level was followed on wheat (variety Leguan) and barley (variety Ferment) during one growth season (2002). One batch of barley seed was treated with fungicide before planting, while a second batch, and the wheat seed, were not treated.

2.4. Fungal Species Associated with Fruit and Cereals

In our laboratory, data on the occurrence of fungal contamination of fresh, stored, processed and mouldy samples of cereals and fruits have been collected, recorded and compiled for more than 15 years. Samples from the field, food factories and local supermarkets have been surveyed. A number of cultivars of apples (e.g. Cox's Orange, Discovery, Gala, Jonagored), cherries (e.g. Bing, Van, Vicky), barley (e.g. Alexis, Chariot, Ferment, Krona) and wheat (e.g. Leguan) have been sampled and the resulting data analysed.

2.5. Fungal Identification and Metabolite Profiling

Alternaria and *Stemphylium* isolates were transferred to DRYES and PCA, while other black fungi, including *Cladosporium* isolates, were transferred to DRYES and V8. Identifications of the black fungi were done according to Andersen et al. (2002), Simmons and Roberts (1993), Samson et al. (2004) and Simmons (1967; 1969; 1986). Metabolite profiling involved extracting nine plugs of 14 day old DRYES cultures in ethyl acetate (1 ml) with formic acid (1%; Andersen et al., 2002) in an ultrasonic bath for 60 min. The ethyl acetate was evaporated and the dried sample redissolved in methanol (500 µl). Samples were then filtered and analysed on a HP1100 HPLC-DAD (Agilent, Germany) (Andersen et al., 2002).

Fusarium isolates were transferred to SNA, PDA and YES and identified according to Gerlach and Nirenberg (1982), Nirenberg (1989), Burgess et al. (1994) and Samson et al. (2004). Metabolites were profiled by extracting nine plugs of 14 day old PDA and YES cultures, each in dichloromethane: ethyl acetate (2:1 vol/vol; 1 ml) with formic acid (1%) in an ultrasonic bath for 60 min. Then the organic phase was evaporated and the dried sample redissolved in methanol (500 µl). Samples were then filtered and analysed on a HP1100 HPLC-DAD (Smedsgaard, 1997).

Penicillium isolates were transferred to CYA, MEA and YES and identified according to Samson et al. (2004) and Samson and Frisvad (2004). Metabolites were profiled by extracting three plugs of 7 day old CYA and YES cultures, then treated as for *Fusarium* extracts.

2.6. Metabolites from Mouldy Fruit and Cereals

Metabolites were extracted from the mouldy samples in the same way as for the pure fungal cultures. Each sample (100 g) was mixed with ethyl acetate (100 ml) containing formic acid (1%). The mixture was shaken regularly over a 2 hour period to extract the metabolites, then held overnight in a freezer. Extracts (14 ml) were decanted from the frozen water and sample matrix, evaporated to dryness, redissolved in methanol (500 µl) and analysed as before.

3. RESULTS

3.1. Fungal Development in Fruit

The dominant fungal genera found in flowers, immature and mature fruit are given in Table 1. In apples, *Cladosporium* and *Alternaria* species constituted the major infection in the flowers. *Botrytis* and *Fusarium* species were also found frequently, whereas *Penicillium* species were isolated only rarely. After the trees were sprayed with fungicide, *Botrytis* was eliminated and the number of *Cladosporium* colonies was halved. The spraying did not seem to effect the numbers of *Alternaria*, *Fusarium* or *Penicillium* colonies enumerated. The numbers of these three genera increased with time and were highest at harvest.

Table 1. Fungal infection (%) in Danish apples and cherries during the growth seasons 2002 and 2001 respectively

	Apples (Jonagored)			Sour cherries (Vicky)			Sweet cherries (Van)		
	Fl.[a]	Imm.[b]	Mat.[c]	Fl.	Imm.	Mat.	Fl.	Imm.	Mat.
Alternaria	77	80	88	68	33	60	66	22	18
Botrytis	35	0	0	4	0	5	78	1	2
Cladosporium	81	73	40	26	0	31	37	5	21
Fusarium	20	53	64	2	0	7	0	1	6
Penicillium	5	13	20	0	0	0	0	1	0

[a]Fl. = flowers; [b] Imm. = immature; [c]Mat. = mature

On the flowers of sweet cherries, *Botrytis* was the dominant genus, whereas it was isolated in very low numbers from the flowers of sour cherries (Table 1). In flowers of both cherry types, *Alternaria* and *Cladosporium* were found in high numbers and *Fusarium* was found in low numbers. After spraying, *Botrytis* was more or less eliminated in both immature and mature cherries. The number of *Cladosporium* colonies seen was greatly reduced in immature cherries after spraying, but the numbers rose again as the cherries matured. Numbers of *Alternaria* isolated were somewhat reduced in immature cherries after spraying, but the numbers rose again in sour cherries while it fell in sweet cherries as they matured. A *Penicillium* species was found in only one sample of immature sweet cherries.

The dominant toxigenic fungi on apples were *Alt. tenuissima* species-group, followed by *Alt. arborescens* species-group, *F. avenaceum*, *F. lateritium*, *P. crustosum* and *P. expansum*. On cherries, the dominant species were *Alt. arborescens* species-group followed by *Alt. tenuissima* species-group, *F. lateritium* and *P. expansum*.

3.2. Fungal Development in Cereals

The dominant genera found in seed, immature (harvested by hand) and mature (machine harvested) wheat and barley kernels are shown in Table 2. In untreated wheat seed, *Penicillium* constituted the major infection in the seed together with *Alternaria*, *Eurotium* and *Aspergillus*. The same composition of fungi was seen in untreated barley seed, except that *Penicillium* counts were less than 50% of those in treated seed. In barley seeds that had been treated with fungicides before sowing, only two out of 500 grains were found to be infected with fungi. The changes in mycobiota in immature kernels compared to the seed were most pronounced in barley, where the numbers of

Table 2. Fungal infection (%) in Danish wheat and barley seeds and kernels without surface disinfection during the growth season 2002

	Wheat (Leguan)			Barley (Ferment)			Barley (Ferment)		
	Seed[a]	Imm.[b]	Mat.[c]	Seed[a]	Imm.	Mat.	Seed[d]	Imm.	Mat.
Alternaria	42	42	67	40	73	64	0	88	58
Botrytis	0	0	9	5	43	37	0	22	31
Eurotium/ Aspergillus	25	0	1	19	1	0	0	1	0
Cladosporium	0	20	2	0	94	0	1	12	18
Fusarium	0	82	90	0	59	83	0	78	98
Penicillium	98	89	98	41	7	98	1	0	98

[a]Seed without fungicide treatment; [b] Imm. = Immature; [c] Mat. = Mature; [d] Fungicide treated seed

Fusarium and *Alternaria* rose markedly and the storage fungi disappeared. The change was less dramatic in wheat. The differences in mycobiota from immature to mature kernels were minor and mostly the number of fungal infected kernels was stable or rose slightly. The number of *Cladosporium* found on immature and mature samples varied a great deal as high number of *Fusarium* and *Penicillium* colonies often obscured the smaller *Cladosporium* colonies. Surface disinfection reduced the numbers of *Penicillium* and *Eurotium* colonies by 80-90 %, and of *Cladosporium* and *Fusarium* by 40-50 %. Only 10-15% of *Alternaria* and *Bipolaris* could be removed, indicating that the grains had internal infections with these genera.

Dominant fungi in common to both wheat and barley kernels were isolates of *Alt. infectoria* species-group, *F. avenaceum, P. aurantiogriseum, P. cyclopium* and *P. polonicum*. However, barley had higher infection rates with *Bipolaris sorokiniana, P. hordei* and *P. verrucosum* compared to wheat.

3.3. Fungal Species Associated with Fruit and Cereals

The frequencies of occurrence of fungal species in several cultivars of apples, cherries, barley and wheat are given in Tables 3 and 4. In our laboratory, such data have been compiled for more than 15 years. The differences in the mycobiota between cultivars were in most cases small. Most often the same fungal species were found and the variation was quantitative only. As can been seen from Tables 3 and 4, only a limited number of fungal species are found in both fruit and cereals.

Newly harvested, undamaged apples and cherries were not usually infected by fungi. When infection was present, fungi mostly belonged to the dematiaceous hyphomycetes (Table 3). The mycobiota of fresh apples consisted of mainly of *Alt. tenuissima* species-group, *Botrytis* and *Cladosporium* spp. In fresh cherries, *Alt. arborescens* species-group and *Stemphylium* spp. constituted the mycobiota, along with *Botrytis* and *Cladosporium* spp. In storage, the mycobiota changed in both types of fruit. In cherries, after only a few weeks in storage, *Botrytis* spp., *P. expansum* and Zygomycetes dominated the mycobiota and they often spoiled the cherries. In stored apples the same fungal species were found after months of storage together with *P. solitum*. Fungi were rarely found in juice made from apples or cherries, but when fungal growth was detected, *Byssochlamys* spp. and *P. expansum* were found in pasteurised and untreated juices, respectively.

In contrast to fruit, samples of newly harvested, sound cereal grains always had some fungal infections after surface disinfection. Colonies of *Alt. infectoria* species-group, *Cladosporium* spp., *F. avenaceum* and *F. tricinctum* were always isolated from fresh barley and wheat. *Epicoccum nigrum*, *F. culmorum*, *F, equiseti* and *F. poae* were often seen also. As in fruit, the mycobiota changed in cereals in storage in favour of *Aspergillus* and *Penicillium* species. In dry barley and wheat samples that had been stored for one year or more, *Eurotium* spp. and

Table 3. Fungal occurrence (frequency) in apples and cherries from the north temperate zone: data accumulated over a 15 year period [a]

	Apples			Cherries		
	Fresh	Stored	Juice	Fresh	Stored	Juice
Alternaria arborescens sp.-grp.	+	–	–	++	+	–
Alt. infectoria sp.-grp.	(+)	–	–	(+)	–	–
Alt. tenuissima sp.-grp.	++	+	–	+	–	–
Botrytis spp.	++	++	–	++	++	–
Byssochlamys spp.	–	–	+	–	–	+
Cladosporium spp.	++	+	–	++	+	–
Fusarium avenaceum	+	+	–	–	–	–
F. lateritium	+	(+)	–	+	–	–
Monilia spp.	(+)	+	–	(+)	+	–
Penicillium carneum	(+)	+	–	–	–	–
P. crustosum	+	+	–	–	–	–
P. expansum	+	++	(+)	+	++	(+)
P. polonicum	–	(+)	–	–	–	–
P. solitum	(+)	++	–	–	–	–
Stemphylium spp.	(+)	–	–	++	–	–
Zygomycetes	–	++	–	–	++	–

[a]+++: Always present; ++: often present; +: sometimes present; (+): rarely present; –: never detected or found only once. Fresh indicates direct plated, surface disinfected sound samples; stored, direct plated disinfected sound or visibly mouldy samples

P. aurantiogriseum always dominated the mycobiota. However, *Alt. infectoria* species-group could still be isolated from barley that had been stored for two years. *Aspergillus candidus*, *Asp. flavus*, *P. cyclopium*, *P. hordei*, *P. melanoconidium*, *P. polonicum*, *P. verrucosum* and *P. viridicatum* were also often found in stored barley and in lower numbers in stored wheat.

3.4. Production of Toxic Metabolites in Pure Culture

Six genera out of the twelve listed in Tables 3 and 4 are regarded as being non-toxigenic, namely *Botrytis*, *Cladosporium*, *Epicoccum,*

Table 4. Fungal occurrence (frequency) in barley and wheat from the north temperate zone: data accumulated over a 15 year period [a]

	Barley		Wheat	
	Fresh	Stored	Fresh	Stored
Alternaria arborescens sp.-grp.	(+)	–	(+)	–
Alt. infectoria sp.-grp.	+++	++	+++	++
Alt. tenuissima sp.-grp.	+	–	+	–
Aspergillus candidus	–	++	–	++
Asp. flavus	–	++	–	++
Asp. niger	–	+	–	+
Bipolaris sorokiniana	+	(+)	(+)	–
Botrytis spp.	(+)	–	(+)	–
Cladosporium spp.	+++	(+)	+++	(+)
Epicoccum nigrum	++	–	++	–
Eurotium spp.	–	+++	–	+++
Fusarium avenaceum	+++	+	+++	+
F. culmorum	++	(+)	++	(+)
F. equiseti	++	–	++	–
F. graminearum	+	–	+	–
F. langsethiae	(+)	–	–	–
F. lateritium	(+)	–	–	–
F. poae	++	+	++	+
F. sporotrichioides	+	–	+	–
F. tricinctum	+++	+	+++	+
Penicillium aurantiogriseum	(+)	++	(+)	++
P. cyclopium	(+)	++	(+)	++
P. freii	(+)	+	(+)	++
P. hordei	+	++	(+)	+
P. melanoconidium	–	++	–	++
P. polonicum	–	++	–	++
P. verrucosum	+	++	+	++
P. viridicatum	–	++	–	++
Stemphylium spp.	(+)	–	(+)	–

[a]See footnote to Table 3

Eurotium, Monilia and *Stemphylium*, together with the Zygomycetes (Pitt and Hocking, 1997). Two genera which include toxigenic species, *Bipolaris* and *Byssochlamys*, were found only relatively rarely in cereals and fruit, respectively. *Alternaria*, *Aspergillus*, *Fusarium* and *Penicillium* species were much more commonly isolated. These four genera include 25 toxigenic species found on a regular basis in either fruit or cereals. Only *Alt. tenuissima* species-group and *F. avenaceum* were regularly found in both fresh fruit and fresh cereals. In Table 5 the mycotoxins that are produced in pure culture by the fungal species listed in Tables 3 and 4 are given. Of all of the species in Tables 3 and 4, only *Alt. infectoria* species-group and *P. solitum* are regarded as non-toxigenic. As can been seen from Table 5, several fungal species within the same genus and found in the same product can produce the same mycotoxins (e.g. roquefortine C and penitrem A by Penicillia in fruit or culmorins and trichothecenes by Fusaria in cereals). In fruit that has either been damaged in the orchard and/or stored poorly, one or more of the following toxic metabolites might theoretically be found: altenuene, alternariols, chaetoglobosins, citrinin, patulin, roquefortine C and tenuazonic acid. In fresh cereals that have been harvested during a rainy period, the following toxic metabolites would be relevant: antibiotic Y, beauvericin, culmorins, enniatins, fusarin C, fusarochromanone, moniliformin, trichothecenes and zearalenone. In cereals that have been stored poorly and/or not dried down after harvest, the following toxic metabolites should be considered: aflatoxins, aspergillic acid, citrinin, cyclopiazonic acid, nephrotoxic glycopeptides, ochratoxin A, penicillic acid, terphenyllin, verrucosidin, viomellein, vioxanthin, viridic acid, xanthoascin and xanthomegnin. However, it should be noted that mycotoxins produced in the field or at an early stage of storage always should be taken into consideration, as mycotoxins in general are persistent through storage and processes for food and feed production.

3.5. Production of Toxic Metabolites in Fruit and Cereals

Analyses of extracts from pure fungal cultures can indicate which fungal metabolites should be analysed, but the mycobiota of the actual sample needs to be determined to make realistic recommendations. Mycotoxins that have been detected in naturally moulded apples, cherries, barley and wheat samples examined in our laboratory are given in Table 6. Samples that were mouldy when received were extracted immediately, while sound samples were incubated for a week

Table 5. Fungal species found in apples, cherries wheat and barley from the north temperate zone and some mycotoxins they are known to produce

Toxigenic fungal species	Mycotoxins produced in pure culture
Alternaria arborescens sp.-grp.	Altenuene, alternariols, altertoxins, tenuazonic acid
Alt. tenuissima sp.-grp.	Altenuene, alternariols, altertoxins, tenuazonic acid
Aspergillus candidus	Terphenyllin, xanthoascin
Asp. flavus	Aflatoxin, aspergillic acid, cyclopiazonic acid
Asp. niger	Malformins, naphtho-γ-pyrones
Bipolaris sorokiniana	Sterigmatocystin
Byssochlamys spp.	Byssochlamic acid, patulin
Fusarium avenaceum	Antibiotic Y, aurofusarin, enniatins, fusarin C, moniliformin
F. culmorum	Aurofusarin, culmorin, fusarin C, trichothecenes, zearalenone
F. equiseti	Fusarochromanone, trichothecenes, zearalenone
F. graminearum	Aurofusarin, culmorin, fusarin C, trichothecenes, zearalenone
F. langsethiae	Culmorin, enniatins, trichothecenes
F. lateritium	Antibiotic Y, enniatins,
F. poae	Beauvericins, culmorins, fusarin C, trichothecenes
F. sporotrichioides	Aurofusarin, beauvericins, culmorins, fusarin C, trichothecenes
F. tricinctum	Antibiotic Y, aurofusarin, enniatins, fusarin C, moniliformin
Penicillium aurantiogriseum	Penicillic acid, verrucosidin, nephrotoxic glycopeptides
P. carneum	Patulin, isofumigaclavin, penitrem A, roquefortine C
P. crustosum	Penitrem A, roquefortine C
P. cyclopium	Penicillic acid, xanthomegnin, viomellein, vioxanthin
P. expansum	Citrinin, chaetoglobosins, communesins, patulin, roquefortine C
P. freii	Penicillic acid, xanthomegnin, viomellein, vioxanthin
P. hordei	Roquefortine C
P. melanoconidium	Penicillic acid, penitrem A, xanthomegnin
P. polonicum	Penicillic acid, verrucosidin, nephrotoxic glycopeptides
P. verrucosum	Citrinin, ochratoxin A
P. viridicatum	Penicillic acid, viridic acid, xanthomegnin, viomellein

Table 6. Mycotoxins detected in naturally infected samples

Sample	Metabolites detected
Immature apples with mouldy core (1)	Alternariols, antibiotic Y, aurofusarin
Immature apples with mouldy core (2)	Altenuene, alternariols
Mature apples with mouldy core (3)	Alternariols, patulin
Mouldy apple pulp (4)	Alternariols, antibiotic Y, aurofusarin, ascladiol,
Mouldy apple on tree (5)	Chaetoglobosin A
Mouldy sweet cherries 'June drop' (6)	Alternariols, antibiotic Y
Mouldy, mature sweet cherries (7)	Alternariols
Mouldy cherry juice (8)	Chaetoglobosins, communesins, roquefortine C
Mouldy barley (9)	Ochratoxin A
Mouldy barley 'hot spot' (10)	Ochratoxin A
Mouldy wheat (11)	Antibiotic Y, ochratoxin A, zearalenone
Mouldy wheat (12)	Aurofusarin, fusarin C, zearalenone

or until mould could be seen to simulate a worst case. *Fusarium* and/or *Alternaria* metabolites were detected in samples of immature apples from the field with visible fungal growth (samples 1 and 2, worst cases), while the mature, fresh apple samples (samples 3 and 4, worst cases) also contained *Penicillium* metabolites. One mature, mouldy apple (sample 5) still hanging on the tree contained only chaetoglobosin A. The cherry sample (sample 6), which consisted of immature cherries that had been shed by the trees, had 100 % infection with *Alternaria* species. Extraction showed that it contained high amounts of both *Alternaria* and *Fusarium* metabolites, whereas the mature cherries (sample 7, worst case) only contained *Alternaria* metabolites. The mouldy cherry juice (sample 8), on the other hand, contained only *Penicillium* metabolites. The mouldy barley samples (samples 9 and 10), which had been stored without drying after harvest, contained ochratoxin A. Sample 10, sampled in the 'green hot spot' of the mouldy lot, contained approximately 1000 times the amount of ochratoxin A as sample 9, which had little visible fungal growth. In the mouldy wheat sample (sample 11) *Fusarium* metabolites as well as ochratoxin A were detected, while sample 12 only contained *Fusarium* metabolites.

4. DISCUSSION

Analyses of seasonal changes in apple mycobiota showed that many fungal species found in mature fruit already were present in the flowers and later colonized the immature fruit (Table 1). Spraying with fungicides decreased *Botrytis* infection and had some effect on *Cladosporium* numbers, but no effect on *Alternaria, Fusarium* and *Penicillium*. The number of *Alternaria, Fusarium* and *Penicillium* infected apples also increased after spraying and continued to increase until apples were picked. In cherries, the same fungal species were seen in mature cherries as in flowers (Table 1). Application of fungicides had an effect on all the fungal genera found in flowers; *Alternaria* infections decreased by 50-65%. *Alternaria* and *Cladosporium* numbers, however, increased again before harvest in sour cherries, while the numbers of *Alternaria* remained constant in sweet cherries, probably due to the early drop of immature cherries, which had 100% infection with *Alternaria*. Our results show that the mycobiota of apples and cherries are similar at genus level, but different in species composition. *Alternaria tenuissima* species-group, *P. expansum* and *P. solitum* dominate in apples, whereas *A. arborescens* species-group, and *Stemphylium* spp. dominate in cherries (Table 3).

Analyses of the mycobiota in cereals from sowing to harvest showed, in contrast to fruit, that the initial mycobiota present in the untreated seed played only a small role in the subsequent mycobiota on mature kernels, though it may play a great role in the viability of the seed (Table 2). The two untreated seed samples contained a high number of *Alternaria*, *Penicillium* and *Eurotium* species. Surface disinfection removed 80-90% of the *Penicillium* and *Eurotium* numbers, while the same only could be done for 40-45% of the *Alternaria*. Furthermore, the *Penicillium* species in the seed and in the immature kernels were different. In the seed *P. chrysogenum, P. cyclopium* and *P. freii* were found, whereas *P. aurantiogriseum, P. polonicum* and *P. verrucosum* were found in mature kernels growing from the untreated seed. The only fungi that were found in larger amounts in seed and recovered in more than 50% of the harvested cereal samples belonged to *A. infectoria* species-group. The numbers of *Alternaria* and *Fusarium* found in the three mature samples were low (less than 60%) and high (more than 80%), respectively, compared with other reports (Andersen et al., 1996; Kosiak et al., 2004). A very wet period in June and July 2002, during the growing season in Western Denmark was probably responsible. Comparisons of the mycobiota from two mature barley samples grown from fungicide treated and untreated seed showed few differences.

As fruits mature and are harvested, fungi such as *Botrytis, Monilia* and Zygomycetes are known to cause fruit spoilage in orchards as well as in storage, whereas *Cladosporium* and *Epicoccum* are known for their discolouration of cereals in the field. These fungi cause economical losses, but none of them are associated with production of mycotoxins.

Different species of *Alternaria*, *Fusarium* and *Penicillium,* on the other hand, all spoil fruit and cereals, but produce species specific mycotoxins (Table 5) and hundred mycotoxins and other biologically active metabolites from these three genera have been characterised within recent years (Nielsen and Smedsgaard, 2003) and it is reasonable to expect that more than the few included in the legislation (aflatoxin, ochratoxin, deoxynivalenol, fumonisins in cereals and patulin in fruits) can be produced in mouldy foods.

The results presented in this study show that *Alternaria* and *Fusarium* in fruit and cereals may pose a mycotoxin risk. During spoilage of apples and cherries, *P. expansum* is known to produce patulin, which has been incorporated in the legislation on fruit produce. However, both *Alternaria* and *Fusarium* were able to produce additional metabolites in mouldy fruit samples (Table 6, sample 4): alternariols, antibiotic Y and aurofusarin. In cereals, *P. verrucosum* is known to produce ochratoxin A, which has also been incorporated in the legislation on raw cereal grain. However, *Fusarium* was able to produce antibiotic Y and zearalenone in addition to ochratoxin A from *P. verrucosum* in mouldy wheat (Table 6, sample 11). For these lesser known metabolites no or very limited data are available on the toxicity on co-produced metabolites and their possible synergistic effects, which make risk assessment in food and food production systems difficult. In conclusion, we see the co-occurrence of these specific *Alternaria* and *Fusarium* metabolites and their potential toxicities as the major future challenge in food mycology.

5. ACKNOWLEDGEMENTS

The authors are grateful to Dr. Jens C. Frisvad for discussion of manuscript and identification of some of the *Penicillium* cultures. This work was partly supported by the Danish Ministry of Food, Agriculture and Fisheries through the program "Food Quality with a focus on Food Safety", by LMC Centre for Advanced Food Studies and by the Danish Technical Research Council through 'Program for Predictive Biotechnology'.

6. REFERENCES

Abildgren, M. P., Lund, F., Thrane, U., and Elmholt, S., 1987, Czapek-Dox agar containing iprodione and dicloran as a selective medium for the isolation of *Fusarium* species, *Lett. Appl. Microbiol.* **5**:83-86.

Andersen, B., Thrane, U., Svendsen, A., Rasmussen, I. A., 199, Associated field mycobiota on malting barley, *Can. J. Bot.* **74**:854-858.

Andersen, B., Krøger, E., and Roberts, R.G., 2002, Chemical and morphological segregation of *Alternaria arborescens*, *A. infectoria* and *A. tenuissima* species-groups, *Mycol. Res.* **106**:170-182.

Andersen, B., Smedsgaard, J., and Frisvad, J. C., 2004, *Penicillium expansum*: consistent production of patulin, chaetoglobosins and other secondary metabolites in culture and their natural occurrence in fruit products, *J. Agric. Food Chem.* **52**:2421-2428.

Burgess, L. W., Summerell, B. A., Bullock, S., Gott, K. P., and Backhouse, D., 1994, *Laboratory Manual for Fusarium Research*. 3rd Edition, University of Sydney, Sydney, Australia.

EC (European Commission), 2004, European Commission Regulation 455/2004 of 11 March, 2004. European Commission, http://europa.eu.int/eur-lex/pri/en/oj/dat/2004/l_074/l_07420040312en00110011.pdf

Filtenborg, O., Frisvad, J. C., and Thrane, U., 1996, Moulds in food spoilage, *Int. J. Food Microbiol.* **33**:85-102.

Frisvad, J. C., 1983, A selective and indicative medium for groups of *Penicillium viridicatum* producing different mycotoxins on cereals, *J. Appl. Bacteriol.* **54**:409-416.

Frisvad, J. C., Thrane, U., and Filtenborg, O., 1998, Role and use of secondary metabolites in fungal taxonomy, in: *Chemical Fungal Taxonomy*, J. C. Frisvad, P. D. Bridge, and D. K. Arora, (eds), Marcel Dekker, New York, pp. 289-319.

Gerlach, W., and Nirenberg, H., 1982, *The genus* Fusarium *-A Pictorial Atlas,* Mitteilungen aus der Biologische Bundesanstalt für Land-und Forstwirtschaft, Berlin-Dahlem **209**:1-406.

Hocking, A. D., and Pitt, J. I., 1980, Dichloran-glycerol medium for enumeration of xerophilic fungi from low moisture foods, *Appl. Environ. Microbiol.* **39**:488-492.

Kosiak, B., Torp, M., Skjerve, E., and Andersen, B., 2004, *Alternaria* and *Fusarium* in Norwegian grains of reduced quality -a matched pair sample study, *Int. J. Food Microbiol.* **93**:51-62.

Marasas, W. F. O., Nelson, P. E., and Toussoun, T. A., 1984, *Toxigenic* Fusarium *Species. Identity and Mycotoxicology,* The Pennsylvania State University Press, University Park, pp. 1-328.

Nielsen, K. F., and Smedsgaard, J., 2003, Fungal metabolite screening: database of 474 mycotoxins and fungal metabolites for de-replication by standardised liquid chromatography-UV detection-mass spectrometry methodology, *J. Chromatogr. A.* **1002**:111-136.

Nirenberg, H., 1976, Untersuchungen über die morphologische und biologische Differenzierung in der *Fusarium*-Sektion *Liseola,* Mitteilungen aus der Biologische Bundesanstalt für Land-und Forstwirtschaft. Berlin-Dahlem **169**:1-117.

Nirenberg, H. I., 1989, Identification of Fusaria occurring in Europe on cereals and potatoes, in: *Fusarium: Mycotoxins, Taxonomy and Pathogenicity*, J. Chelkowski (ed.), Elsevier Science Publishers B.V., Amsterdam, pp. 179-193.

Pitt, J. I., and Hocking, A. D., 1997, *Fungi and Food Spoilage*, Blackie Academic and Professional, London, pp. 1-593.

Samson, R. A., Frisvad, J. C., 2004, *Penicillium* subgenus *Penicillium*: new taxonomic schemes, mycotoxins and other extrolites, *Stud. Mycol.* **49**:1-257.

Samson, R. A., Hoekstra, E. S., Frisvad, J. C., (eds), 2004, *Introduction to Food- and Airborne Fungi*. 7th Edition. Centraalbureau voor Schimmelcultures, Utrecht, pp. 1-389.

Simmons, E. G., 1967, Typification of *Alternaria, Stemphylium* and *Ulocladium, Mycologia* **59**:67-92.

Simmons, E. G., 1969, Perfect states of *Stemphylium, Mycologia* **61**:1-26.

Simmons, E. G., 1986, *Alternaria* themes and variations (22-26), *Mycotaxon* **25**:287-308.

Simmons, E. G., 1992, *Alternaria* taxonomy: Current Status, viewpoint, challenge, in: *Alternaria Biology, Plant Diseases and Metabolites*, J. Chelkowski and A. Visconti, eds, Elsevier, Amsterdam, pp. 1-35.

Simmons, E. G., and Roberts, R. G., 1993, *Alternaria* themes and variations (73), *Mycotaxon* **48**:109-140.

Smedsgaard, J., 1997, Micro-scale extraction procedure for standardized screening of fungal metabolites production in cultures, *J. Chromatogr. A* **760**:264-270.

Thrane, U., 2001, Developments in the taxonomy of *Fusarium* species based on secondary metabolites, in: *Fusarium*. Paul E. Nelson Memorial Symposium. B. A. Summerell, J. F. Leslie, D. Backhouse, W. L. Bryden, and L. W. Burgess (eds), APS Press, St. Paul, Minnesota, pp. 29-49.

BLACK *ASPERGILLUS* SPECIES IN AUSTRALIAN VINEYARDS: FROM SOIL TO OCHRATOXIN A IN WINE

Su-lin L. Leong,[*‡] *Ailsa D. Hocking,*[*] *John I. Pitt,*[*] *Benozir A. Kazi,*[†] *Robert W. Emmett*[†] *and Eileen S. Scott*[‡§]

1. INTRODUCTION

Fungi classified in *Aspergillus* Section *Nigri* (the black Aspergilli) are ubiquitous saprophytes in soils around the world, particularly in tropical and subtropical regions (Klich and Pitt, 1988; Pitt and Hocking, 1997). Several species in this Section are common in vineyards and are often associated with bunch rots (Amerine et al., 1980). *A. niger* is reported to be the primary cause of Aspergillus rot in grapes before harvest (Nair, 1985; Snowdon, 1990), while *A. aculeatus* (Jarvis and Traquair, 1984) and *A. carbonarius* (Gupta, 1956) have also been reported. The development of fungal bunch rots has been correlated with the splitting of grape berries (Barbetti, 1980), and *Aspergillus* counts on grapes grown for drying were greater during seasons when rain before harvest caused the berries to split (Figure 1) (Leong et al., 2004). Spores of black *Aspergillus* spp. are resistant to UV light (Rotem and Aust, 1991), which may account for their

* Su-lin L. Leong, Ailsa D. Hocking and John I. Pitt, CSIRO Food Science Australia, North Ryde, NSW 2113, Australia.
[†]Benozir A. Kazi and Robert W. Emmett, Department of Primary Industries, Mildura, Victoria 3502, Australia.
[‡]Su-lin L. Leong and Eileen S. Scott, University of Adelaide, Glen Osmond, South Australia 5064, Australia.
[§]All authors, Cooperative Research Centre for Viticulture, Glen Osmond, South Australia 5064, Australia. Correspondence to: ailsa.hocking@csiro.au

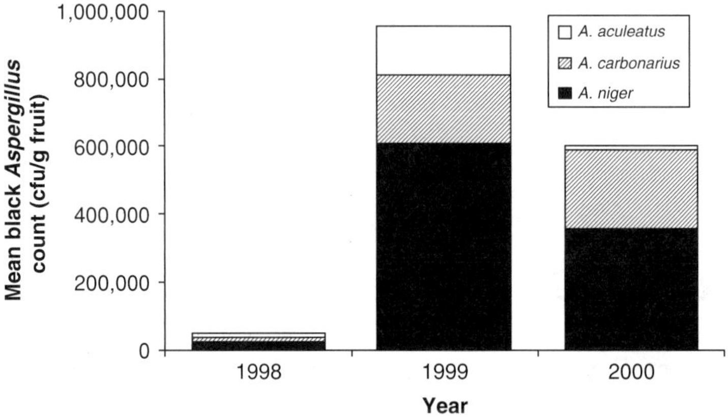

Figure 1. Mean severity of infection of fresh and drying grapes (combined) by each of three black *Aspergillus* species over three successive harvest seasons. Rain occurred before harvest in 1999 and 2000. The mean count was derived by summing the counts for each species of black *Aspergillus* from all the fruit samples, and dividing that sum by the total number of samples. Reproduced from Leong, S. L., Hocking. A. D., and Pitt, J. I., 2004, *Australian Journal of Grape and Wine Research* **10**: 83-88 (with permission from the Australian Society of Viticulture and Oenology).

persistence in vineyards and on grape berries even after drying (King et al., 1981; Abdel-Sater and Saber, 1999; Abarca et al., 2003).

Within Section *Nigri*, *A. carbonarius* and *A. niger* have been shown to produce the mycotoxin, ochratoxin A (OA) (Abarca et al., 1994; Téren et al., 1996; Heenan et al., 1998; reviewed in Abarca et al., 2001). OA is a demonstrated nephrotoxin, which may also be carcinogenic, teratogenic, immunogenic and genotoxic. It has been classified as Group 2B, a "possible human carcinogen" (Castegnaro and Wild, 1995).

OA has been detected in grapes and grape products including juice, wine, dried vine fruit and wine vinegars (Zimmerli and Dick, 1996; MacDonald et al., 1999; Majerus et al., 2000; Markarki et al., 2001; Da Rocha Rosa et al., 2002; Sage et al., 2002; OA in wine and grape juice reviewed by Bellí et al., 2002). A survey of 600 Australian wines showed that OA was present only at low levels. Only 15% of samples had levels > 0.05 μg/l and 85% of these were < 0.2 μg/l. The maximum level found was 0.61 μg/l (Hocking et al., 2003). In Europe, a limit of 2 μg/kg in table wines is under discussion (Anon., 2003) and a limit of 10 μg/kg in dried vine fruits has been set (European Commission, 2002).

OA in grapes and grape products is produced by toxigenic *A. carbonarius* and *A. niger* species which have been isolated from grapes in

France (Sage et al., 2002), South America (Da Rocha Rosa et al., 2002), Spain (Cabañes et al., 2002), Italy (Battilani et al., 2003b), Portugal (Serra et al., 2003) and Greece (Tjamos et al., 2004). In an extensive study of Australian dried vine fruit, strains of *A. carbonarius* were commonly isolated from semi-dried and dried vine fruit in the field, and all were capable of producing OA in the laboratory (Leong et al., 2004). Hence *A. carbonarius* is thought to be the primary species responsible for OA production in grapes in Australia.

Assuming that OA production in grapes ceases at the commencement of processing, typically a sterilisation step in industrial juice and wine production (Roset, 2003), the concentration of OA in the final product is a function of the initial concentration in the grapes and the effect of processing. Processes which reduce OA can be classified into two groups, physical removal and degradation.

Physical removal of OA first involves removing the site where OA has been produced, for example, the removal of visibly mouldy berries from table grapes. It is not well understood if OA is primarily associated with the skin, pulp or juice of grape berries. However, a strong association with the skin or pulp would suggest that a relatively small proportion of OA remains in the finished beverage. The high water content of grape berries may lead to the migration of OA from the zone of fungal growth to other parts of the berry (Engelhardt et al., 1999).

A second aspect of physical removal of OA is the partitioning of the toxin between solid and liquid phases during processing. Fernandes et al. (2003) conducted microvinification trials on crushed grapes spiked with OA, and reported reductions in OA of 50-95%. The most significant reductions resulted from solid-liquid separation steps, such as pressing the juice or wine from the skins, and decanting the wine from precipitated solids. Many of the solids present in grape juice have an affinity for OA and will loosely bind and precipitate the toxin from solution (Roset, 2003), as do some fining agents added during winemaking, such as activated charcoal (Dumeau and Trione, 2000; Castellari et al., 2001; Silva et al., 2003).

Little is known about the degradation of OA by wine yeasts during fermentation, though this has been demonstrated during beer fermentation (Baxter et al., 2001). Silva et al. (2003) reported reduction in OA by lactic acid bacteria during malolactic fermentation which follows the completion of primary (yeast) fermentation. However, Fernandes et al. (2003) argued that this is not a true degradation, rather, bacterial biomass binding OA that later settles out of the wine. The addition of sulphur dioxide and the pasteurisation of juice by heating have no effect on OA (Roset, 2003).

This paper presents original data on OA contamination of wine, covering the source of *A. carbonarius* in Australian vineyards, the survival of *A. carbonarius* spores on the surface of bunches, and the passage of OA throughout vinification of grapes inoculated with *A. carbonarius*.

2. MATERIALS AND METHODS

2.1. *Aspergillus carbonarius* in the Vineyard Environment

Substrates were collected from vineyards in the grape-growing region centred around Mildura, Victoria, Australia. Substrates collected included parts of vines [green, yellow (senescing) and dead leaf tissue, green and dead berries, dead bunch remnants (dried rachides), tendrils, canes and bark] and materials from the vineyard floor [green cover crop plants, dead cover crop trash, vine trash and soil]. Collections were made over three growing seasons, from six vineyards in 2000-01 and from three vineyards in 2001-02 and 2002-03. In 2000-01, samples of substrates were collected from three sites along a diagonal transect in each vineyard 2 weeks after veraison and at harvest. In the latter two seasons, samples of substrates were collected from five sites along a diagonal transect in each vineyard when berries were pea size, at 2 weeks after veraison and at harvest. To quantify *A. carbonarius* present on the surface of these substrates, samples were washed for 2 min in sterile water containing Citowett® (BASF Australia Ltd, Victoria, Australia) as a wetting agent, and aliquots of the solution were plated in duplicate onto Dichloran Rose Bengal Chloramphenicol Agar (DRBC) (Pitt and Hocking, 1997). Serial dilutions were performed on soil samples, followed by plating onto DRBC. After plates were incubated at 25°C for 5-7 days, colonies of *A. carbonarius* were identified and enumerated.

The presence of *A. carbonarius* in vineyard soils was also compared among four vineyards in 2002-03 by dilution plating as described above. Thirty soil samples were collected from each vineyard, ten samples at each stage of vine growth, i.e. when berries were pea size, at 2 weeks after veraison and at harvest. The tillage practices of the vineyards were noted. Soil was sampled directly under vines and also between vine rows. In one vineyard, five soil sampling cores of 0.2 m^2 were taken from under vines. For each soil core, *A. carbonarius* was enumerated in soil at the surface, 5 cm and 15 cm below the surface. *A. carbonarius* was also enumerated in the air of this vineyard. Colonies from 20 l of air sampled using a MAS-100® air sampler

(Merck KGaA, Darmstadt, Germany) were enumerated on DRBC. Samples were taken on nine occasions from air in the vineyard at 10 cm, 100 cm and 180 cm above the vineyard floor.

Statistical analyses were performed using Genstat (6th edition, Lawes Agricultural Trust, Rothamsted, UK).

2.2. Survival of *Aspergillus carbonarius* Spores on the Surface of Bunches Preharvest

A trial was conducted in the Hunter Valley, New South Wales, Australia to examine the survival of *A. carbonarius* spores on the surface of Chardonnay and Shiraz grapes (three replicate rows) during the growing season in 2002-03. The vines were over 25 years old, trained onto horizontal wires and under drip irrigation.

Spores of 7-14 day old cultures of *A. carbonarius* on Czapek Yeast Agar (CYA) (Pitt and Hocking, 1997) plates were harvested into sterile water containing Tween-80® (0.05% w/v; Merck, Victoria, Australia), and diluted to $2\text{-}4 \times 10^5$ colony forming units per ml (cfu/ml). Bunches were inoculated by immersion in 1 l of suspension contained within a plastic bag. The same inoculum was used for up to 40 bunches without decrease in the spore concentration. Twelve bunches in each row were inoculated at pre-bunch closure (berries green and pea size), veraison and 11-16 days preharvest. Two bunches were combined into a single sample, resulting in six samples per replicate at each sampling stage. Inoculated bunches were sampled after the inoculum had dried to give an initial value, at each of the subsequent stages and at harvest.

Bunches were homogenised for 3 min in a stomacher (BagMixer, Interscience, France) with the addition of sterile distilled water, and serial dilutions of the suspension were plated onto DRBC. After incubation for 3 days at 25°C, colonies of *A. carbonarius* were enumerated. The average berry weight at each growth stage was calculated, and the number of *A. carbonarius* colonies was expressed as cfu per berry, in order to compare the number of viable spores present during each stage.

2.3. Winemaking

2.3.1. Inoculation and Vinification of Grapes

Berries were inoculated preharvest with a spore suspension of *A. carbonarius* (prepared as described in 2.2) at approximately 10^7 cfu/ml.

Strains selected for inoculation were local to the region of the experimental vineyard, and were strong producers of OA when screened on Coconut Cream Agar (CCA) (Heenan et al., 1998). A variety of inoculation techniques was employed, all involving puncture damage to the berry skin and subsequent contact with the spore suspension. In addition to the primary inoculation, a supplementary inoculation of additional fruit was often performed towards harvest to ensure sufficient fruit for vinification. At harvest, inoculated and uninoculated fruit were mixed to simulate high, intermediate and low or absent levels of OA in fruit. Table 1 summarises the inoculation, incubation and harvest details.

Table 1. Preparation of OA-contaminated grapes for winemaking

Location, vintage	Mildura, Victoria, 2002		Hunter Valley, New South Wales, 2003	
***A. carbonarius* strains**	FRR 5374[a], FRR 5573, FRR 5574		FRR 5682, FRR 5683	
Grape variety	Chardonnay	Shiraz	Semillon	Shiraz
Method of inoculation	Berries injected using syringe {berries injected using syringe}[b]	Berries injected using syringe {skin scored using grater and sprayed with spore suspension}	Berries punctured with a bed of pins dipped in spore suspension {berries injected using syringe}	Berries injected using syringe
Period from primary inoc. until harvest	21 days {4 days}[b]	14 days {13 days}	9 days {3 days}	8 days
High OA wine: mass of grapes	53 kg inoculated	120 kg inoculated	25 kg inoculated	28 kg inoculated
Intermediate OA wine: mass of grapes	34 kg inoculated + 23 kg uninoculated	46 kg inoculated + 73 kg uninoculated	15 kg inoculated + 13 kg uninoculated	11 kg inoculated + 16 kg uninoculated
Control wine: mass of grapes	51 kg uninoculated	118 kg uninoculated	32 kg uninoculated	27 kg uninoculated
Size of ferment	4 l	16 l including skins	2 l	4 l including skins

[a]FRR numbers are from the culture collection of Food Science Australia, North Ryde, NSW, Australia.
[b]{} bracketed text refers to supplementary inoculation of additional fruit.

Harvested bunches were chilled at 4°C prior to crushing. After crushing, eight samples of must from each toxin level were collected in order to establish the initial total OA present in the berries. Samples were also collected throughout the vinification process as described below.

During white vinification, the must was pressed, after which potassium metabisulphite was added to give 50 ppm SO_2 in the juice. Pectinase was added in the form of Pomolase AC50 (0.05 ml/l juice; Enzyme Solutions, Victoria, Australia) or Pectinase (0.5 g/l juice; Fermtech, Queensland, Australia). The juice was overlaid with nitrogen or carbon dioxide, and refrigerated at 4°C for at least 24 h to precipitate solids. In 2002, the juice was divided into four replicate ferments at each toxin level before clarification; this division occurred after clarification in 2003. The pH was adjusted to approximately 3.3 by the addition of tartaric acid to give a titratable acidity of 6.5-7.0 g/l. The clarified juice was siphoned into bottles filled with nitrogen or carbon dioxide and fitted with stoppers and air traps. The yeast QA23 (Lallemand, Toulouse, France) was rehydrated and added at a rate equivalent to 0.2 g dry yeast/l juice. Diammonium phosphate was added at 0.5 g/l juice. The fermentation temperature was 19°C in 2002 and 15°C in 2003. Diammonium phosphate was added during fermentation as required and after fermentation was completed, the wine was racked. Potassium metabisulphite was added at a rate equivalent to 50 ppm SO_2 to stabilise the wine and prevent further fermentation. Bentonite (0.5 g/l; Fermtech, Queensland, Australia) and Liquifine (2002: 1 ml/l; 2003: 0.6 ml/l; Winery Supplies, Victoria, Australia) were added, and the bottles placed at 19°C (2002) or 15°C (2003) to allow precipitation of solids. A second racking was performed for all bottles, and potassium metabisulphite was added to bring the free SO_2 to 20 ppm. The bottles were placed at < 4°C for cold stabilisation for > 30 days. The wine was filtered into 375 ml glass bottles with cork closures.

During red vinification, the must was divided into 4 replicate fermentations at each toxin level. Potassium metabisulphite was added to give 50 ppm SO_2 in the must, diammonium phosphate was added at 0.5 g/l must, and tartaric acid was added to bring the titratable acidity to 6.5 g/l. The yeast D254 (Lallemand, Toulouse, France) was rehydrated and added at approximately 0.3 g/l must. The cap was plunged 2-3 times daily. The must was pressed after 4 days of fermentation at room temperature in 2002, and after 6 days of fermentation at approximately 20°C in 2003. Fermentation was finished in bottles at room temperature. During the first racking, 50 ppm SO_2 was added, after which the wine was held at 19°C (2002) or 15°C (2003) to precipitate yeast cells and other solids. At the second racking, SO_2 was added to maintain a final concentration of 50 ppm. The wine was held at < 2°C for cold stabilisation, after which the pH was adjusted and the wine bottled through a filtration line.

Bottles were cellared at room temperature (approximately 22°C) in 2002, and at 15°C in 2003.

2.3.2. OA Assays

A new method was developed for the rapid analysis of OA in grape matrices. Samples were standardised by weight.

Grape musts were homogenised and a 10 g subsample weighed into a centrifuge tube. Methanol (10 ml), Milli-Q water (10 ml) and 10N HCl (≈ 0.15 ml) were added, and mixed thoroughly with the sample. For liquid samples, 10 g of sample was mixed with methanol (1.5 ml) and 10N HCl (≈ 0.15 ml). The mixture was centrifuged at 2500 rpm for 15 min. A 900 mg C18 solid phase extraction cartridge (Maxi-Clean, Alltech, Deerfield, USA) was conditioned with 5 ml acetonitrile followed by 5 ml water, and the supernatant was passed dropwise through this cartridge under vacuum. The pellet was resuspended in 10 ml 10% methanol, then centrifuged for a further 15 min at 2500 rpm. This supernatant was also passed through the C18 cartridge. For must samples, an additional 10 ml water was washed through the C18 cartridge at this stage.

A 200 mg aminopropyl cartridge (4 ml Extract-Clean, Alltech, Deerfield, USA) was conditioned with 3 ml methanol. The C18 and aminopropyl cartridges were attached in series, and the sample was eluted from the C18 cartridge onto the aminopropyl cartridge with the addition of 10 ml methanol. The sample was eluted from the aminopropyl cartridge with 10 ml 35% ethyl acetate in cyclohexane containing 0.75% formic acid.

The eluate was dried under reduced pressure at 45°C and was resuspended in 1 ml mobile phase (35% acetonitrile containing 0.1% acetic acid) for analysis by HPLC (Hocking et al., 2003).

Aliquots of the wine extracts were chromatographed on an Ultracarb (30) C18 4.6 × 250 mm, 5 μm column (Phenomenex, Torrance, USA). The mobile phase consisted of acetonitrile: water:acetic acid (50:49:1, v/v), and was delivered through the heated column (40°C) at a flow of 1.3 ml/min using a Shimadzu 10A VP high pressure binary gradient solvent delivery system. Detection of OA was achieved by post column addition of ammonia (12.5% w/w,

0.2 ml/min) and monitoring the natural fluorescence of OA at 435 nm after excitation at 385 nm (Shimadzu, RF-10AXL). Sample injections were performed using a Shimadzu SIL-10Advp autosampler, and the typical injection volume was 30 µl. OA in the wine extracts was quantified by comparison with a calibration curve. Typical recoveries ranged from 80-100%. The results presented have not been corrected for recovery.

3. RESULTS AND DISCUSSION

3.1. *Aspergillus carbonarius* in the Vineyard Environment

Counts of *A. carbonarius* were high in soil and in vine trash on soil and relatively low on other substrates (Table 2). Hence, soil is likely to be the primary source of *A. carbonarius* in vineyards.

In the four vineyards surveyed, soil beneath vines contained more *A. carbonarius* than soil between rows ($P < 0.05$, Figure 2). This association is likely to be due to damaged and dead berries falling onto the soil and providing a sugar-rich medium for the growth of indigenous saprophytic *Aspergillus* species. The concentration of *A. carbonarius* propagules was highest at the surface of soil, where this debris is found, and decreased deeper within the soil profile of an untilled vineyard (Figure 3). The vineyard in which the soil profile was regularly disturbed by tilling contained a higher concentration of *A. carbonarius* in the soil than vineyards in which the soil was less disturbed ($P < 0.05$, Figure 2). In vineyards with minimal tillage, *A. carbonarius* may be one member of a complex and stable microbial community associated with the cover crop and other flora on the soil surface. One potential effect of regular tillage is to allow the increase of a dominant species (Marfenina and Mirchink, 1989).

Table 2. Aspergillus carbonarius on vineyard materials. Results expressed as cfu/ml surface wash unless otherwise indicated.

Substrate	2000-01	2001-02	2002-03
Canes (dead)	8	1	not recorded
Vine bark	31	9	not recorded
Bunch remnants	47	15	not recorded
Cover trash (dead)	8	10	2
Vine trash on soil	669	45	20
Soil (cfu/g)	4,987	1,219	1,342

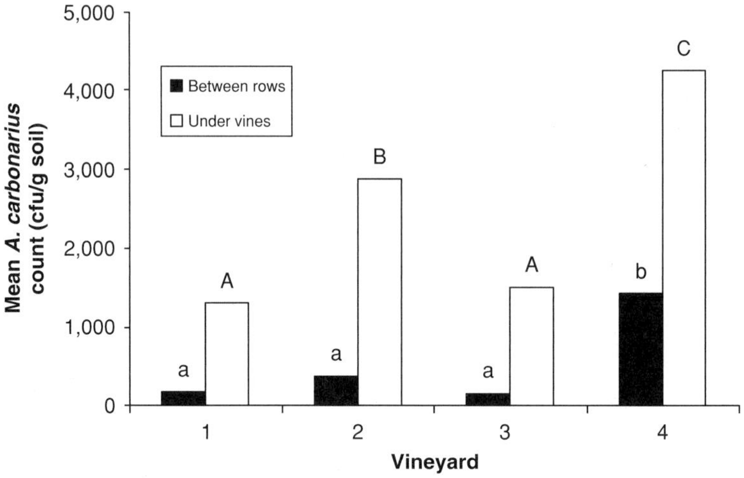

Figure 2. *Aspergillus carbonarius* in soil under vines and between vine rows in vineyards with minimal and continual tillage, n = 30. Vineyard 1 and 3: No tillage for the last 3 and 4 years, respectively. Vineyard floors were covered with wild grasses and/or weeds. Vineyard 2: Tilled once each year before sowing a rye grass cover crop. Vineyard 4: Tilled monthly after a cover crop was established. Vertical bars with different letters differ significantly (LSD, $P < 0.05$).

A. carbonarius may compete more effectively for sugars from fallen berries, and tillage would also distribute propagules throughout the soil profile. Other authors have observed higher levels of some fungi in tilled soil than in untilled soil. An example is the primary

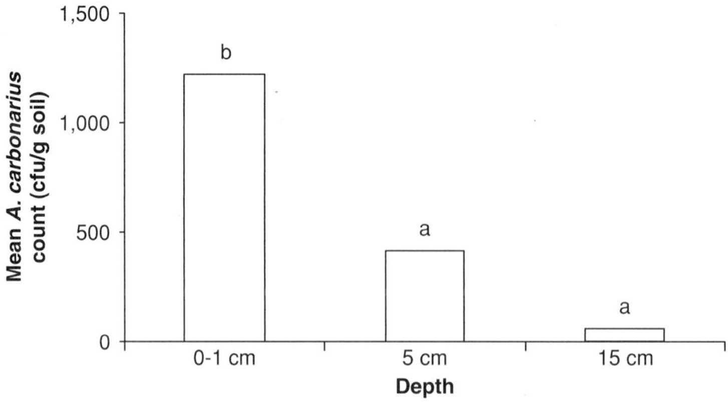

Figure 3. *Aspergillus carbonarius* in untilled vineyard soil at different depths in 2001-02, n = 5. Vertical bars with different letters differ significantly (LSD, $P < 0.05$).

pathogen of root rot of wheat, *Cochliobolus sativus* (Diehl, 1979; Duczek, 1981; Diehl et al., 1982).

A. carbonarius was present in vineyard air, though the concentration of conidia decreased with height from the vineyard floor (Figure 4). This suggests that wind may distribute conidia of *A. carbonarius* from the soil onto berry surfaces.

During the early stages of berry development from pre-bunch closure until veraison, *A. carbonarius* spores survived poorly and a nine fold decrease in the number of viable propagules was observed (Figure 5). This suggests that the surface of green berries is a hostile environment for the survival of spores. The vine canopy during the early part of the season in 2002-2003 was sparse due to drought. Thus, the berries were not shielded from UV light by overhanging leaves. A sparse canopy would also allow greater penetration of routine fungicide sprays, which, together with the UV light may have contributed to the death of the spores (Rotem and Aust, 1991).

3.2. Survival of *Aspergillus carbonarius* Spores on Bunch Surfaces Preharvest

For Shiraz bunches inoculated at pre-bunch closure, there was a consistent decrease in the counts of *A. carbonarius* between the inoculation time and veraison. This decrease continued in some bunches between veraison and harvest (Figure 6). However, in other bunches, the count increased due to infection of berries and subsequent

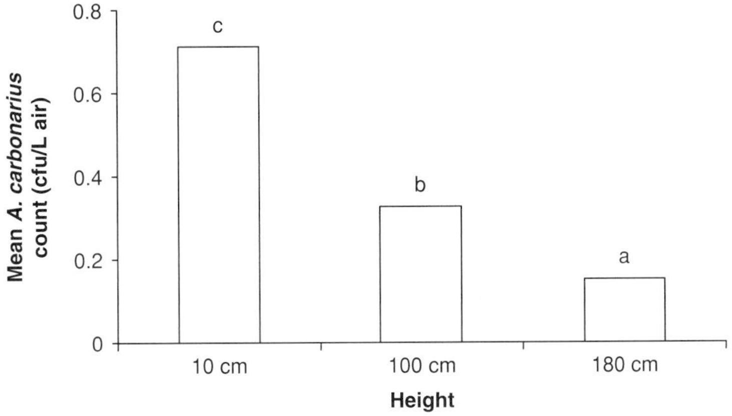

Figure 4. Aspergillus carbonarius in air at different levels above the vineyard floor, n = 9. Vertical bars with different letters differ significantly (LSD, $P < 0.05$).

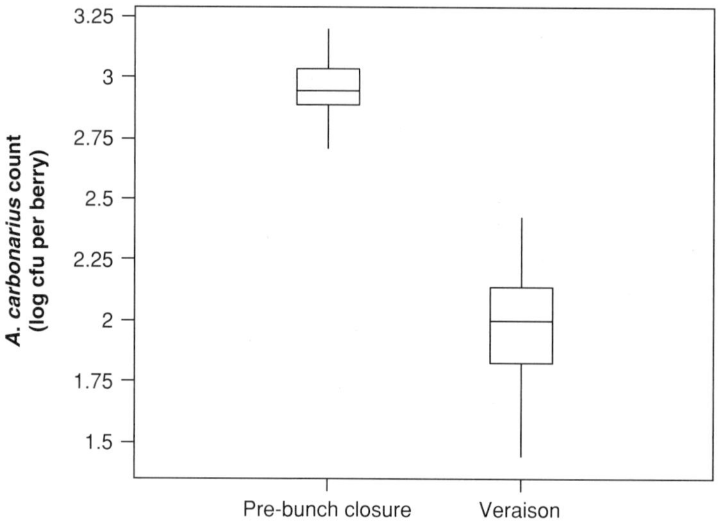

Figure 5. Spore death on berry surfaces between pre-bunch closure and veraison; n = 32, comprising both Chardonnay and Shiraz varieties. 50% of the data are contained within the boxes, while the bar within the box plot shows the median value. Maximum and minimum values are indicated by vertical lines. The difference between *Aspergillus carbonarius* spore count per berry at pre-bunch closure and veraison was significant at $P < 0.001$.

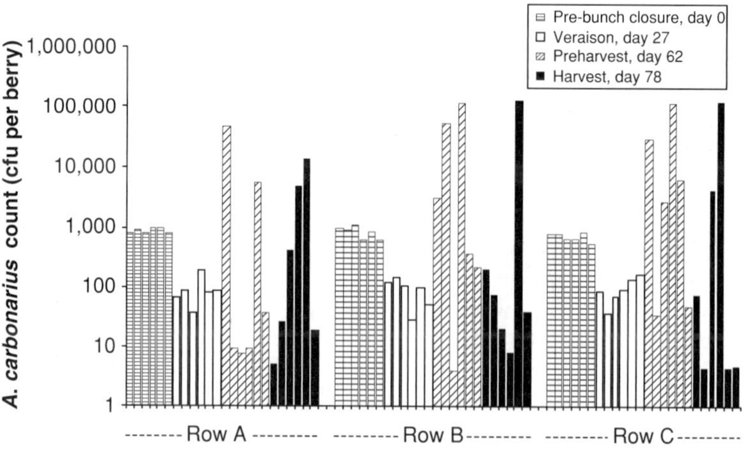

Figure 6. Survival of *Aspergillus carbonarius* inoculated on the surface of Shiraz berries at pre-bunch closure.

sporulation by the mould. A slight but statistically significant decrease in the counts of *A. carbonarius* was also observed on Chardonnay bunches inoculated 11 days before harvest (Figure 7). There were no fungicide sprays applied during this period, hence death of spores could be attributed to residual fungicide activity and/or exposure to UV light. Samples with increased *A. carbonarius* counts had foci of infection and sporulation on some berries.

These results suggest that the critical time for the development of Aspergillus rots occurs from veraison onwards. Battilani et al. (2003a) and Serra et al. (2003) both noted black *Aspergillus* spp. were more frequently isolated from berries after veraison.

3.3. Winemaking

HPLC analysis of samples taken throughout vinification showed that the greatest reduction in OA was observed at pressing, as the mean concentration of OA in white juice was 26% of the concentration in crushed grapes (must); the corresponding figure for Shiraz was 28% (Figure 8). This suggests that there is a strong association of OA with the skins and seeds (marc) trapped during pressing. Clarification of white juice with pectinase and precipitation of solids overnight resulted in a mean reduction in OA of an additional 12%, within the range observed for precipitation of must sediments during industrial

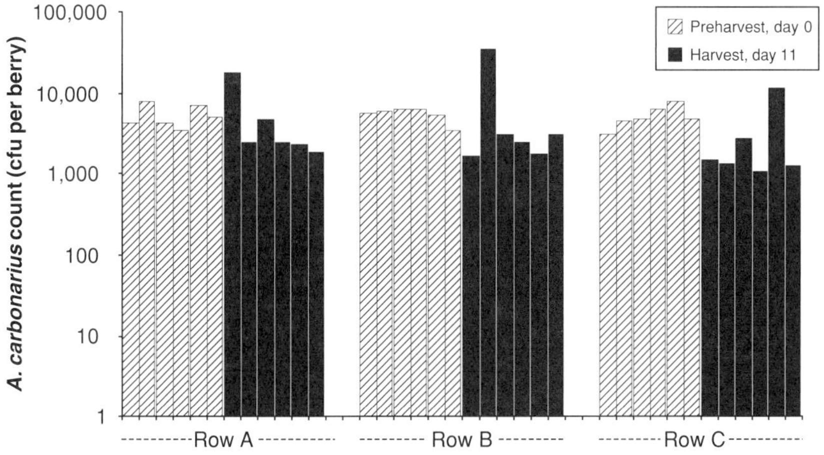

Figure 7. Survival of *Aspergillus carbonarius* inoculated on the surface of Chardonnay berries preharvest. The difference between *A. carbonarius* spore count per berry at preharvest and harvest was significant at $P < 0.05$.

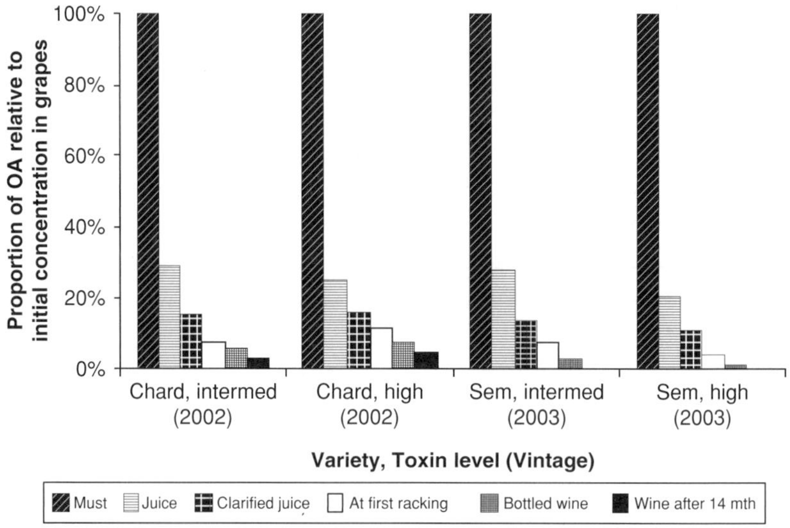

(a)

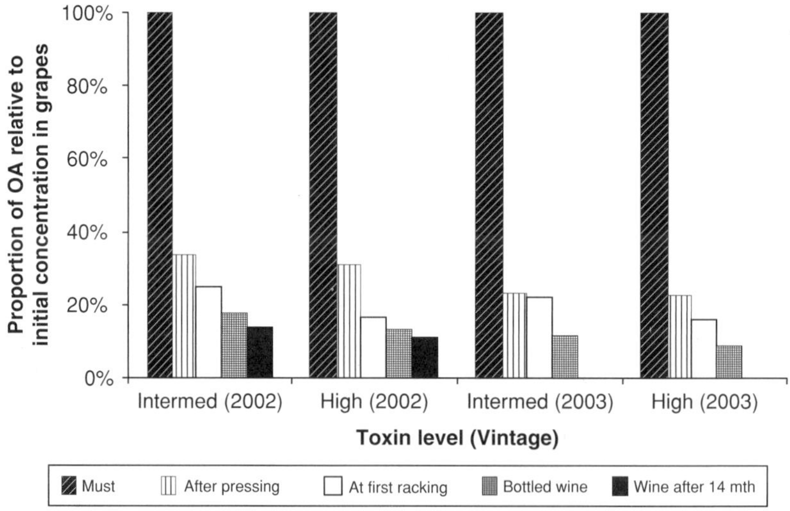

(b)

Figure 8. Reduction in ochratoxin A during **a.** white and **b.** red wine production. The white varieties used were Chardonnay (Chard) and Semillon (Sem) and the red variety was Shiraz. Berries which had been crushed and destemmed were termed "must", and for this study, were deemed to contain the total ochratoxin A initially present. Only wines from 2002 vintage had been stored for 14 months at the time of analysis, hence these results are absent for the 2003 vintage.

storage of grape juice (Roset, 2003). In most white and red ferments, racking (decanting wine after fermentation) also reduced OA content. These reductions may be due to loose interactions between OA and the solids settling out of solution. The mean OA concentration of white and red wines, respectively, at bottling was 4% and 13% of the initial concentration in grapes. Slight reductions in OA were observed in red and white wines after 14 months of storage. These trends in reduction in OA were apparently unaffected by the initial concentration of OA in the grapes, which varied in the two vintages. Initial OA concentrations fell within the range 2-66 ng/g for white grapes and 2-114 ng/g for red grapes.

The reduction in OA concentration at pressing was similar for both white and red wines (approximately 70%), however, at bottling, red wines retained three fold more of the initial OA concentration than white wines. This difference may be inherent to the vinification process. In white vinification, the juice after pressing does not contain alcohol, and OA may readily bind to proteins and other solids in the juice, to be removed during clarification. In red vinification, the wine after pressing contains alcohol and hence OA present is less bound to solids and more soluble in the liquid phase. Thus racking of red wines after fermentation does not result in the same reduction in OA as clarification of white juice. Also, during fermentation of red must on skins, there may be increased partitioning of OA from the pulp and skins into the alcohol produced during fermentation.

Fernandes et al. (2003) reported the opposite effect, with white wines retaining a greater proportion of initial OA than red wines (8-14% *cf.* 6%). This difference can be explained by noting that OA that has been spiked into crushed grapes may interact differently with grape solids compared with OA exuded directly in the berries from fungal hyphae. However, the importance of solid-liquid separations in the removal of OA during vinification is clear, regardless of the means of OA contamination.

3.4. Future Directions

An understanding of the source of *A. carbonarius* and other members of Section *Nigri* in vineyards may aid in the development of management strategies to minimise dispersal from the soil to bunches. Reducing the frequency of tillage may be one strategy to minimise *Aspergillus* in the soil and air. The presence of *A. carbonarius* spores on bunches during the early stages of berry development does not necessarily lead to the development of Aspergillus rots and subsequent production of OA, as the spores do not survive well on the surface of green berries. Berry softening from veraison onwards appears to increase berry susceptibility to Aspergillus rots. Fungicide sprays are the primary tool for management of Aspergillus rots in maturing bunches, however, Australia has strict guidelines governing their use in the weeks before harvest to minimise chemical residues in the wine. The efficacy and timing of these sprays is under investigation, as part of an overall strategy to reduce the incidence of *Aspergillus* in vineyards. Studies of the fate of OA during vinification are also continuing, to increase understanding of the relative proportion of OA remaining in wine after vinification under Australian conditions. This has implications for setting acceptable limits for OA in winegrapes at harvest, and also for further processing of waste streams from vinification.

4. ACKNOWLEDGEMENTS

This research was supported by the Australian Government and Australian grapegrowers and winemakers through their investment in the Cooperative Research Centre for Viticulture and Horticulture Australia Ltd. Support from grape growers in the Sunraysia district, Victoria, and Glen Howard of Somerset Vineyard, Pokolbin (Hunter Valley, New South Wales) is gratefully acknowledged, as is the contribution from Syngenta Crop Protection Pty Ltd. Narelle Nancarrow, Kathy Clarke and Margaret Leong are thanked for their help with the field trials. Winemaking was carried out with assistance from Mark Krstic, Glenda Kelly and Fred Hancock at the Victorian Department of Primary Industries, Mildura, Victoria, and Stephen W. White, Nai Tran-Dinh and Nick Charley at Food Science Australia. The key role of Peter Varelis, Kylie McClelland, Shane Cameron, Kathy Schneebeli (Analytical Chemistry, Food Science Australia) in development of the OA assays is most gratefully acknowledged. Advice on the statistical analyses was provided by John Reynolds (formerly Senior Biometrician, Victorian Department of Primary Industries, Attwood, Victoria).

5. REFERENCES

Abarca, M. L., Accensi, F., Bragulat, M. R., and Cabañes, F. J., 2001, Current importance of ochratoxin A-producing *Aspergillus* spp., *J. Food Prot.* **64**:903-906.

Abarca, M. L., Accensi, F., Bragulat, M. R., Castellà, G., and Cabañes, F. J., 2003, *Aspergillus carbonarius* as the main source of ochratoxin A contamination in dried vine fruits from the Spanish market, *J. Food Prot.* **66**:504-506.

Abarca, M. L., Bragulat, M. R., Castellá, G., and Cabañes, F. J., 1994, Ochratoxin A production by strains of *Aspergillus niger* var. *niger*, *Appl. Environ. Microbiol.* **60**:2650-2652.

Abdel-Sater, M. A., and Saber, S. M., 1999, Mycoflora and mycotoxins of some Egyptian dried fruits, *Bull. Fac. Sci., Assiut Univ.* **28**:91-107.

Amerine, M. A., Berg, H. W., Kunkee, R. E., Ough, C. S., Singleton, V. L., and Webb, A. D., 1980, *The Technology of Wine Making*, AVI Publishing Company, Westport, CT, pp. 154-185.

Anon., 2003, (March, 2004) EU food law news, FSA Letter, 29 July; http://www.foodlaw.rdg.ac.uk/news/eu-03068.htm.

Barbetti, M. J., 1980, Bunch rot of Rhine Riesling grapes in the lower south-west of Western Australia, *Aust. J. Expt. Agric. Anim. Husb.* **20**:247-251.

Battilani, P., Giorni, P., and Pietri, A., 2003a, Epidemiology of toxin-producing fungi and ochratoxin A occurrence in grape, *Eur. J. Plant. Pathol.* **109**:715-722.

Battilani, P., Pietri, A., Bertuzzi, T., Languasco, L., Giorni, P., and Kozakiewicz, Z., 2003b, Occurrence of ochratoxin A-producing fungi in grapes grown in Italy, *J. Food Prot.* **66**:633-636.

Baxter, D. E., Shielding, I. R., and Kelly, B., 2001, Behaviour of ochratoxin A in brewing, *J. Am. Soc. Brew. Chem.* **59**:98-100.

Bellí, N., Marín, S., Sanchis, V., and Ramos, A. J., 2002, Ochratoxin A (OTA) in wines, musts and grape juices: occurrence, regulations and methods of analysis, *Food Sci. Technol. Int.* **8**:325-335.

Cabañes, F. J., Accensi, F., Bragulat, M. R., Abarca, M. L., Castellá, G., Minguez, S., and Pons, A., 2002, What is the source of ochratoxin A in wine?, *Int. J. Food. Microbiol.* **79**:213-215.

Castegnaro, M. and Wild, C. P., 1995, IARC activities in mycotoxin research, *Nat. Toxins* **3**:327-331.

Castellari, M., Versari, A., Fabiani, A., Parpinello, G. P., and Galassi, S., 2001, Removal of ochratoxin A in red wines by means of adsorption treatments with commercial fining agents, *J. Agric. Food Chem.* **49**:3917-3921.

Da Rocha Rosa, C. A., Palacios, V., Combina, M., Fraga, M. E., De Oliveira Rekson, A., Magnoli, C. E., and Dalcero, A. M., 2002, Potential ochratoxin A producers from wine grapes in Argentina and Brazil, *Food Addit. Contam.* **19**:408-414.

Diehl, J. A., 1979, Influencia de sistemas de cultivo sobre podridoes de raizes de trigo. *Summa Phytopathol.* **5**:134-139.

Diehl, J. A., Tinline, R. D., Kochhann, R. A., Shipton, P. J., and Rovira, A. D., 1982, The effects of fallow periods on common root rot of wheat in Rio Grande do Sul, Brazil, *Phytopathology* **72**:1297-1301.

Duczek, L. J., 1981, Number and viability of conidia of *Cochliobolus sativus* in soil profiles in summer fallow fields in Saskatchewan, *Can. J. Plant Path.* **3**:12-14.

Dumeau, F., and Trione, D., 2000, Trattamenti e "ochratossina A" nei vini, *Vignevini* **9**:79-81.

Engelhardt, G., Ruhland, M., and Wallnöffer, P. R., 1999, Occurrence of ochratoxin A in moldy vegetables and fruits analysed after removal of rotten tissue parts, *Adv. Food Sci.* **21**:88-92.

European Commission, 2002, Commission regulation (EC) No 472/2002 of 12 March 2002 amending regulation (EC) No 466/2001 setting maximum levels for certain contaminants in foodstuffs, *Off. J. Eur. Comm.* **75**:18-20, 42.

Fernandes, A. Venâncio, A., Moura, F., Garrido, J., and Cerdeira, A., 2003, Fate of ochratoxin A during a vinification trial, *Aspect. Appl. Biol.* **68**:73-80.

Gupta, S. L., 1956, Occurrence of *Aspergillus carbonarius* (Bainier) Thom causing grape rot in India, *Sci. Cult.* **22**:167-168.

Heenan, C. N., Shaw, K. J., and Pitt, J. I., 1998, Ochratoxin A production by *Aspergillus carbonarius* and *A. niger* and detection using coconut cream agar, *J. Food Mycol.* **1**:67-72.

Hocking, A. D., Varelis, P., Pitt, J. I., Cameron, S., and Leong, S., 2003, Occurrence of ochratoxin A in Australian wine, *Aust. J. Grape Wine Res.* **9**:72-78.

Jarvis, W. R., and Traquair, J. A., 1984, Bunch rot of grapes caused by *Aspergillus aculeatus*, *Plant Dis.* **68**:718-719.

King, A. D., Hocking, A. D., and Pitt, J. I., 1981, The mycoflora of some Australian foods, *Food Technol. Aust.* **33**:55-60.

Klich, M. A., and Pitt, J. I., 1988, *A Laboratory Guide to Common* Aspergillus *species and their Teleomorphs*, CSIRO, Division of Food Processing, North Ryde, NSW, pp. 26-27, 48-51, 58-59.

Leong, S. L., Hocking, A. D., and Pitt, J. I., 2004, Occurrence of fruit rot fungi (*Aspergillus* Section *Nigri*) on some drying varieties of irrigated grapes, *Aust. J. Grape Wine Res.* **10**:83-88.

MacDonald, S., Wilson, P., Barnes, K., Damant, A., Massey, R., Mortby, E., and Shepherd, M. J., 1999, Ochratoxin A in dried vine fruit: method development and survey, *Food Addit. Contam.* **16**:253-260.

Majerus, P., Bresch, H., and Otteneder, H., 2000, Ochratoxin in wines, fruit juices and seasonings, *Arch. Lebensmittelhyg.* **51**:81-128.

Marfenina, O. E., and Mirchink, T. G., 1989, Effect of human activity on soil microfungi, *Sov. Soil Sci.* **21**:40-47.

Markarki, P., Delpont-Binet, C., Grosso, F., and Dragacci, S., 2001, Determination of ochratoxin A in red wine and vinegar by immunoaffinity high-pressure liquid chromatography, *J. Food Prot.* **64**:533-537.

Nair, N. G., 1985, Fungi associated wtih bunch rot of grapes in the Hunter Valley, *Aust. J. Agric. Res.* **36**:435-442.

Pitt, J. I., and Hocking, A. D., 1997, *Fungi and Food Spoilage*, 2nd edition, Blackie Academic and Professional, London, pp. 385-388, 510-511.

Roset, M., 2003, Survey on ochratoxin A in grape juice, *Fruit Process.* **13**:167-172.

Rotem, J., and Aust, H. J., 1991, The effect of ultraviolet and solar radiation and temperature on survival of fungal propagules, *J. Phytopathol.* **133**:76-84.

Sage, L., Krivobok, S., Delbos, É., Seigle-Murandi, F., and Creppy, E. E., 2002, Fungal flora and ochratoxin A production in grapes and musts from France, *J. Agric. Food Chem.* **50**:1306-1311.

Serra, R., Abrunhosa, L., Kozakiewicz, Z., and Venâncio, A., 2003, Black *Aspergillus* species as ochratoxin A producers in Portuguese wine grapes, *Int. J. Food Microbiol.* **88**:63-68.

Silva, A., Galli, R., Grazioli, B., and Fumi, M. D., 2003, Metodi di riduzione di residui di ocratossina A nei vini, *Ind. Bevande* **32**:467-472.

Snowdon, A. L., 1990, *A Colour Atlas of Post-Harvest Diseases and Disorders of Fruit and Vegetables. I. General Introduction and Fruits*, Wolfe Scientific, London, p. 256.

Téren, J., Varga, J., Hamari, Z., Rinyu, E., and Kevei, F., 1996, Immunochemical detection of ochratoxin A in black *Aspergillus* strains, *Mycopathologia* **134**: 171-176.

Tjamos, S. E., Antoniou, P. P., Kazantzidou, A., Antonopoulos, D. F., Papageorgiou, I., and Tjamos, E. C., 2004, *Aspergillus niger* and *Aspergillus carbonarius* in Corinth raisin and wine-producing vineyards in Greece: population composition, ochratoxin A production and chemical control, *J. Phytopathol.* **152**:250-255.

Zimmerli, B., and Dick, R., 1996, Ochratoxin A in table wine and grape-juice: occurrence and risk assessment, *Food Addit. Contam.* **13**:655-668.

OCHRATOXIN A PRODUCING FUNGI FROM SPANISH VINEYARDS

Marta Bau, M. Rosa Bragulat, M. Lourdes Abarca, Santiago Minguez and F. Javier Cabañes[*]

1. INTRODUCTION

Ochratoxin A (OA) is a nephrotoxic mycotoxin naturally occurring in a wide range of food commodities. It has been classified by IARC as a possible human renal carcinogen (group 2B) (Castegnaro and Wild, 1995) and among other toxic effects, is teratogenic, immunotoxic, genotoxic, mutagenic and carcinogenic (Creppy, 1999). Wine is considered the second major source of OA in Europe, with cereals being the primary source. Since the first report on the occurrence of OA in wine (Zimmerli and Dick, 1996) its presence in wine and grape juice have been reported in a broad variety of wines from different origins. Maximum OA levels have been established for cereals and dried vine fruits in the European Union, and it is possible that other commodities such as wine and grape juices will be regulated before the end of 2003 (Anonymous, 2002).

Until recently, *Aspergillus ochraceus* and *Penicillium verrucosum* were considered the main OA-producing species. *P. verrucosum* is usually found in cool temperate regions and has been reported almost exclusively in cereal and cereal products while *A. ochraceus* is found

*Marta Bau, M. Rosa Bragulat, M. Lourdes Abarca and F. Javier Cabañes: Departament de Sanitat i d'Anatomia Animals, Universitat Autònoma de Barcelona, 08193 Bellaterra, Barcelona, Spain. Santiago Minguez: Institut Català de la Vinya i el Vi (INCAVI), Generalitat de Catalunya, Vilafranca del Penedés, Barcelona, Spain. Correspondence to: Javier.Cabanes@uab.es

sporadically in different commodities in warmer and tropical climates (Pitt and Hocking, 1997). The production of OA by species in *Aspergillus* section *Nigri* has received considerable attention since the first description of OA production by *Aspergillus niger* var. *niger* (Abarca et al., 1994) and by *Aspergillus carbonarius* (Horie, 1995). Recently, *A. carbonarius* and other black aspergilli belonging to the *A. Niger* aggregate have been described as a main possible sources of OA contamination in grapes (Da Rocha Rosa et al., 2002; Sage et al., 2002; Battilani et al., 2003; Magnoli et al., 2003; Serra et al., 2003), wine (Cabañes et al., 2002), and also in dried vine fruits (Heenan et al., 1998; Abarca et al., 2003).

The objective of this study was to identify the ochratoxigenic mycobiota of grapes from vineyards mainly located along the Mediterranean coast of Spain.

2. MATERIALS AND METHODS

2.1. Samples

During the 2001 and 2002 seasons, fungi capable of producing ochratoxin A were isolated from the grapes from seven Spanish vineyards. The vineyards were located mainly along the Mediterranean coast and belonged to five winemaking regions: Barcelona (two vineyards), Tarragona (two vineyards), Valencia (one vineyard), Murcia (one vineyard) and Cádiz (one vineyard). In each vineyard, from May to October, samplings were made at four different times, coinciding with the following developmental stages of the grape: setting, one month after berry-set, veraison and harvesting. At each sampling time, 10 bunches were collected from 10 different plants located approximately along two crossing diagonals of the vineyard. Every bunch was collected in a separate paper bag and analyzed in the laboratory within 24-48 h of collection.

2.2. Mycological study

Ten berries from each bunch were randomly selected, with five plated directly onto dichloran rose bengal chloramphenicol agar (DRBC) (Pitt and Hocking, 1997) and five onto malt extract agar (MEA) (Pitt and Hocking, 1997) supplemented with 100 ppm of chloramphenicol and 50 ppm of streptomycin. In total, 5,600 berries were analyzed. Plates were incubated at 25°C for 7 days. All fungi belonging to *Aspergillus* and *Penicillium* genera were isolated for identification to species level. (Raper and Fennell, 1965; Pitt, 1979; Klich and Pitt, 1988; Pitt and Hocking, 1997).

2.3. Ability of Fungal Isolates to Produce Ochratoxin

Isolates belonging to *Aspergillus* spp. and *Penicillium* spp. were evaluated using a previously described HPLC screening method (Bragulat et al., 2001). Briefly, the isolates were grown on Czapek Yeast extract Agar (CYA) and on Yeast extract Sucrose agar (YES) (Pitt and Hocking, 1997) and incubated at 25°C for 7 days. Isolates identified as *A. carbonarius* were grown on CYA for 10 days at 30°C because these incubation conditions have been cited as optimal for detecting OA production in this species (Cabañes et al., 2002; Abarca et al., 2003). From each isolate, three agar plugs were removed from different points of the colony and extracted with 0.5 ml of methanol. The extracts were filtered and injected into the HPLC.

2.4. Data analysis

Data obtained were analyzed statistically by means of one-way analysis of variance test and Student's test. All statistical analyses were performed using SPSS software (version 10.0).

3. RESULTS AND DISCUSSION

The occurrence of *Aspergillus* spp. in the 5,600 berries plated on the two culture media used are shown in Table 1. Although the number of isolates recovered on DRBC medium was higher than on MEA, the differences were not statistically significant. A total of 1,061 isolates belonging to twenty *Aspergillus* spp. (including *Emericella* spp. and *Eurotium amstelodami*) were identified. Isolates of *A. carbonarius* and *A. niger* aggregate constituted 88.7% of the total *Aspergillus* isolates (Figure 1). *Aspergillus niger* aggregate were isolated from 14.2% of plated berries, and *A. carbonarius* from 2.6%. The occurrence of the remaining *Aspergillus* spp. ranged from 0.02% to 0.5%.

The distribution of the *A. niger* aggregate and *A. carbonarius* isolates in 2001 and 2002 seasons during the development of berries is shown in Figure 2. Although they were recovered in all the stages

Table 1. Occurrence of *Aspergillus* spp. in grapes from Spanish vineyards examined during the 2001 and 2002 seasons

	No. (%) of positive berries		
Species	Total (n = 5,600)	DRBC[a] (n = 2,800)	MEA[a] (n = 2,800)
Aspergillus niger aggregate	797 (14.23)	438	359
A. carbonarius	144 (2.57)	83	61
A. ustus	26 (0.46)	15	11
A. fumigatus	19 (0.34)	10	9
A. flavus	11 (0.20)	6	5
A. tamarii	9 (0.16)	4	5
A. japonicus var. *aculeatus*	8 (0.14)	3	5
A. ochraceus	8 (0.14)	8	0
Emericella nidulans	6 (0.11)	4	2
A. alliaceus	5 (0.09)	4	1
A. terreus	5 (0.09)	4	1
A. wentii	5 (0.09)	3	2
A. melleus	4 (0.07)	4	0
A. flavipes	3 (0.05)	2	1
A. ostianus	3 (0.05)	3	0
A. parasiticus	3 (0.05)	2	1
Eurotium amstelodami	2 (0.04)	1	1
Emericella astellata	1 (0.02)	0	1
Emericella variecolor	1 (0.02)	0	1
A. versicolor	1 (0.02)	0	1
Total *Aspergillus* spp.	1061	594	467

[a]DRBC: dichloran rose bengal chloramphenicol agar; MEA: malt extract agar.

sampled, there was a statistically significant increase at harvesting. The number of isolates recovered in 2002 was lower than in 2001, probably due to different climatic conditions. Nevertheless in both seasons black Aspergilli showed the same tendency, with the highest levels of isolation at harvesting.

A total of 165 isolates belonging to genus *Penicillium* were identified. The most frequent species were *P. glabrum, P. brevicompactum, P. sclerotiorum, P. citrinum, P. chrysogenum* and *P. thomii*. The occurrence of the remaining *Penicillium* spp. was lower than 0.12%. OA production was not detected by any of the 165 *Penicillium* isolates. Only one isolate of *P. verrucosum* was identified. This isolates was able to produce citrinin but did not produce OA.

The ability of *Aspergillus* isolates to produce ochratoxin A is shown in Table 2. All the *A. carbonarius* isolates (n = 144) were able to produce OA whereas only eight isolates of *A. niger* aggregate (n = 797) were toxigenic. None of the *A. japonicus* var. *aculeatus* (n = 8) produced OA.

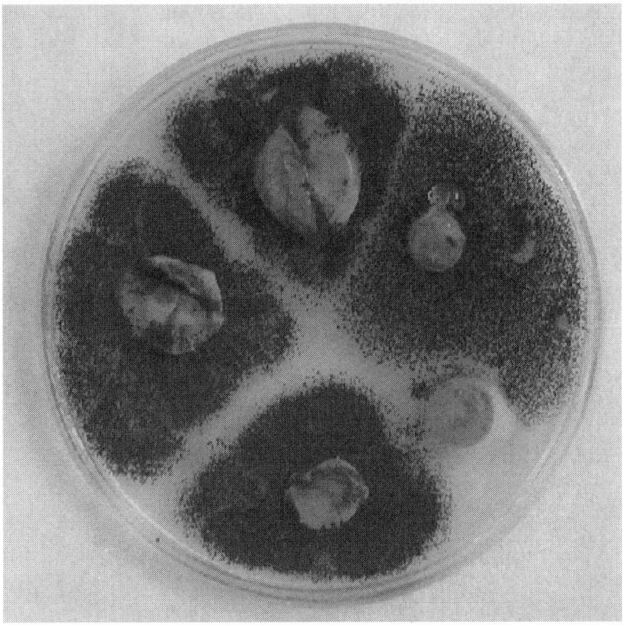

Figure 1. Black aspergilli growing on plated berries from harvesting time. (Note their high occurrence at this sampling time).

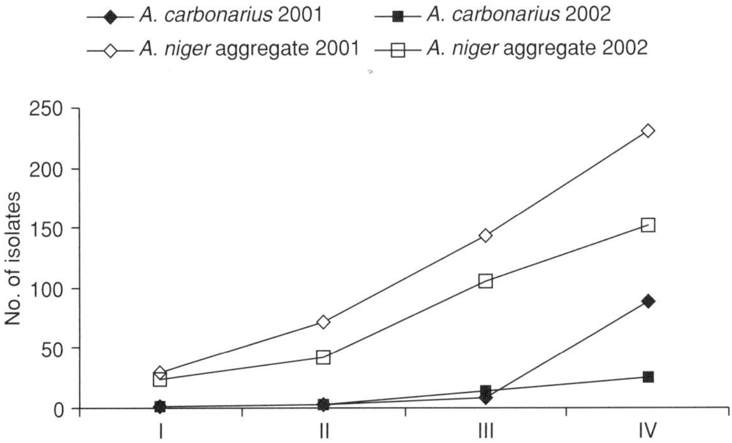

Figure 2. Distribution of *A. carbonarius* and *A. niger* aggregate isolates at each developmental stages of the berries: I, berry set; II, one month after berry set; III, veraison; IV, harvest

Table 2. Ochratoxin A production (μg/g of culture medium) by *Aspergillus* spp. isolated from Spanish grapes

Species	No. of positive isolates /Total	Mean concentration	Range
A. carbonarius	144 / 144	24.6	0.1 – 378.5
A. niger aggregate	8 / 797	29.2	0.05 – 230.9
A. alliaceus	5 / 5	351.4	197.6 – 715.4
A. ochraceus	4 / 8	440.8	1.3 – 1026.7
A. ostianus	3 / 3	1273.3	245.9 – 2514.1
A. melleus	2 / 4	19.7	7.3 – 32.2

OA production was also detected by other *Aspergillus* species outside section *Nigri*. Four of the eight isolates of *A. ochraceus* and two of the four isolates of *A. melleus* produced OA. All isolates classified as *A. ostianus* (n=3) and *A. alliaceus* (n=5) were able to produce OA. Some of these species were able to produce OA in large quantities in pure culture, but due to their low occurrence, they are probably a relatively unimportant source of this mycotoxin in grapes.

Although the possible participation of different OA producing species may occur, our results are strong evidence of the contribution of *A. carbonarius* to OA contamination in grapes, mainly at the final developmental stage of the berries, and consequently in wine.

4. ACKNOWLEDGEMENTS

This research was supported by the European Union project QLK1-CT-2001-01761 (Quality of Life and Management of Living Resources Programme (QoL), Key Action 1 on Food, Nutrition and Health). The financial support of the Ministerio de Ciencia y Tecnología of the Spanish Government (AGL01-2974-C05-03) is also acknowledged.

5. REFERENCES

Abarca, M. L., Bragulat, M. R., Castellá, G., and Cabañes, F. J., 1994, Ochratoxin A production by strains of *Aspergillus niger* var. *niger*, *Appl. Environ. Microbiol.* **60:**2650-2652.

Abarca, M. L., Accensi, F., Bragulat, M. R., Castellá. G., and Cabañes, F. J., 2003, *Aspergillus carbonarius* as the main source of ochratoxin A contamination in dried vine fruits from the Spanish market, *J. Food Prot.* **66:**504-506.

Anonymous, 2002. Commission regulation (EC) no 472/2002 of 12 March 2002 amending regulation (EC) no 466/2001 setting maximum levels for certain contaminants in foodstuffs, *Off. J. Eur. Communities* L75, 18-20.

Battilani, P., Pietri, A., Bertuzzi, T., Languasco, L., Giorni, P., and Kozakiewicz, Z., 2003, Occurrence of ochratoxin A-producing fungi in grapes grown in Italy, *J. Food Prot.* **66**:633-636.

Bragulat, M. R., Abarca, M. L., and Cabañes, F. J., 2001, An easy screening method for fungi producing ochratoxin A in pure culture, *Int J. Food Microbiol.* **71**: 139-144.

Cabañes, F. J., Accensi, F., Bragulat, M. R., Abarca, M. L., Castellá, G., Minguez, S., and Pons, A., 2002, What is the source of ochratoxin A in wine?, *Int J. Food Microbiol.* **79**:213-215.

Castegnaro, M., and Wild, C. P., 1995, IARC activities in mycotoxin research, *Natural Toxins* **3**:327-331.

Creppy, E. E., 1999. Human ochratoxicoses. *J. Toxicol. – Toxin Rev.* **18**:273-293.

Da Rocha Rosa, C. A., Palacios, V., Combina, M., Fraga, M. E., De Oliveira Rekson, A., Magnoli, C. E., amd Dalcero, A. M., 2002, Potential ochratoxin A producers from wine grapes in Argentina and Brazil. *Food Addit. Contam.* **19**:408-414.

Heenan, C. N., Shaw, K. J., and Pitt, J. I., 1998, Ochratoxin A production by *Aspergillus carbonarius* and *A. niger* isolates and detection using coconut cream agar, *J. Food Mycol.* **1**:67-72.

Horie, Y., 1995, Productivity of ochratoxin A of *Aspergillus carbonarius* in *Aspergillus* section *Nigri. Nippon Kingakukai Kaiho* **36**:73-76.

Klich, M. A., and Pitt, J. I., 1988, *A Laboratory Guide to Common* Aspergillus *Species and their Teleomorphs*. CSIRO Division of Food Processing, North Ryde, NSW.

Magnoli, C., Violante, M., Combina, M., Palacio, G., and Dalcero, A., 2003, Mycoflora and ochratoxin-producing strains of *Aspergillus* section *Nigri* in wine grapes in Argentina, *Lett. Appl. Microbiol* **37**:179-184.

Pitt, J. I., 1979, *The Genus* Penicillium *and its Teleomorphic States* Eupenicillium *and* Talaromyces, Academic Press, London.

Pitt, J. I., and Hocking, A. D., 1997, *Fungi and Food Spoilage*, Blackie Academic and Professional, London.

Raper, K. B., and Fennell, D. I., 1965, *The Genus* Aspergillus, The William and Wilkins Co., Baltimore.

Sage, L., Krivobok, S., Delbos, E., Seigle-Murandi, F., and Creppy, E. E., 2002, Fungal flora and ochratoxin A production in grapes and musts from France, *J. Agric. Food Chem.* **50:**1306-1311.

Serra, R., Abrunhosa, L., Kozakiewicz, Z., and Venancio, A., 2003, Black *Aspergillus* species as ochratoxin A producers in Portuguese wine grapes, *Int. J. Food Microbiol.* **88**:63-68.

Zimmerli, B., and Dick, R., 1996, Ochratoxin A in table wine and grape-juice: occurrence and risk assessment, *Food Addit. Contam.* **13**:655-668.

FUNGI PRODUCING OCHRATOXIN IN DRIED FRUITS

*Beatriz T. Iamanaka, Marta H. Taniwaki, E. Vicente and Hilary C. Menezes**

1. INTRODUCTION

Ochratoxin A (OA) has been shown to be a potent nephrotoxin in animal species and has been found in agricultural products. OA is believed to be produced in nature by three main species of fungi, *Aspergillus ochraceus, Aspergillus carbonarius* and *Penicillium verrucosum*, with a minor contribution from *A. niger* and several species closely related to *A. ochraceus* (JECFA, 2001). *P. verrucosum* occur mainly in cool temperate climates, and is usually associated with cereals (Pitt, 1987; Pitt and Hocking, 1997).

Studies carried out in Europe have reported the presence of the ochratoxigenic fungi *A. ochraceus*, *A. niger* and *A. carbonarius* and sometimes OA in dried fruits (MAFF, 2002). *A. carbonarius* and *A. niger* were described as sources of OA in maturing and drying grapes in Spain and Australia (Abarca et al., 1994; Heenan et al., 1998).

Grape juice and wines from southern regions of Europe have been reported to contain detectable levels of OA (Zimmerli and Dick, 1996). Detectable concentrations of OA have been found in sultanas imported into the United Kingdom: a survey of 20 samples of dried fruit found more than 80% were positive for OA (MAFF, 1997; MacDonald et al., 1999).

*Beatriz T. Iamanaka, Marta H. Taniwaki and E. Vicente: Food Technology Institute, ITAL C.P 139 CEP13.073-001 Campinas-SP, Brazil; Hilary C. Menezes: Food Engineering Faculty-Unicamp, Campinas-SP, Brazil. Correspondence to: mtaniwak@ital.sp.gov.br

The aim of this work was to investigate the incidence of toxigenic fungi and OA in dried fruits from different countries of origin sold on the Brazilian market.

2. MATERIALS AND METHODS

2.1. Sampling

A total of 119 samples (500 g each) of dried fruits were purchased in Campinas and São Paulo markets in Brazil in 2002-2003, comprising black sultanas (24), white sultanas (19), apricots (14), figs (19), dates (22) and plums(21). The dried fruit samples originated from Turkey, Spain, Mexico, Tunisia, USA, Argentina and Chile.

2.2. Mycological Analyses

Larger fruit (apricots, figs, dates and plums) were cut aseptically into small pieces, whereas smaller fruit (sultanas) were analysed whole. Whole fruit or fruit pieces were surface disinfected with 0.4% chlorine solution for 1 min. Fifty sultanas or fruit pieces were plated onto Dichloran 18% Glycerol agar (DG18; Pitt and Hocking, 1997). The plates were incubated at 25°C for 5-7 days. Colonies with the appearance of *A. niger*, *A. carbonarius* and *A. ochraceus* were isolated onto Malt Extract agar (Pitt and Hocking, 1997) and identified according to Klich and Pitt (1988). The percentage infection of the fruit or pieces was calculated.

2.3. Ochratoxin A Production

Ochratoxin A production from each isolate was analysed qualitatively using the agar plug technique of Filtenborg et al. (1983) or extracted with chloroform as described below. The isolates were inoculated onto Yeast Extract (0.1%) Sucrose (15%) agar and incubated at 25°C for 7 days. For the agar plug technique, a small plug was cut from the colony using a cork borer and tested by TLC as described by Filtenborg et al. (1983). If isolates were found to be negative for OA production, the whole colony was extracted with chloroform. The whole colony from the Petri dish was placed in chloroform in a Stomacher and homogenised for 3 min, filtered and concentrated in a water bath at 60°C to near dryness then dried under a stream of nitrogen. The residue was resuspended in chloroform and spotted onto TLC plates which were developed in toluene: ethyl acetate: formic acid

(5:4:1) and visualized under UV light at 365nm. An OA standard (Sigma Chemicals, St Louis, USA) was used for comparison.

2.4. Analysis of Ochratoxin A from Fruit Samples

The fruit samples were analysed for OA using the Ochratest™ HPLC Procedure for Currants and Raisins (Vicam, 1999). The method was validated for this study. Samples were extracted with a solution of methanol: sodium bicarbonate, 1% (70:30) in a blender at high speed for 1 min. The extracts were filtered onto qualitative paper and an aliquot diluted with phosphate buffered saline containing 0.01% Tween 20. This solution was filtered through a microfibre filter and an aliquot applied to an immunoaffinity column (Vicam, Watertown MA, USA) containing a monoclonal antibody specific for OA. The column was washed with phosphate buffered saline with 0.01% Tween 20, followed by purified water. The OA was eluted with HPLC methanol and the eluate was added to purified water, mixed and injected in the HPLC with a fluorescence detector. The mobile phase was acetonitrile:water:acetic acid (99:99:2) and the flow rate was 0.8 ml/min. The HPLC equipment was a Shimadzu LC-10VP system (Shimadzu, Japan) set at 333 nm excitation and 477 nm emission. The HPLC was fitted with a Shimadzu CLC G-ODS(4) (4 × 10 mm) guard column and Shimadzu Shimpack CLC-ODS (4.6 × 250 mm) column.

2.5. Confirmation of Ochratoxin A

Ochratoxin A was confirmed by methyl ester formation (Pittet et al., 1996). Aliquots (200 µl) of both the sample and standard were evaporated under a stream of nitrogen and the residue redissolved in 300 µl of boron trifluoride-methanol complex (20% solution in methanol). The solution was heated at 80°C for 10 min and allowed to cool to room temperature. The solution was evaporated and taken up in mobile phase (1 ml) and injected into the HPLC. A positive confirmation of identity was provided by the disappearance of the OA peak at retention time 16.4 min and the appearance of a new peak (OA methyl ester) at a retention time of 41.0 min.

2.6. Water Activity

The water activity (a_w) was measured using an Aqualab T3 device (Decagon, Pullman, WA, USA), at a constant temperature of 25°C. The samples were analysed in triplicate.

3. RESULTS AND DISCUSSION

The water activities of the samples examined are summarised in Table 1. Samples of all the fruit except plums and apricots were below 0.75 a_w and microbiologically stable. Some samples of apricots and plums were higher than 0.75 a_w, but contained preservatives and were also shelf stable.

Table 2 shows the mean and range of percentage infection by *A. niger* plus *A. carbonarius* and *A. ochraceus* in dried fruits. The predominance of black Aspergilli can be explained by their black spores which possess resistance to ultraviolet light. The high sugar concentration and low water activity in dried fruits also assist the development of these fungi because they are xerophilic.

Table 3 shows the results of the concentrations of ochratoxin A in dried fruits. The average level of contamination by OA in dried fruits was low except for one sample each of black sultanas and figs with more than 30 and 20 μg/kg OA and mean values of 4.73 and 4.10 μg/kg respectively. OA was detected at levels ranging from 0.13 to 5.0 μg/kg in most samples (88.2%). Although date samples were contaminated with a high level of toxigenic species of black Aspergilli

Table 1. Average and range of water activity (a_w) in dried fruits

Dried Fruits	a_w Mean	Range
Black sultanas	0.629	0.527–0.765
White sultanas	0.567	0.473–0.638
Dates	0.629	0.549–0.712
Plums	0.796	0.712–0.863
Apricots	0.694	0.638–0.782
Figs	0.682	0.638–0.751

Table 2. Percentage infection of dried fruits by *A. niger* plus *A. carbonarius* and *A. ochraceus*

		% Infection			
	No.	*A. niger + A. carbonarius*		*A. ochraceus*	
Dried fruits	Samples	Mean	Range	Mean	Range
Black sultanas	24	22	0 – 90	0.8	0 –18
White sultanas	19	0.5	0 – 8	-	-
Dates	22	8.6	0 – 86	1.1	0 –24
Plums	21	8.0	0 – 60	0.5	0 –10
Apricots	14	-	-	-	-
Figs	19	4.5	0 – 38	-	-

Table 3. Ochratoxin A levels in dried fruits

Ochratoxin A (μg/kg)	Black sultanas	White sultanas	Dates	Plums	Apricots	Figs
<0.13	5	9	20	19	14	1
0.13 –5.0	11	10	2	1	-	13
5.1 – 10.0	4	-	-	-	-	2
10.1 – 20.0	3		-	-	-	2
20.1 – 30.0	-		-	-	-	1
>30.0	1		-	-	-	-
Mean	4.73	0.52	<0.13	<0.13	N.D*	4.10

*ND-not detected (limit of detection: 0.13 μg/kg)

(Figure 1), the levels of OA were not so high. Most of the white sultanas, apricots, plums and dates were contaminated with OA at levels below 5 μg/kg.

From a total of 119 samples of dried fruits, 518 strains of potentially ochratoxigenic fungi were isolated. Black Aspergilli were the dominant fungi with 491 (95% of isolates), the other isolates belonged to *A. ochraceus*. Among the black Aspergilli, 123 (25%) were able to produce OA under the conditions tested. Only 5 isolates were identified as *A. carbonarius* and only three of these produced OA. The remaining 120 isolates were identified as *A. niger*. OA production by such a high proportion of *A niger* isolates (25%) is unexpected.

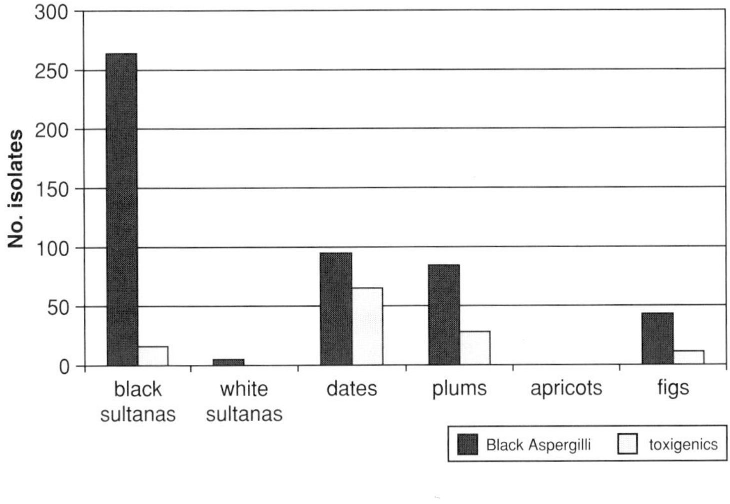

Figure 1. Number of isolates of black Aspergilli from dried fruits and production of ochratoxin A

Figure 1 shows the total numbers of black *Aspergillus* species isolated from each type of sample and the numbers of isolates producing OA. Of 264 isolates from black sultanas, only 6.1% were toxigenic, whereas in dates 68.4% of black Aspergilli were toxigenic. Plums and figs yielded 84 and 43 isolates, of which 28 (33.3%) and 11 (25.6%) were toxigenic, respectively. Studies on dried fruits have shown that these products are commonly contaminated with black Aspergilli such as *A. carbonarius, A. niger* and related species (King et al., 1981; Abarca et al., 2003). Apricots were not contaminated with ochratoxigenic fungi because the use of high levels of sulphur dioxide to preserve colour renders dried apricots essentially sterile.

Abarca et al. (2003) isolated black Aspergilli from 98% of dried fruit samples (currants, raisins and sultanas). They found that 96.7% of *A. carbonarius* and 0.6% of *A niger* isolates produced OA, indicating that among black Aspergilli, *A. carbonarius* was the most probable source of this toxin in these substrates in Spain. However, Da Rocha Rosa et al. (2002) found a higher percentage of *A. niger* producing OA in grapes from Argentina (17%) and Brazil (30%). Only *A. carbonarius* was found in Brazilian grapes and only eight isolates (25%) were able to produce this toxin.

In the present study, the high incidence of toxigenic black Aspergilli in dates, plums and figs is of concern, but only a few isolates were identified as *A. carbonarius*. Compared with studies that have been carried out in several parts of the world (Heenan et al., 1998; Da Rocha Rosa et al., 2002; Abarca et al., 2003; Battilani et al., 2003), the results presented here differ significantly because they show a higher percentage of black Aspergilli other than *A. carbonarius* producing OA.

Figure 2 shows the numbers of *A. ochraceus* found in dried fruits. *A. ochraceus* contamination was not high, only 27 isolates were found in black sultanas, dates and plums, however 52% of the isolates were toxigenic. Most of isolates from black sultanas and plums produce OA; 80% and 100%, respectively. From dates, only one of 12 isolates was toxigenic. *A. ochraceus* was not found in white sultanas, apricots or figs.

In general, the presence of OA in dried fruits was related to contamination by black Aspergilli and the major incidence occurred in black sultanas, figs and dates.

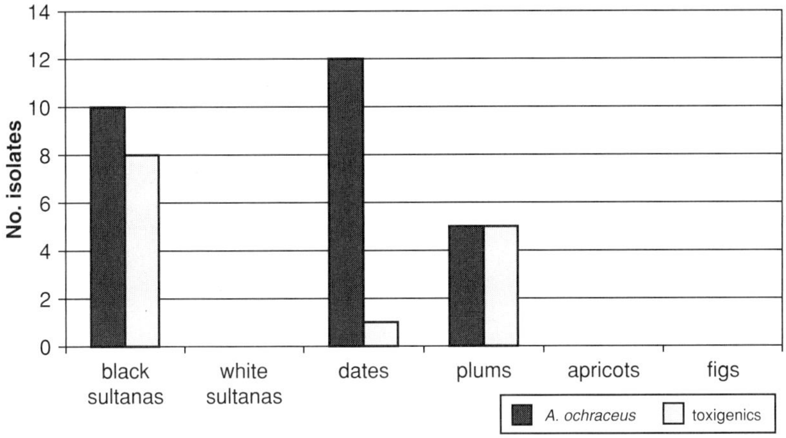

Figure 2. Number of isolates of *Aspergillus ochraceus* from dried fruits and production of ochratoxin A

4. REFERENCES

Abarca, M. L., Bragulat, M. R., Castella, G., and Cabanes, F. J., 1994, Ochratoxin A production by strains of *Aspergillus niger* var. *niger*, *Appl. Environ. Microbiol.* **60**:2650-2652.

Abarca, M. L., Accensi, F., Bragulat, M. R., Castella, G., and Cabanes, F. J., 2003, *Aspergillus carbonarius* as the main source of ochratoxin A contamination in dried vine fruits from the Spanish market, *J. Food Prot.* **66**:504-506.

Battilani, P., Pietri, A., Bertuzzi, T., Languasco, L., Giorni, P. and Kozakiewicz, Z., 2003, Occurrence of ochratoxin A-producing fungi in grapes grown in Italy, *J. Food Prot.* **66**:633-636.

Da Rocha Rosa, C. A., Palacios, V., Combina, M., Fraga, M. E., De Oliveira Rekson, A., Magnoli, C. E., and Dalcero, A. M., 2002, Potential ochratoxin A producers from wine grapes in Argentina and Brazil, *Food Addit. Contam.* **19**:408-414.

Filtenborg, O., Frisvad, J. C., and Svendensen, J. A., 1983, Simple screening method for molds producing intracellular mycotoxins in pure cultures, *Appl. Environ. Microbiol.* **45**:581-585.

Heenan, C. N., Shaw, K. J. and Pitt, J. I., 1998, Ochratoxin A production by *Aspergillus carbonarius* and *A. niger* isolates and detection using coconut cream agar, *J. Food Mycol.* **1**:67-72.

JECFA (Joint FAO/WHO Expert Committee on Food Additives), 2001, Ochratoxin A, in: *Safety Evaluation of Certain Mycotoxins in Food.* Prepared by the Fifty-sixth meeting of the JECFA. FAO Food and Nutrition Paper 74, Food and Agriculture Organization of the United Nations, Rome, Italy.

King, A. D., Hocking, A. D., and Pitt, J. I., 1981, The mycoflora of some Australian foods, *Food Technol. Aust.* **33**:55-60.

Klich, M. A., and Pitt, J. I., 1988, *A Laboratory Guide to Common* Aspergillus *Species and their Teleomorphs,* CSIRO Division of Food Science and Technology, Sydney, Australia.

MacDonald, S., Wilson, P., Barnes, K., Damant, A., Massey, R., Mortby, E., and Shepherd, M. J., 1999, Ochratoxin A in dried vine fruit: method development and survey, *Food Addit. Contam.* **6**:253-260.

MAFF (Ministry of Agriculture, Fisheries and Food), 1997, Survey of aflatoxins and ochratoxin A in cereals and retail products, Food Surveillance Information Sheet 130, MAFF, UK.

MAFF (Ministry of Agriculture, Fisheries and Food), 2002, Survey of nuts, nut products and dried tree fruits for mycotoxin, http://www.food.gov.uk/multimedia/pdfs/21nuts.pdf Food Surveillance Information, MAFF, UK.

Pitt J. I., 1987, *Penicillium viridicatum, Penicillium verrucosum* and production of ochratoxin A. *Appl. Environ. Microbiol.* **53**,:266-269.

Pitt, J. I., and Hocking, A. D., 1997, *Fungi and Food Spoilage*, Blackie Academic and Professional, London.

Pittet, A., Tornare, D., Huggett, A., and Viani, R., 1996, Liquid chromatographic determination of ochratoxin A in pure and adulterated soluble coffee using an immunoaffinity column cleanup procedure, *J. Agric. Food Chem.* **44**:3564-3569.

VICAM, 1999, Ochratest™ Instruction Manual, VICAM Pty Ltd, Watertown, MA. CD-Rom.

Zimmerli, B., and Dick, R., 1996, Ochratoxin A in table wine and grape juice: occurrence and risk assessment. *Food Addit. Contam.* **13**:655-668.

AN UPDATE ON OCHRATOXIGENIC FUNGI AND OCHRATOXIN A IN COFFEE

*Marta H. Taniwaki**

1. INTRODUCTION

The occurrence of ochratoxin A (OA) in raw coffee was first reported by Levi et al. (1974). Initial data suggested that almost complete destruction of mycotoxins would occur during the roasting process. However, occurrence of ochratoxin A in market samples of roasted coffee, as well as in coffee beverage, was reported after an improved detection method. Consequently, the European Commission's Scientific Committee for Food (SCF) considered that there was real potential for OA contamination of coffee. Since then, surveys on the presence of OA in coffee have been undertaken in several countries. These surveys looked first at raw coffee imported into Europe from different origins, then on roasted and soluble coffee produced and sold in Europe (MAFF, 1996; Patel et al., 1997; van der Stegen et al., 1997).

Some countries have stated limits of ochratoxin A for raw coffee: Italy 8 μg/kg; Finland 10 μg/kg; Greece 20 μg/kg; or for coffee products: Switzerland 5 μg/kg. Pressure from European authorities has prompted coffee importers, and the coffee-producing countries, to begin surveys for the presence of OA in coffee products, and as well as investigation of the fate of OA during the production, handling and manufacture of raw coffee.

*Food Technology Institute (ITAL), Av. Brasil, 2880, Campinas, SP, CEP 13.073-001, Brazil. Correspondence to: mtaniwak@ital.sp.gov.br

2. INCIDENCE OF OCHRATOXIN A IN COFFEE SAMPLES WORLDWIDE

Surveys all over the world have confirmed the presence of OA in commercial raw, roasted and soluble coffee (Tables 1-3). Extensive sampling of raw coffee from all origins and both types of coffee (Arabica, Robusta) has shown that OA contamination may be more frequent in some areas, but that no producing country is entirely free from contamination (Table 1). Similarly it has been shown that, while the initial contamination may occur at farm level, OA formation may happen throughout the entire chain, at every stage of production.

What happens to OA during processing of coffee beans is not yet fully understood, but the results of surveys for OA in retail roasted and soluble coffees all over the world indicate that coffee is not a major source of OA in the diet, with estimated intakes being well within safety limits. The low levels of OA contamination found in roasted and soluble coffee (Tables 2 and 3) support this conclusion.

3. FUNGI PRODUCING OCHRATOXIN A IN COFFEE

Aspergillus ochraceus has been isolated from coffee samples by several authors (Urbano et al., 2001b; Taniwaki et al., 2003; Martins et al., 2003; Batista et al., 2003; Suárez-Quiroz et al., 2004) and was proposed as the major cause of OA in coffee beans (Frank, 1999). However, *A. carbonarius* and *A. niger,* capable of producing OA, have also been isolated from coffee (Téren et al., 1997; Nakajima et al., 1997; Joosten et al., 2001; Urbano et al., 2001b; Taniwaki et al., 2003; Pardo et al., 2004; Suárez-Quiroz et al., 2004), indicating that these two species are also potential sources of OA in coffee. These findings have led to the consensus viewpoint that the three Aspergilli: *A. ochraceus, A. carbonarius* and *A. niger* are responsible for the formation of OA on coffee.

In a study of fungi with the potential to produce OA in Brazilian coffee (Taniwaki et al., 2003), the major fungus responsible was found to be *A. ochraceus*. Of 269 isolates of this species, 75% were found to be capable of toxin production, a higher percentage than had been reported previously. *A. carbonarius* was also found, though much less commonly, and it also had the potential to produce OA in coffee. Few

Table 1. Incidence of ochratoxin A (OA) in commercial raw coffee worldwide

Origin	No. positive/ No. samples	Range of OA (µg/kg)	Coffee type	Reference
Angola	0/4	< 20[a]	N.S.[b]	Levi et al. (1974)
Brazil	3/7	Trace – 360	"	"
Colombia	17/139	Trace – 50	"	"
Cameroon	0/1	< 20[a]	"	"
Ivory Coast	1/12	Trace	"	"
Uganda	1/2	Trace	"	"
Unknown	7/102	Trace	"	"
Brazil	10/14	0.2 – 3.7	Arabica	Micco et al. (1989)
Cameroon	3/3	Traces – 2.2	Robusta	"
Colombia	1/2	3.3	Arabica	"
Costa Rica	1/2	Traces	Arabica	"
Ivory Coast	1/2	1.3	Robusta	"
Kenya	0/2	< 0.01[a]	Arabica	"
Mexico	1/2	1.4	Arabica	"
Zaire	2/2	8.4 – 15.0	Robusta	"
Brazil	3/5	2.0 – 7.4	N.S.[b]	Studer-Rhor et al. (1995)
Colombia	3/5	1.2 – 9.8	"	"
Central America	0/1	< 0.5[a]	N.S.[b]	"
Costa Rica	0/1	< 0.5[a]	"	"
Guatemala	0/1	< 0.5[a]	"	"
Ivory Coast	2/2	9.9 – 56.0	"	"
Kenya	0/3	< 0.5[a]	"	"
New Guinea	0/1	< 0.5[a]	"	"
Tanzania	1/1	2.2	"	"
Zaire	1/1	17.3	"	"
Unknown	2/4	2.2 – 11.8		
America, Africa, Papua New Guinea	31/153	0.2 – 9.0	Arabica	MAFF, 1996
America, Africa, Asia	55/75	0.2 – 27.3	Robusta	"
Yemen	7/10	0.7 – 17.4	Arabica	Nakajima et al. (1997)
Tanzania	5/9	0.1 – 7.2	Arabica	"
Indonesia	2/9	0.2 – 1.0	Robusta	"
Ethiopia	0/1	< 0.1[a]	Arabica	"
Central America	0/6	< 0.1[a]	Arabica	"
South America	0/12	< 0.1[a]	Arabica	"
East Africa	42/33	0.2 – 62.0	N.S.[b]	Heilmann et al. (1999)
West Africa	9/9	0.3 – 5.0	N.S.[b]	Heilmann et al. (1999)

Table 1. Incidence of ochratoxin A (OA) in commercial raw coffee worldwide—*cont'd*

Origin	No. positive/ No. samples	Range of OA (μg/kg)	Coffee type	Reference
Asia	20/29	0.2 – 4.9	N.S.[b]	"
Central America	6/15	0.2 – 0.8	"	"
South America	5/17	0.2 – 1.0	"	"
South America	9/19	0.1 – 4.9	N.S.[b]	Trucksess et al. (1999)
Africa	76/84	0.5 – 48.0	N.S.[b]	Romani et al. (2000)
Latin America	19/60	0.1 – 7.7	"	"
Asia	11/18	0.2 – 4.9	"	"
Brazil	17/37	0.2 – 6.2	Arabica	Gollücke et al. (2001)
Brazil	27/132	0.7 – 47.8	Arabica	Leoni et al. (2000)
Brazil	5/40	0.6 – 4.4	Arabica	Batista et al. (2003)
Brazil	20/60	0.2 – 7.3	Arabica	Martins et al. (2003)
Africa	12/12	2.4 – 23.3	Robusta	Pardo et al. (2004)
America	31/31	1.3 – 27.7	Arabica	"
Asia	14/14	1.6 – 31.5	Arabica and Robusta	"

[a]Corresponds to the detection limit of the method; [b]Not Specified.

Table 2. Incidence of ochratoxin A in commercial roasted coffee worldwide

Retail country	No. positive/ No. samples	Range of AO (μg/kg)	Reference
Japan	5/68	3.2 – 17.0	Tsubouchi et al. (1988)
United Kingdom	17/20	0.2 – 2.1	Patel et al. (1997)
Europe	?/484	< 0.5[a] – 8.2	Van der Stegen et al. (1997)
Denmark	11/11	0.1 – 3.2	Jorgensen (1998)
Spain	29/29	0.22 – 5.64	Burdespal & Legarda (1998)
United States	9/13	0.1 – 1.2	Trucksess et al. (1999)
Brazil	23/34	0.3 – 6.5	Leoni et al. (2000)
Brazil	41/47	0.99 – 5.87	Prado et al. (2000)
Germany	22/67	0.3 – 3.3	Wolff (2000)
Germany	273/490	0.21 – 12.1	Otteneder & Majerus (2001)
Canada	42/71	0.1 – 2.3	Lombaert et al. (2002)
Hungary	22/38	0.17 – 1.3	Fazekas et al. (2002)

[a]Corresponds to the detection limit of the method.

Table 3. Incidence of ochratoxin A in commercial soluble coffee worldwide

Retail country	No. positive/ No. samples	Range of OA (μg/kg)	Reference
Australia	7/22	0.2 – 4.0	Pittet et al. (1996)
United States	3/6	1.5 – 2.1	"
Germany	5/9	0.3 – 2.2	"
United Kingdom	64/80	0.1 – 8.0	Patel at al. (1997)
Europe	?/149	< 0.5[a] – 27.2	Van der Stegen et al. (1997)
Spain	9/9	0.19 – 1.08	Burdaspal & Legarda (1998)
Brazil	8/10	0.31 – 1.78	Prado et al. (2000)
Brazil	16/16	0.5 – 5.1	Leoni et al. (2000)
Germany	23/52	0.3 – 9.5	Wolff (2000)
Germany	12/41	0.28 – 4.8	Otteneder & Majerus (2001)
Canada	20/30	0.1 – 3.1	Lombaert et al. (2002)

[a]Corresponds to the detection limit of the method.

cherries on the coffee trees were infected with these species, indicating that infection mostly occurred after harvest, and the fungal sources were likely to be soil, equipment and drying yard surfaces. Variability in infection rates of these toxigenic species was reflected in a wide range of ochratoxin A levels in samples from the drying yard and storage. *A. niger* was more common than *A. ochraceus* or *A. carbonarius,* but only 3% of isolates were capable of producing OA, so this species is probably a relatively unimportant source of OA in coffee.

4. COFFEE PRODUCTION

The formation of OA during coffee production can be assessed in three stages: 1) pre-harvest; 2) post-harvest to storage and transportation of raw coffee to the processing plant, and 3) raw coffee processing to roasted and ground coffee and soluble coffee.

4.1. Pre-harvest

The issue of time of invasion of coffee by toxigenic fungi is of great importance in understanding the problem of OA in coffee and in developing control strategies. Experience with other crops has shown that if invasion occurs pre-harvest, control will be much more difficult than if it occurs post-harvest, i.e. during drying and storage. Post-harvest problems can be expected to relate to unfavourable climates for drying, poor drying practice or quality control, or inadequate storage conditions. The major risk factors and processing steps that can lead to contamination of raw coffee with ochratoxin A have been reviewed (Bucheli and Taniwaki, 2002).

The suggestion by Mantle (2000) that OA in coffee beans may result from uptake of OA in soil by the roots of the coffee tree and then translocated is conjectural. Taniwaki et al. (2003) found an average of *A. ochraceus* infection of less than 0.6% of fruit on the tree. The highest percentage of *A. ochraceus* infection in coffee fruit sampled from trees was 4%, but the figure increased to 16% in fruit harvested from the ground, and to 35% in fruit during drying and storage. No evidence has been found that either *A. ochraceus* or *A. carbonarius* invades coffee beans before harvest or has an association with the coffee tree. These facts indicate that infection mostly occurs after harvest, and the fungal sources are likely to be soil, equipment and drying yard surfaces.

4.2. Post-harvest, storage and transport

Cherries contain sufficient amounts of water to support mould growth and OA formation on the outer part of the cherries during the initial 3-5 days of drying. Sun drying of coffee cherries if done incorrectly can potentially result in OA contamination. In general, drying of fruits appears to be a particularly risky step for the natural occurrence of black Aspergilli. This may be linked to the resistance of the black spores of this species to UV irradiation. Drying is also the most favourable time for development of *A. ochraceus*, the main problem being the time it takes for the berries to dry below a critical water activity (a_w) of about 0.80. Berries should spend no more that 4 days between 0.97 to 0.80 a_w. Palacios-Cabrera et al. (2004) showed that *A. ochraceus* produced little OA (0.15 µg/kg) in coffee beans at a_w of 0.80 and temperature of 25:C, but at 0.86 and 0.90 a_w the production was 2500 and over 7000 µg/kg, respectively.

4.2.1. Drying

The formation of OA during coffee drying was studied in Thailand by Bucheli et al. (2000), who showed that the toxin was normally produced during the sun drying of coffee; overripe cherries were more susceptible than green ones. They also noted that during sun drying, OA was formed in the coffee cherry pericarp (pulp and parchment), the part of the cherry that is removed as husk in the dehulling process. The results obtained by Bucheli et al. (2000) demonstrated that the initial raw material quality, weather conditions during drying, drying management, presence of OA producing fungi, and local farm conditions, undoubtedly played a more important role in OA contamination in raw coffee than the drying methodology used, on bamboo tables, bare ground or concrete. Taniwaki et al. (2003) agreed with Bucheli et al. (2000) that if drying is rapid and effective, OA will not be produced. Good sun drying or a combination of sun drying and mechanical dehydration provide effective control.

4.2.2. Dehulling

Dried coffee cherries are subjected to a dehulling process to separate the raw coffee beans from the husks. This is often a rather dusty procedure and it is possible that part of the OA contained in the husks will be transferred as dust particles or husk fragments to the raw coffee. After dehulling, husks can be highly contaminated with OA.

Bucheli et al. (2000) studied Robusta coffee drying in Thailand over three seasons. They found that sun drying of cherries consistently resulted in OA formation in the pulp and husks. The coffee beans, on the other hand, had only about 1% of the OA found in husks. OA contamination of raw coffee depended on cherry maturation, with overripe cherries being the most susceptible. Inclusion of defective berries and husks was found to be the most important source of OA contamination. The main fungal source of OA in this case was *A. carbonarius*.

The largest proportion (more than 90%) of OA contamination in coffee was usually concentrated in the husks. Cleaning, grading and hygienic storing of raw coffee is, therefore, of paramount importance. An indirect confirmation of the importance of husk removal to reduce OA contamination is the observation that husk addition, a fraudulent practice sometimes encountered in the manufacture of soluble coffee, may lead to the presence of relatively high levels of OA contamination in adulterated soluble coffee (Pittet et al., 1996). More recently, Suárez-Quiroz et al. (2004) reported that toxin was not totally eliminated after hulling coffee containing OA. The content of the toxin in coffee beans, parchment or husk was similar. However, these authors analysed only seven samples and the contamination level was between a trace and 0.3 μg/kg.

4.2.3. Wet processing

Bucheli and Taniwaki (2002), reviewing the impact of wet processing on the presence of OA in coffee, noted that there is indirect evidence that the depulping process must reduce significantly the risk of OA contamination, as the fruit pulp is an excellent substrate for the growth of OA producing fungi. Suárez-Quiroz et al. (2004) evaluated the effect on growth of fungi producing OA of three methods of coffee processing (wet, mechanical and dry) at different stages from harvest to storage. These authors concluded that for *A. ochraceus* there was no great difference between the three processes, but the dry method seemed to promote the presence of *A. niger*. This may be due to the black spores which give protection from sunlight and UV light, providing a competitive advantage in such habitats. Black Aspergilli have been frequently isolated from sun dried products, such as dried vine fruits, dried fish and spices (Pitt and Hocking, 1997).

Whether the wet or dry process is preferred, some measures must be taken to avoid contamination, by having a good initial quality of harvested coffee, and well controlled processing conditions.

4.2.4. Storage

Bucheli et al. (1998) reported on the impact of storage on mould growth on industrial green Robusta coffee and consequent OA formation. Under the storage conditions tested (bag storage, and silo storage under air-conditioning, aeration and non-aeration), neither growth and presence of OA producing fungi, nor consistent OA production, was observed. On average, an 18-fold decrease of fungal counts was found. This storage study demonstrated that safe storage of green Robusta coffee under humid tropical conditions can be achieved, even over a rainy period of several months, without finding OA formation and bean damage during storage. However, the initial a_w of the beans stored in bags was 0.72 and did not exceed 0.75 even in the rainy season. Improper storage and transportation do not appear to be a major route for OA contamination of coffee, unless coffee is remoistened.

4.2.5. Transportation

Condensation can occur during transportation of raw coffee to the consuming countries and lead to mould growth. Blanc et al. (2001) reported on transportation of raw coffee in bulk or bags in shipping containers and showed that condensation sometimes occurred at the top of the container, for example during winter in a European harbour. The development of areas with high moisture can favour mould growth and potentially OA formation.

Storage and transportation trials have shown that risk of condensation and wetting of coffee occur mainly during transport overland to the harbour for shipping and/or arrival at the destination.

4.3. Effect of roasting on ochratoxin levels

The roasting process subjects raw coffee to temperatures of 180-250:C for 5 to 15 min. During roasting, chemical changes occur with the development of aromas and the formation of dark coloured compounds, while physical changes include loss of water and dry material, mainly CO_2 and other volatiles (Clarke, 1987).

Conflicting data with respect to the influence of roasting (Table 4), grinding and beverage preparation on the residual levels of OA are found in the literature. Blanc et al. (1998) investigated the behaviour of OA during roasting and the production of soluble coffee. In this study, a small proportion of the OA was eliminated during the initial cleaning process, due to the discarding of defective and black beans, but the most significant reduction occurred during the roasting process. The ground roasted coffee contained only 16% of the original

Table 4. Effect of roasting on ochratoxin A reduction

No samples	Toxin origin	Roasting condition	% reduction	References
4	Inoculation[a]	200°C/10-20 min	0 –12	Tsubouchi et al. (1988)
2	Natural[b]	5 – 6 min/dark roasting	90 – 100	Micco et al. (1989)
3	Natural[b]	252°C/100-190 seg	14 – 62	Studer-Rohr et al. (1995)
2	Inoculation[a]	252°C/100-190 seg	2 – 28	”
6	Natural[b]	223°C / 14 min	84	Blanc et al. (1998)
3	Inoculation[a]	200:C/10 min (medium roasting)	22.5	Urbano et al. (2001a)
3	”	200:C/15 min (medium roasting)	48.1	”
3	”	210:C/10 min (medium dark)	39.2	”
3	”	210:C/15 min (medium dark)	65.6	”
3	”	220:C/10 min (dark)	88.4	”
3	”	220:C/15 min (dark)	93.9	”

[a]Coffee beans inoculated with toxigenic spores of *Aspergillus ochraceus*; [b]Naturally contaminated beans.

OA present in the raw coffee. Cleaning and thermal degradation were the most important factors in the elimination of OA.

Heilmann et al. (1999), studied OA reduction in raw coffee beans roasted industrially, and showed that levels of OA were significantly reduced, especially in coffee decaffeinated by solvent extraction. Leoni et al. (2000) studied 34 samples of ground roasted coffee, 14 instant coffee and 2 decaffeinated coffee. They found an average of 2.2 μg/kg of OA in the instant coffee, and in 23 of the ground roasted samples, values between 0.3 and 6.5 μg/kg of OA were found. When coffee beverage was prepared from the ground, roasted samples, the OA content of the resultant coffee was 74 to 86% of that in the original ground samples.

Urbano et al. (2001a) analysed 18 samples of coffee inoculated with a strain of *Aspergillus ochraceus* that produced OA. The samples were subjected to temperatures of 200°, 210° and 220°C for 10 to 15 min as shown in Table 4. The level of OA destruction varied from 22% to 94% depending on the time and temperature combination. In practice, however, a treatment of 220°C for 10 to 15 min may not produce a sensorially acceptable beverage.

The data in the literature show evidence that the roasting process is efficient in reducing OA, but there is a lack of more conclusive research on the effects of the stages of roasting, grinding and beverage preparation on the stability of the toxin. The effect on OA destruction of coffee production processes such as roasting, drink preparation, soluble coffee and decaffeination have been reviewed elsewhere (Gollucke et al., 2004). It is important that to maintain good quality coffee, it is advisable to avoid coffee with a high OA content. Even though roasting or other processes may reduce this toxin, the quality of the coffee can be affected.

5. OCHRATOXIN A CONSUMPTION FROM COFFEE

A great deal of interest has been focused on the possible role of coffee in ochratoxin A consumption. JECFA (2001) has set a Provisional Tolerable Weekly Intake for OA of 100 ng/kg bw/week which corresponds to a Provisional Tolerable Daily Intake (PTDI) of 14 ng/kg bw/day. From a survey of coffee drinkers in the United Kingdom and the data shown in Table 2, Patel et al. (1997) developed an Estimated Weekly Intake of OA from soluble coffee. Based on a consumption of 4.5 g of soluble coffee per day, the average coffee drinker ingested 0.4 ng/kg body weight/week, while the heavy consumer (97.5% percentile) consumed nearly 20 g soluble coffee per day to give an OA intake of 1.9 ng/kg body weight/week. Those figures translate into 3.5 ng and 17 ng intake of OA per day and 0.4% or 2% of the PTDI respectively (Patel et al., 1997). Comparing these data with the data obtained by Leoni et al. (2000) in Brazil, the average of OA concentration in roasted coffee was 0.9 μg/kg (Table 2). Ground and roasted coffee is the type of coffee most used by Brazilians and many other coffee producing countries. According to Leoni et al. (2000), the average Brazilian adult drinks five cups of coffee per day and this would correspond to 30 g of roast and ground coffee to brew five cups of the beverage (each 60 ml) prepared according to common household procedures. The probable daily intake of OA by a 70 kg adult would be 0.4 ng/kg bw/day and this falls far below the JECFA PTDI. Examining a worst case situation of a heavy coffee drinker who may drink up to 33 cups of coffee a day (Camargo et al. 1999) this would mean an ingestion of 2.5 ng/kg bw/day which is still well below JECFA PTDI. These results indicate that coffee is not a major dietary source of ochratoxin A in the UK or Brazil and the situation should not be very different in other coffee drinking countries.

6. CONCLUSION

As the evidence shows that ochratoxin A is formed in raw coffee beans after harvest, OA contamination can be minimized by following good agricultural practice and post-harvest handling consisting of appropriate techniques for drying, grading, transportation and storage of raw coffee. Moreover, better quality raw material, appropriate dehulling procedures and reduction of defects using colour sorting can substantially reduce the concentration of OA in raw coffee. These procedures are well established. In June 2002, the European coffee associations and bodies published the Code of Practice "Enhancement of coffee quality through prevention of mould formation" (www.ecf-coffee.org). The objective of this code of practice is to assist operators to apply Good Agricultural Practice, Good Practices in Transport and Storage and Good Manufacturing Practices preventing OA contamination and formation throughout the coffee chain. Preventive measures taken by all participants in the chain from tree to cup are the best way to prevent OA contamination in coffee.

7. REFERENCES

Batista, L. R., Chalfoun, S. M., Prado, G., Schwan, R. F., and Wheals, A. E., 2003, Toxigenic fungi associated with processed (green) coffee beans (*Coffea arabica* L.), *Int. J. Food Microbiol.* **85**:293-300.

Blanc, M., Pittet, A., Muñoz-Box, R., and Viani, R., 1998, Behavior of ochratoxin A during green coffee roasting and soluble coffee manufacture, *J. Agric. Food. Chem.* **46**:673-675.

Blanc, M., Vuataz, G., and Hilckmann, L., 2001, Green coffee transport trials, *Proc. 19th ASIC Coffee Conf.*, Trieste, Italy, 14-18 May.

Bucheli, P., Kanchanomal, C., Meyer, I., and Pittet, A., 2000, Development of ochratoxin A during Robusta (*Coffee canephora*) coffee cherry drying, *J. Agric. Food. Chem.* **48**:1358-1362.

Bucheli, P., Meyer, I., Vuataz, A., and Viani, R., 1998, Industrial storage of green Robusta coffee under tropical conditions and its impact on raw material quality and ochratoxin A content, *J. Agric. Food. Chem.* **46**: 4507-4511.

Bucheli, P., and Taniwaki, M. H., 2002, Research on the origin, and the impact of postharvest handling and manufacturing on the presence of ochratoxin A in coffee, *Food Addit. Contam.* **19**:655-665.

Burdaspal, P. A., and Legarda, T. M., 1998, Ochratoxin A in roasted and soluble coffee marketed in Spain, *Alimentaria* **296**:31-35.

Camargo, M. C. R., Toledo, M. C. F., and Farah, H. G., 1999, Caffeine daily intake from dietary sources in Brazil, *Food Addit. Contam.* **16**:79-87.

Clarke, R. J., 1987, Roasting and grinding, in: *Coffee Technology,* R. J. Clarke, and R. Macrae, eds, Elsevier Applied Science, Essex, pp. 73-107.

Fazekas, B., Tar, A. K., and Zomborszky-Kovács, M., 2002, Ochratoxin A contamination of cereal grains and coffee in Hungary in the year 2001, *Acta Vet. Hungarica* **50**:177-188.

Frank, J. M., 1999, HACCP and its mycotoxin control potential: ochratoxin A (OTA) in coffee production, *Proc. 7th Int. Committee Food Microbiol. Hyg.* The Netherlands, Veldhoven, pp. 1222-1225.

Gollucke, A. P. B., Taniwaki, M. H., and Tavares, D. Q., 2001, Occurrence of ochratoxin A in raw coffee for export from several producing regions in Brazil, *Proc. 19th ASIC Coffee Conf.*, Trieste, Italy, 14-18 May.

Gollucke, A. P. B., Tavares, D. Q., and Taniwaki, M. H., 2004, Efeito do processamento sobre a ocratoxina A, em café. *Higiene Alimentar*. **18**:38-48.

Heilmann, W., Rehfeldt, A. G., and Rotzoll, F., 1999, Behaviour and reduction of ochratoxin A in green beans in response to various processing methods, *Eur. Food Res. Technol.* **209**:297-300.

JECFA (Joint FAO/WHO Expert Committee on Food Additives), 2001, Ochratoxin A, in: *Safety Evaluation of Certain Mycotoxins in Food*, prepared by the Fifty-sixth meeting of the JECFA, FAO Food and Nutrition Paper 74, Food and Agriculture Organization of the United Nations, Rome, Italy.

Joosten, H. M. L. J., Goetz, J., Pittet, A., Schellenberg, M., and Bucheli, P., 2001, Production of ochratoxin A by *Aspergillus carbonarius* on coffee cherries, *Int. J. Food Microbiol.* **65**:39-44.

Jorgensen, K., 1998, Survey on pork, poultry, coffee, beer and pulses for ochratoxin A, *Food Addit. Contam.* **15**:550-554.

Leoni, L. A. B., Valente Soares, L. M., and Oliveira, P. L. C., 2000, Ochratoxin A in Brazilian roasted and instant coffees, *Food Addit. Contam.* **17**:867-870.

Levi, C. P., Trenk, H. L. and Mohr, H. K., 1974, Study of the occurrence of ochratoxin A in green coffee beans, *J. Assoc. Off. Anal. Chem.* **57**:866-870.

Lombaert, G. A., Pellaers, P., Chettiar, M., Lavalce, D., Scott, P. M., and Lau, B. P. Y., 2002, Survey of Canadian retail coffees for ochratoxin A, *Food Addit. Contam.* **19**:869-877.

MAFF (Ministry of Agriculture, Fisheries and Food), 1996, Surveillance of ochratoxin A in green (unroasted) coffee beans, Food Surveillance Information Sheet 80.

Mantle, P. G., 2000, Uptake of radiolabelled ochratoxin A from soil by coffee plants, *Phytochemistry* **53**:377-378

Martins, M. L., Martins, H. M., and Gimeno, A., 2003, Incidence of microflora and of ochratoxin A in green coffee benas (*Coffee arabica*), *Food Addit. Contam.* **20**:1127-1131.

Micco, C., Grossi, M., Miraglia, M., and Brera, C., 1989, A study of the contamination by ochratoxin A of green and roasted coffee beans, *Food Addit. Contam.* **6**:333-339.

Nakajima, M., Tsubouchi, H., Miyabe, M., and Ueno, Y., 1997, Survey of aflatoxin B_1 and ochratoxin A in commercial green coffee beans by high-performance liquid chromatography linked with immunoaffinity chromatography, *Food Agric. Immunol.* **9**:77-83.

Otteneder, H. and Majerus, P., 2001, Ochratoxin A (OTA) in coffee: nation wide evaluation of data collected by German food control 1995-1999, *Food Addit. Contam.* **18**:431-435.

Palacios-Cabrera, H., Taniwaki, M. H., Menezes, H. C., and Iamanaka, B. T., 2004, The production of ochratoxin A by *Aspergillus ochraceus* in raw coffee at different equilibrium relative humidity and under alternating temperatures, *Food Control* **15**:531-535.

Pardo, E., Marin, S., Ramos, A. J. and Sanchis, V., 2004, Occurrence of ochratoxigenic fungi and ochratoxin A in green coffee from different origins, *Food Sci. Tech. Int.* **10**:45-49.

Patel, S., Hazel, C. M., Winterton, A. G. M., and Gleadle, A. E., 1997, Survey of ochratoxin A in UK retail coffees. *Food Addit. Contam.* **14**:217-222.

Pitt, J. I., and Hocking, A. D., 1997, *Fungi and Food Spoilage,* 2nd edition, Blackie Academic and Professional, London.

Pittet, A., Tornare, D., Huggett, A., and Viani, R., 1996, Liquid chromatographic determination of ochratoxin A in pure and adulterated soluble coffee using an immunoaffinity column cleanup procedure, *J. Agric. Food Chem.* **44**:3564-3569.

Prado, G., Oliveira, M. S., Abrantes, F. M. Santos, L. G., Veloso, T., and Barroso, R. E. S., 2000, Incidência de ocratoxina A em café torrado e moído e café solúvel consumido na cidade de Belo Horizonte, MG, *Ciência e Tecnologia de Alimentos* **20**:192-196.

Romani, S., Sacchetti, G., Chaves López, C., Pinnavaia, G. G., and Dalla Rosa, M., 2000, Screening on the occurrence of ochratoxin A in green coffee beans of different origins and types, *J. Agric. Food Chem.* **48**: 3616-3619.

Studer-Rohr, I., Dietrich, D.R., Schlatter, J., and Schlatter, C., 1995, The occurrence of ochratoxin A in coffee, *Food Chem. Toxicol.* **33**:341-355.

Suárez-Quiroz, M., González-Rios, O., Barel, M., Guyot, B., Schorr-Galindo, S. and Guiraud, J. P., 2004, Study of ochratoxin A producing strains in coffee processing, *Int. J. Food Sci. Technol.* **39**:501-507.

Taniwaki, M. H., Pitt, J. I., Teixeira, A. A., and Iamanaka, B. T., 2003, The source of ochratoxin A in Brazilian coffee and its formation in relation to processing methods, *Int. J. Food Microbiol.* **82**:173-179.

Téren, J., Palágyi, A. and Varga, J., 1997, Isolation of ochratoxin producing aspergilli from green coffee beans of different origin, *Cereal Res.*, **25**:303-304.

Trucksess, M. W., Giler, J., Young, K., White, K. D., and Page, S. W., 1999, Determination and survey of ochratoxin A in wheat, barley and coffee-1997, *J. AOAC. Int.*, **82**:85-89.

Tsubouchi, H., Terada, H., Yamamoto, K., Hisada, K., and Sakabe, Y., 1988, Ochratoxin A found in commercial roast coffee, *J. Agric. Food Chem.* **36**:540-542.

Urbano, G. R., Freitas Leitão, M. F., Vicentini, M. C. and Taniwaki, M. H., 2001a, Preliminary studies on the destruction of ochratoxin A in coffee during roasting, *Proc. 19th ASIC Coffee Conf.*, Trieste, Italy, 14-18 May, 2001.

Urbano, G.R., Taniwaki, M. H., Leitão, M. F. and Vicentini, M. C. 2001b, Occurrence of ochratoxin A producing fungi in raw Brazilian coffee, *J. Food Prot.* **64**:1226-1230.

Van der Stegen, G., Jorissen, U., Pittet, A., Saccon, M., Steiner, W., Vincenzi, M., Winkler, M., Zapp, J., and Schlatter, C., 1997, Screening of European coffee final products for occurrence of ochratoxin A (OTA), *Food Addit. Contam.* **14**:211-216.

Wolff, J., 2000, Forschungsbericht: Belastung des Verbrauchers und der Lebensmittel mit Ochratoxin A, study funded by German Federal Ministry of Health (BMG vom 03.02.2000. Gesch.Z. 415-6080-1/54).

MYCOBIOTA, MYCOTOXIGENIC FUNGI, AND CITRININ PRODUCTION IN BLACK OLIVES

Dilek Heperkan, Burcak E. Meric, Gülcin Sismanoglu, Gözde Dalkiliç and Funda K. Güler[*]

1. INTRODUCTION

Olives have been reported to be a poor substrate for the production of aflatoxins (Mahjoub and Bullerman, 1987; Eltem, 1996; Leontopoulos et al., 2003). However Yassa et al., (1994) isolated *A. flavus* and *A. parasiticus* from black table olives in Egypt, from which nine strains of *A. flavus* and five strains of *A. parasiticus* were found to produce aflatoxin B_1 on olive paste. Toussaint (1997) and Daradimos et al. (2000) found aflatoxin B_1 in olive oil samples at levels of 5-10 μg/kg and 3-46 ng/kg respectively.

Mahjoub and Bullerman (1987) found that mixing olive paste with yeast extract sucrose agar caused a moderate increase of aflatoxin B_1 production compared with yeast extract sucrose medium alone, and that heat and sodium hydroxide treatments stimulated growth and aflatoxin production in olives. However, Leontopoulos et al. (2003) found higher levels of aflatoxin produced in olives sanitized with chlorine than in those sterilized by autoclaving.

* Dilek Heperkan, Burcak E. Meric, Gülcin Sismanoglu, Gözde Dalkiliç, and Funda K. Güler, Istanbul Technical University, Dept. of Food Engineering Istanbul, Turkey, 34469 Maslak. Correspondence to: heperkan @itu.edu.tr

Daradimos et al. (2000) highlighted the importance of methods for aflatoxin analysis. Applying two different methods for determination of aflatoxin B_1 in olive oil, with one method they found 72% of the samples were contaminated with aflatoxin B_1, whereas no aflatoxin B_1 was detected with the other method.

Mycotoxins such as ochratoxin and patulin (Gourama and Bullerman, 1987) were not produced by *A. ochraceus*, *A. petrakii* and *P. expansum* isolated from olives. However citrinin production by *Penicillium citrinum* has been reported in olives (Oral and Heperkan, 1999).

Citrinin contamination of foods is important because citrinin is nephrotoxic (Krogh et al., 1973; Frank, 1992) and genotoxic (Föllmann, et al., 1998) and enhances the effect of ochratoxin A in induction of renal tumors in animals such as rats and mice (Pfohl-Leszkowicz et al., 2002). The LD_{50} of citrinin varies from 20 mg/kg b.w. by subcutaneous injection in the rabbit to 112 mg/kg b.w. in mice after intraperitoneal administration (Pfohl-Leszkowicz et al., 2002).

Citrinin is produced by some *Penicillium*, *Aspergillus* and *Monascus* species. Pitt (2002) indicated that production of citrinin has been reported from at least 22 *Penicillium* species, but the true number of producers was less than five. Citrinin producing strains include *P. citrinum*, *P. verrucosum* (Frisvad and Thrane, 2002; Pitt and Hocking, 1997; Pitt, 2002), *P. expansum* (Vinas et al., 1993) *A. terreus* (Frisvad and Thrane, 2002), *Monascus ruber* and *M. purpureus* (Blanc et al., 1995; Hajjaj et al., 1999; Xu et al., 1999).

Penicillium and *Aspergillus* commonly occur on black olives (Fernandez et al., 1997; Sahin et al., 1999). Olives are grown mainly in Spain, Italy, Greece and Turkey and are used to produce olive oil or consumed directly. Fermented olives are an important product worldwide: table olive production and total olive production are approximately 350,000 and 1,150,000 tonnes respectively per year in Turkey. Table olives comprise approximately 30% of Turkey's total olive production. They have high economic value for the producing country as well as being an important source of nutrition.

The process, by which olives are produced varies from country to country and table olives often take their name from the country of origin, such as Turkish style, Greek style and Californian style, etc. During conventional olive production, the surface of the brine may be covered with a thick layer of mould. Mould growth may also be visible on one end of black olives sold in transparent packages in the market. Mould growth can cause softening of the olive tissue, a mouldy taste and appearance and thus reduce the acceptable quality of olives.

Moulds may also shorten the shelf life and produce mycotoxins. *P. citrinum* and *P. crustosum* have been isolated from the surface of olives during fermentation (Oral and Heperkan, 1999).

This study provides information on the mycoflora and citrinin production in black table olives obtained from a survey conducted in Turkey.

2. MATERIALS AND METHODS

2.1. Samples

Whole black table olives were surveyed. A total of 69 samples were randomly collected from markets in the Marmara and Aegean Regions, Turkey in 2000-2001. Each sample comprised approximately 2 kg and was maintained at 4°C until analyzed.

2.2. Mycobiota

Dilution plate techniques were used to determine the mycobiota. For the enumeration and isolation of moulds Malt Extract Agar with chloramphenicol was used. Plates were incubated at 25°C for 7 days. Moulds were identified according to Pitt and Hocking (1997) and Samson et al. (1996).

2.3. Citrinin Analysis

Citrinin standards were purchased from Sigma Chemical, St Louis, MO. Pre-coated silica gel TLC plates (Merck) were used for citrinin analyses. The citrinin standard was dissolved in chloroform and the UV absorption was read at 322 nm. The concentration of the solution was calculated using the formula for aflatoxins (Trucksess, 2000). The extraction and separation method of Comerio et al. (1998) was modified. Olive samples (25 g) were blended with acetonitrile (180 ml), 4% KCl (20 ml) and 20% H_2SO_4 (2 ml) for two min at high speed and filtered through Whatman No 4 filter paper. After filtration, hexane (50 ml) was added and the mixture was shaken for 15 min in a separating funnel. The oil from the olives was extracted into the hexane (upper) phase preventing it from interfering with the assay. Citrinin partitioned into the lower phase. The first 100 ml of the lower phase was transferred to a second separating funnel to which chloroform (50 ml) and distilled water (25 ml) were added. After extraction, the toxin contained in the lower phase was collected in a beaker and was evaporated to dryness under stream of nitrogen in water bath at approximately 55°C. The dry extract was taken up in 1 ml chloroform in a tube and the chloroform removed under a stream of nitrogen. Before extracts were spotted, TLC plates were dipped into 10% glycolic acid solution in ethanol for two minutes and then dried for 10 min at 110°C.

The dried toxin extracts were dissolved in chloroform (100 μl) and were spotted onto a TLC plate using a micropipette. The plate was developed in a tank containing toluene:ethyl acetate:chloroform:90% formic acid (70:50:50:20), dried and treated with ammonia vapour for 10-15 sec (Martins et al., 2002).

Citrinin was detected under UV light at 366 nm and the amount determined visually by comparing the fluorescence of the sample with the citrinin standard. The recovery of citrinin was determined by spiking a certain amount of citrinin standard e.g 100 μl into 25 g of olive sample. Four parallel experiments were carried out and the mean amount of citrinin was determined. In these experiments the calculated average recovery value was 76.3%.

3. RESULTS AND DISCUSSION

Of the olive samples examined, 17 of 42 (40%) and 15 of 27 (55.5%) from the Marmara and Aegean Regions respectively were found to be contaminated with moulds. The mould species isolated from the black olive samples are given in Table 1.

As can be seen from Table 1, almost all the isolates were *Penicillium* species, with *P. crustosum* and *P. viridicatum* being the most common. No *Aspergillus* species were isolated in the Marmara Region. In the Aegean Region, *A. versicolor* was isolated from only 2 samples; the flora was mainly *Penicillium* spp. with *P. roqueforti* and *P. viridicatum* being most frequently detected. The citrinin content of samples is shown in Table 2.

Citrinin was detected in 34 of the 42 samples (81%) from the Marmara Region; the amount of citrinin varied from a minimum of 75 and a maximum of 350 μg/kg. In the Aegean Region 20 of the 27 (74%) samples contained citrinin with the highest value being 100 μg/kg. There was a large difference between the citrinin amounts in the two regions, with the Marmara Region having much higher levels.

Table 1. Mould flora in black olives

	Number of positive samples	
Mould species	Marmara Region	Aegean Region
Aspergillus versicolor	–	2
Cladosporium spp.	2	3
Penicillium citrinum	2	–
P. crustosum	10	2
P. digitatum	2	2
P. roqueforti	2	14
P. viridicatum	5	4
P. solitum	–	2
P. brevicompactum	–	3
Total strains isolated	23	32

Table 2. Distribution of citrinin in black olives

Concentration of	Number of positive samples	
citrinin (µg/kg)	Marmara Region	Aegean Region
<25	–	12
25–50	–	5
50–75	–	1
75–100	9	2
100–125	6	–
125–150	14	–
150–175	2	–
350–375	3	–
Total	34	20

In another survey (Korukluoğlu et al., 2000), twenty olive samples were studied and ochratoxin was not found, however aflatoxin (7/20), patulin (3/20), citrinin (1/20) and penicillic acid (7/20) were detected in black olives from the Marmara Region.

Although the samples showed low levels of viable mould contamination, they were highly contaminated with citrinin. The results indicate that an investigation of only the mould flora can be misleading: the presence of toxin should also be determined. In this study, *P. citrinum* and other citrinin producing moulds were detected in only two samples taken from the market. In a previous study however, mould growth was observed on the surface during ripening in vats and *P. citrinum* and *P. crustosum* were isolated (Oral and Heperkan, 1999). Sahin et al. (1999) have also detected *P. citrinum, P. verrucosum* and low levels of *P. expansum* in a similar study. Because a study on the mould flora of olives after harvest has not been made, citrinin production was assumed to have taken place during the residence in the concrete vats under adverse conditions. Studies on other products in Turkey also show the frequent occurrence of *P. citrinum*. *P. citrinum* was isolated from 23% of pistachio nuts (Heperkan et al., 1994), 27.5% of stored corn (Özay and Heperkan, 1989) and 14.3% of animal feed (Heperkan and Alperden, 1988).

Although citrinin is considered a minor mycotoxin compared with aflatoxins, ochratoxin A, fumonisins, patulin, zearalenone and deoxynivalenol (Miller, 1995) it is moderately toxic: in humans kidney damage appears to be a likely result of prolonged ingestion (Pitt, 2002). Citrinin co-occurs with ochratoxin A (Vrabcheva et al., 2000) and patulin (Martins et al., 2002) and is nephrotoxic and teratogenic in test animals (Abramson, 1999).

4. CONCLUSIONS

The results of our study and the literature survey indicate that olives, which are widely consumed in Turkey, could be a significant source of mycotoxins in the human diet: our survey found 77% of the samples contained citrinin, with some relatively high amounts detected. As contamination occurs mainly during production, olive production methods should be reviewed. Residence in vats should be eliminated or mould growth prevented during this stage of production. Storage and marketing stages should also be examined, and procedures to prevent mould contamination should be developed.

Although there are no limits for citrinin set by either the Turkish government or the European Union, some countries may choose to set more stringent levels based on dietary intake of their populations (Park and Troxell, 2002). The presence of mycotoxins in food can cause health problems and endanger the commercial value of the product. Therefore sustainable strategies such as Good Agricultural Practices, (GAP) and Hazard Analyses Critical Control Point (HACCP) systems should be established.

5. REFERENCES

Abramson, D., 1999, Rapid determination of citrinin in corn by fluorescence liquid chromatography and enzyme immunoassay, *J. AOAC Int.* **82**:1353-1355.

Blanc, P. J., Laussac, J. P., Le Bars, J., Le Bars, P., Loret, M. O., Pareilleux, A., Prome, D., Prome, J. C., Santerre, A. L., and Goma, G., 1995, Characterization of monascidin A from *Monascus* as citrinin, *Int. J. Food Microbiol.* **27**:201-213.

Comerio, R., Pinto, V. E. F., and Vamonde, G., 1998, Influence of water activity on *Penicillium citrinum* growth and kinetics of citrinin accumulation in wheat, *Int. J. Food Microbiol.* **42**:219-223.

Daradimos, E., Markaki, P., and Koupparis, M., 2000, Evaluation and validation of two fluorometric HPLC methods for the determination of aflatoxin B_1 in olive oil, *Food Addit. Contam.* **17**:65-63.

Eltem, R., 1996, Growth and aflatoxin B_1 production on olives and olive paste by moulds isolated from 'Turkish style' natural black olives in brine, *Int. J. Food Microbiol.* **32**:217-223.

Fernandez, G. A., Diez, M. J. F., and Adams, M. R., 1997, *Table Olives: Production and Processing,* Chapman and Hall, London.

Föllmann, W., Dorrenhaus, A., and Bolt, H. M., 1998, Induktion von mutationen im HPRT-Gen, DNA-reparatursynthese und schwesterchromatidaustausch durch ochratoxin A und citrinin *in vitro*, in: *Proceedings Mycotoxin Workshop*, J. Wolff, T. Betsche., eds, Detmold, Germany, June 8-10.

Frank, H. K., 1992, Citrinin, *Zeits. Ernahrungswiss.* **31**:164-177.

Frisvad, J. C., and Thrane, U., 2002, Mycotoxin production by food-borne fungi, in: *Introduction to Food Borne Fungi*, R. A. Samson, E. S. Hoekstra, J. C. Frisvad, and O. Filtenborg, eds, 6th edition, Centraalbureau voor Schimmelcultures, CBS, Delft, pp. 251-261.

Gourama, H., and Bullerman, L. B., 1987, Mycotoxin production by moulds isolated from Greek-style black olives, *Int. J. Food Microbiol.* **1**:81-90.

Hajjaj, H., Blanc, P. J., Groussac, E., Goma, G., Uribellarrea, J. I. and Loubiere, P., 1999, Improvement of red pigment citrinin production ratio as a function of environmental conditions by *Monascus ruber*, *Biotechnol. Bioeng.* **64**:497-501.

Heperkan, D., and Alperden, I., 1988, Mycological survey of chicken feed and feed ingredients in Turkey, *J. Food Prot.* **51**:807-810.

Heperkan, D., Aran, N., and Ayfer, M., 1994, Mycoflora and aflatoxin contamination in shelled pistachio nuts, *J. Sci. Food Agric.* **66**:273-278.

Korukluoğlu, M., Gürbüz, O., Uylaşer, V., Yildirim, A., and Sahin, I., 2000, Gemlik tipi zeytinlerde mikotoksin kirliliginin arastirilmasi, *Türkiye 1. Zeytincilik Sempozyumu.* Uludağ Üniversitesi 6-9 Haziran, Bursa. Bildiri Kitabi, pp. 218-218.

Krogh, P., Hald, B., and Pedersen, E.J., 1973, Occurrence of ochratoxin A and citrinin in cereals associated with mycotoxic porcine nephropathy, *Acta. Path. Microbiol. Scand. Sect. B.* **81**:689-695.

Leontopoulos, D., Siafaka, A., Markaki, P., 2003, Black olives as substrate for *Aspergillus parasiticus* growth and aflatoxin B_1 production, *Food Microbiol.* **20**:119-126.

Mahjoub, A., and Bullerman, L. B., 1987, Effects of nutrients and inhibitors in olives on aflatoxigenic molds, *J. Food Prot.* **50**:959-963.

Martins, M. L., Gimeno, A., Martins, H. M., and Bernardo, F., 2002, Co-occurrence of patulin and citrinin in Portuguese apples with rotten spots, *Food Addit. Contam.* **19**:568-574.

Miller, J. D., 1995, Fungi and mycotoxins in grain: implications for stored product research, *J. Stored Prod. Res.* **31**:1-16.

Oral, J., and Heperkan, D., 1999, Penicillic acid and citrinin production in olives, in: *Food Microbiology and Food Safety into the Next Millenium. Proceedings of the*

17th International ICFMH Conference. A. C. J. Tuijtelaars, R. A. Samson, F. M. Rombouts, S. Notermans, eds, Veldhoven, The Netherlands, pp. 138-140.

Özay, G., and Heperkan, D., 1989, Mould and mycotoxin contamination of stored corn in Turkey, *Mycotoxin Res*. **5**:81-89.

Park, D. L., and Troxell, T. C., 2002, U.S. perspective on mycotoxin regulatory issues, in: *Mycotoxins and Food Safety*. J. W. DeVries, M. W. Trucksess, and L. S. Jackson, eds, Kluwer Academic / Plenum Publishers, New York, pp. 277-285.

Pfohl-Leszkowicz, A., Petkova-Bocharova, T., Chernozemsky, I. N., and Castegnaro, M., 2002, Balkan endemic nephropathy and associated urinary tract tumors: a review on aetiological causes and potential role of mycotoxins, *Food Addit. Contam*. **19**:282-302.

Pitt, J. I., 2002, Biology and ecology of toxigenic *Penicillium* species, in: *Mycotoxins and Food Safety*. J. W. DeVries, M. W. Trucksess, and L. S. Jackson., eds, Kluwer Academic / Plenum Publishers, New York, pp. 29-41.

Pitt, J. I., and Hocking, A. D., 1997, *Penicillium* and related genera, in: *Fungi and Food Spoilage,* 2nd edition, Blackie Academic and Professional, London, pp. 251-255.

Samson, R. A., Hoekstra, E. S., Frisvad, J. C., and Filtenborg, O., 1996, *Introduction to Food Borne Fungi*. 6th edition, Centraalbureau voor Schimmelcultures, Baarn, The Netherlands.

Sahin, I., Basoglu, F., Korukluoglu, M., and Göcmen, D., 1999, Salamura siyah zeytinlerde rastlanan küfler ve mikotoksin riskleri, *Kükem Dergisi* **22**(2):1-8.

Toussaint, G., Lafaverges, F., and Walker, E. A., 1997, The use of high pressure liquid chromatography for determination of aflatoxin in olive oil, *Arch. Inst. Pasteur, Tunis* **3-4**:325-334.

Trucksess, M. W., 2000, Committee on natural toxins, General referee reports, *J. AOAC Int*. **83**:442-448

Vinas, I., Dadon, J., and Sanchis, V., 1993, Citrinin production capacity of *Penicillium expansum* strains from apple packaging houses of Lerida (Spain), *Int. J. Food Microbiol*. **19**:153-156.

Vrabcheva, T., Usleber, E., Dietrich, R., and Martlauber, E., 2000, Co-ocurrence of ochratoxin A and citrinin in cereals from Bulgarian villages with a history of Balkan endemic nephropathy, *J. Agric. Food Chem*. **48**: 2483-2488.

Xu, Y., Li, Y., Sun, H., Lai, W., and Xu, E., 1999, Study on the production of citrinin in submerged cultures of *Monascus* spp., in: *Food Microbiology and Food Safety Into the Next Millenium. Proceedings of the 17th International ICFMH Conference,* A. C. J. Tuijtelaars, R. A. Samson, F. M. Rombouts and S. Notermans, eds, Veldhoven, The Netherlands, pp. 150.

Yassa, I. A., Abdalla, E. A. M., and Aziz, S. Y., 1994, Aflatoxin B_1 production by moulds isolated from black table olives, *Ann. Agric. Sci*. **39**:525-537.

BYSSOCHLAMYS: SIGNIFICANCE OF HEAT RESISTANCE AND MYCOTOXIN PRODUCTION

*Jos Houbraken, Robert A. Samson and Jens C. Frisvad**

1. INTRODUCTION

Byssochlamys species produce ascospores that are very heat-resistant and survive heating above 85°C for considerable periods (Beuchat and Rice, 1981; Splittstoesser, 1987). Besides their heat resistance, *Byssochlamys* species are also able to grow under very low oxygen tensions (Tanawaki, 1995) and are capable of producing pectinolytic enzymes. The combination of these three physiological characteristics make *Byssochlamys* species very important spoilage fungi in pasteurized and canned fruit in which they can cause great economical losses. The natural habitat of *Byssochlamys* is soil. Fruit that grow near the soil, or are harvested from the ground may thus become contaminated with *Byssochlamys* (Olliver and Rendle, 1934; Hull, 1939).

Besides causing spoilage of pasteurized products, some *Byssochlamys* species are also capable of producing mycotoxins, including patulin, byssotoxin A and byssochlamic acid (Kramer et al., 1976; Rice, 1977). An antitumor metabolite, byssochlamysol, a steroid against IGF-1 dependent cancer cells, is also produced by *Byssochlamys nivea* (Mori et al. 2003).

---

* J. Houbraken and R. A. Samson: Centraalbureau voor Schimmelcultures, P.O. Box 85167, 3508 AD, Ultrecht, Netherlands. J. C. Frisvad: Centre for Process Biotechnology, Biocentrum-DTU, Technical University of Denmark, 2800 Kgs. Lyngby, Denmark. Correspondence to: samson@cbs.knaw.nl

In the taxonomic revision of Samson (1974), three *Byssochlamys* species were accepted: *B. fulva*, *B. nivea* and *B. zollerniae. B. verrucosa* was subsequently described (Samson and Tansey, 1975). Only *B. nivea* and *B. fulva* are currently considered important food spoilage fungi or mycotoxin producers.

Recently we have encountered numerous food spoilage problems in which species of *Byssochlamys* and their *Paecilomyces* anamorphs were involved. To elucidate the taxonomic and ecological characteristics of these isolates in relation to heat resistance and mycotoxin production we have investigated *Byssochlamys* and *Paecilomyces* isolates from various origins, including pasteurized fruit, ingredients based on fruit, and soil. Using a polyphasic approach, a classification of food-related heat resistant *Byssochlamys* species is presented.

2. MATERIALS AND METHODS

2.1. Isolates

The 39 isolates of *Byssochlamys* and *Paecilomyces* studied are listed in Table 1. All isolates are maintained in the collection of the Centraalbureau voor Schimmelcultures (CBS), Utrecht, Netherlands.

2.2. Media

The following media were used in this study: Czapek Yeast Autolysate agar (CYA), Malt Extract Agar (MEA), Potato Dextrose Agar (PDA), Yeast Extract Sucrose agar (YES), Hay infusion agar (HAY), Cornmeal agar (CMA), Creatine sucrose agar (CREA) (Samson et al., 2004) and Alkaloid agar (ALK) (Vinokurova et al.).

2.3. Macromorphological characterisation

The macroscopical features used were acid production on CREA, colony diameter on MEA after 72 h. and colony diameter and degree of growth on CYA after 7 days at 30°C.

2.4. Micromorphological characterisation

Isolates were grown on several media for 5-70 d. Micromorphology of the anamorphic *Paecilomyces* states was characterised on MEA,

Table 1. Byssochlamys and *Paecilomyces* isolates used in this study

CBS[a] No.	Species	Source, remarks about culture	Ascospores
132.33	*Byssochlamys fulva*	Bottled fruit, Type of *Paecilomyces fulvus*	+
146.48	*Byssochlamys fulva*	Bottled fruit, Type of *Byssochlamys fulva*	–
135.62	*Byssochlamys fulva*	Fruit juice, Type of *Paecilomyces todicus*	+
604.71	*Byssochlamys fulva*	Unknown source	–
113245	*Byssochlamys fulva*	Pasteurized fruit juice	+
113225	*Byssochlamys fulva*	Multifruit juice	+
100.11	*Byssochlamys nivea*	Unknown source, Type of *Byssochlamys nivea*	+
133.37	*Byssochlamys nivea*	Milk of cow, Type of *Arachniotus trisporus*	–
606.71	*Byssochlamys nivea*	Oat grain, received as *Byssochlamys musticola*	–
546.75	*Byssochlamys nivea*	Unknown source	+
271.95	*Byssochlamys nivea*	Mushroom bed	+
102192	*Byssochlamys nivea*	Pasteurized drink yoghurt	+
113246	*Byssochlamys nivea*	Apple compote	+
373.70	*Byssochlamys lagunculariae*	Wood of *Laguncularia racemosa* (Mangue), Type of *B. nivea* var. *lagunculariae*	+
696.95	*Byssochlamys lagunculariae*	Pasteurized strawberries	+
338.51	*Byssochlamys spectabilis*	Fruit juice	–
102.74	*Byssochlamys spectabilis*	Unknown source, Type of *Paecilomyces variotii*	–
298.93	*Byssochlamys spectabilis*	Man, breast milk of patient	–
101075	*Byssochlamys spectabilis*	Heat processed fruit beverage, Type of *Talaromyces spectabilis*	+
109072	*Byssochlamys spectabilis*	Pectin, teleomorph present	+
109073	*Byssochlamys spectabilis*	Pectin, teleomorph present	+
110431	*Byssochlamys spectabilis*	Rye bread	–
284.48	*Byssochlamys divaricatum*	Mucilage bottle with library paste, Type of *Penicillium divaricatum*	– initials
110428	*Byssochlamys divaricatum*	Pectin	–, initials
110429	*Byssochlamys divaricatum*	Pectin	–, initials

Table 1. *Byssochlamys* and *Paecilomyces* isolates used in this study—*cont'd*

CBS[a] No.	Species	Source, remarks about culture	Ascospores
110430	*Byssochlamys divaricatum*	Pectin	+
604.74	*Byssochlamys verrucosa*	Nesting material of *Leipoa ocellata*	+
605.74	*Byssochlamys verrucosa*	Nesting material of *Leipoa ocellata,* Type	+
374.70	*Byssochlamys zollerniae*	Wood of *Zollernia ilicifolia* and *Protium heptaphyllum*, Type	+
628.66	*Paecilomyces maximus*	Quebracho-tanned sheep leather, France	–
371.70	*Paecilomyces maximus*	Annona squamosa, Brazil, Type of *Paecilomyces maximus*	–
990.73B	*Paecilomyces maximus*	Unknown source, Type of *Monilia Formosa*	–
296.93	*Paecilomyces maximus*	Man, bone marrow of patient	–
297.93	*Paecilomyces maximus*	Man, blood of patient	–
323.34	*Paecilomyces dactylethromorphus*	Unknown source, Type of *Paecilomyces mandshuricus* var. *saturatus*	–
223.52	*Paecilomyces dactylethromorphus*	Leather	–
251.55	*Paecilomyces dactylethromorphus*	Acetic acid, Type of *Paecilomyces dactylethromorphus*	–
990.73A	*Paecilomyces dactylethromorphus*	Unknown source, Type of *Penicillium viniferum*	–
492.84	*Paecilomyces dactylethromorphus*	*Lepidium sativum*	–

[a]CBS is the culture collection of the Centraalbureau voor Schimmelcultures, Utrecht, Netherlands

HAY and YES agars (Samson et al., 2004). The latter was used for the determination of the presence of chlamydospores. For the analyses of the features of the teleomorphic *Byssochlamys* state, OA and PDA (Samson et al., 2004) were used. The microscopical features recorded included shape and size of conidia, size and ornamentation of ascospores and presence and ornamentation of chlamydospores. Ornamentation of the surface of the conidia, chlamydospores and ascospores was determined by light microscopy after prolonged incubation up to 70 days.

2.5. Multivariate analyses

A matrix consisting of 39 objects (fungal isolates) and 10 variables (macro- and microscopical features) was constructed. Cluster analysis by unweighted pair-group method, arithmetic average (UMPGA) was performed on the data matrix using BIOLOMICS™ software (Bioaware S.A., Hannut, Belgium).

2.6. Secondary metabolite analysis

Isolates studied (see Table 1) were three point inoculated on MEA, YES, PDA, OA, CYA and ALK agars. All isolates were analysed for secondary metabolites after two weeks growth at 30°C. The cultures were extracted according to the method of Smedsgaard (1997) and analysed by HPLC with diode array detection (Frisvad and Thrane, 1993). The metabolites found were compared with a spectral UV library made from authentic standards run under the same conditions, and retention indices were compared with those of standards. The maximal similarity was a match of 1000.

2.7. Heat resistance

The strain *Talaromyces spectabilis* CBS 109073 (now considered to be a *Byssochlamys* species, see below) was inoculated at three points on CMA and incubated for 45 days at 30°C. After incubation, the parts of the colony where ascospores were produced were combined and transferred to 10 mM ACES buffer (pH 6.8, N-[2-acetamido]-2-aminoethane-sulfonic acid; Sigma) supplemented with 0.05% Tween 80. The intact asci were ruptured by suction through a 0.9 mm hypodermic needle with a syringe and by agitation with glass beads. The suspension was sonicated briefly (3 times for 30 s) and filtered through sterile glass wool. The ascospores together with conidia and other fungal fragments were centrifuged at 1,100×g (5 min) and washed three times in buffer.

D values in ACES buffer were determined at 85°C in duplicate. The spore suspension was pre-treated by heating at 65°C for 10 min to eliminate conidia, chlamydospores and hyphae. The suspension was then heated at 85°C at various times (see Figure 2). After heating, the suspension was cooled, serially diluted in sterile water, then spread plated onto CYA and incubated for 5-10 days at 30°C. Colonies were counted and D_{85} values calculated.

3. RESULTS

3.1. Morphological analyses

Paecilomyces variotii sensu lato and anamorphs of *Byssochlamys* species share several micromorphological characteristics, including phialides with cylindrical bases tapering abruptly into long cylindrical necks. The conidia are produced in long divergent chains (Samson, 1974). There are characters that are constant at the species level but distinct between species. Microscopical features such as the shape of the conidia, the sizes of the conidia and ascospores and the presence of chlamydospores can be used to group species belonging to *Byssochlamys* and *Paecilomyces*. Acid production on CREA, colony diameters and degree of growth on CYA are useful characters too. Table 2 summarises the results of the micro- and macroscopical analyses.

The classification of the *Byssochlamys* and *Paecilomyces* taxa based on phenotypical characters is also supported by a molecular taxonomic study of this complex (partial β-tubulin gene sequencing) (Samson et al., submitted).

Figure 1 shows that nine clades could be distinguished among the isolates studied (Table 1). Clades 1, 2 4 and 5 separate the four known species of *Byssochlamys*: *B. verrucosa, B. zollerniae, B. fulva* and *B. nivea*. Clade 6 includes the ex-type culture of *B. nivea* var. *lagunculariae* (CBS 373.70) and CBS 696.95, isolated from strawberries. Clades 3, 7, 8 and 9 are isolates have been classified in *Paecilomyces variotii* complex, now seen to include several taxa. Clade 8 includes the ex-type cultures of *Paec. variotii* and *Talaromyces spectabilis*. Reexamination of the ex-type culture of *T. spectabilis* shows that it logically belongs in *Byssochlamys* and the formal combination *Byssochlamys spectabilis* (Udagawa & Suzuki) Samson et al. is proposed (Samson et al., submitted). Strains which produce ascospores are rare and if ascospores are formed, they often develop only after prolonged incubation. In this clade many strains isolated from drinking yoghurt and pectin are accommodated.

Clade 3 contains the ex-type culture of *Penicillium divaricatum. Pen. divaricatum* Thom 1910 was considered to be a synonym of *Paec. variotii* by Thom (1930). In one isolate, CBS 110430, we have observed ascospore production of the *Byssochlamys* type after prolonged incubation (70 d). We have therefore erected the new name *Byssochlamys divaricata* Samson et al. for *Pen. divaricatum* (Samson et al., submitted). The strains examined were isolated from pectin and

Table 2. Macro- and microscopical features of *Byssochlamys* and *Paecilomyces* isolates

Species	Conidial size (µm) and shape	Chlamydo-spores[a]	Ascospore size (µm) and ornamentation	Colony diameter (mm)[b]	Degree of growth[c]	Acid[d]
B. divaricatum	3.2-4.6 × 1.6-2.5; ellipsoidal to cylindrical with truncate ends	– (+)	5.3-7.0 × 3.8-4.9, smooth	10-17	Moderate	–
B. fulva	3.7-7.5 × 1.4-2.5; cylindrical with truncate ends	– (+)	5.3-7.1 × 3.3-4.3, smooth	(50), >80	Good	+
B. lagunculariae	2.7-4.5 × 2.2-3.3; globose with flattened base	+, smooth	3.8-5.0 × 3.0-3.9, smooth	45-55	Good	–
B. nivea	3.0-4.7 × 2.3-4.0; globose to ellipsoidal with flattened base	+, smooth to finely roughened	4.1-5.5 × 2.9-3.9, smooth	(8) 28-50	Weak	– (+)
B. spectabilis	3.3-6.1 × 1.5-4.4; mostly ellipsoidal and ellipsoidal with truncated ends	+, smooth to finely roughened	5.2-6.8 × 3.5-4.5, almost smooth, sl. roughened	25-40 (56)	Good	–
B. verrucosa	6.3-13.1 × 1.6-4.7; cylindrical with truncate ends	–	6.6-8.4 × 4.0-6.1, rough	25-40	Good	–
B. zollerniae	2.5-4.0 × 1.5-3.0; globose to ellipsoidal, apiculate	+, warted	3.0-4.5 × 2.5-3.0, smooth	30-35	Weak	–
P. dactylethro-morphus	2.3-7.0 × 1.7-3.4; mostly cylindrical and ellipsoidal **without** truncated ends	+, smooth	No ascospores detected	22-55	Good	–
P. maximus	3.0-10 × 1.8-3.5; ellipsoidal to cylindrical with truncate ends	+, smooth and often pigmented	No ascospores detected	18 – >80	Good	+

[a] +, chlamydospores present; –, chlamydospores absent, (+) chlamydospores produced by some isolates after prolonged incubation (40 days); [b] Colony diameter on CYA, 72 h, 30°C; [c] Degree of growth on CYA, 7 d, 30°C; [d] Acid production in CREA, 7 d, 30°C

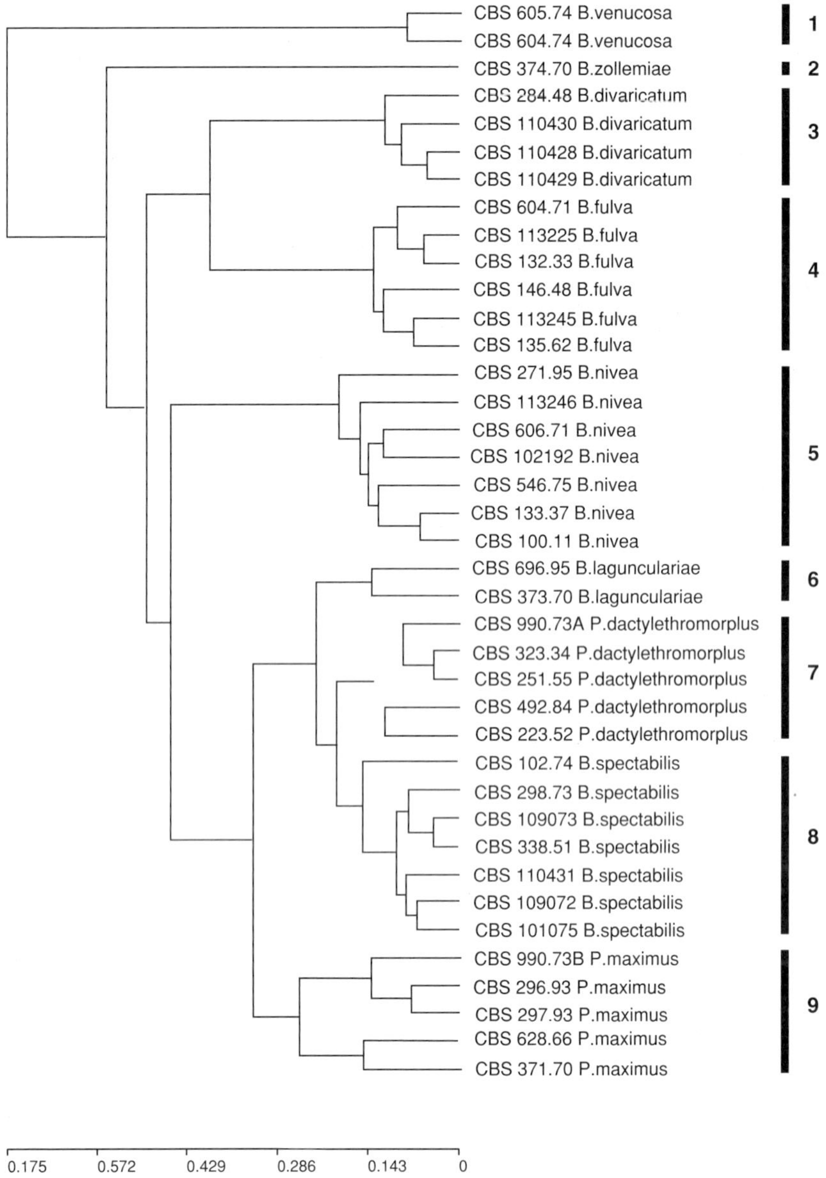

Figure 1. An UPGMA dendrogram based upon micro- and macromorphological characteristics of *Paecilomyces variotii* and *Byssochlamys* isolates.

fruit concentrates, and the ex-type culture came from a mucilage bottle with library paste. Clade 7 includes the ex-type cultures of *Paec. dactylethromorphus* Batista & H. Maia and *Paec. mandshuricus* var. *saturatus* Nakazawa et al. while the ex-type culture *Paec. maximus* C.

Ram is accommodated in Clade 9. These three taxa are now considered to be synonyms of *Paec. variotii*. Both clades contain strictly conidial isolates and no ascospores have been observed.

3.2. Mycotoxin analysis

Species in *Byssochlamys* and related *Paecilomyces* species can also be distinguished by differences in secondary metabolites. The results of the analyses are summarized in Table 3. *B. spectabilis* produced the mycotoxin viriditoxin. Some isolates of *B. nivea* and *P. dactylethromorphus* produced patulin, however, many did not. Mycophenolic acid was produced by *B. nivea* and *B. laguncularіae*, whereas emodin was produced by *B. divaricatum*. Byssotoxin A has been reported to be produced by isolates of *B. fulva* (Kramer, 1976) but because the structure of byssotoxin A was not elucidated, isolates were not screened for the presence of this mycotoxin.

3.3. Heat resistance

Paecilomyces and *Byssochlamys* isolates in the CBS collection were re-identified as described above. The results of the identification were related to the origin of the isolates. In Table 4 the number of strains isolated from heat-treated products or samples is correlated with the total number of isolates. The Table shows that species with a teleomorph are more often found in heat treated products, with the exception of *B. verrucosa* and *B. zollerniae* which do not occur in foods. The presence of *P. variotii sensu stricto* (the anamorph of *B. spectabilis*) in heat treated products could be explained by the production of heat resistant ascospores.

Table 3. Mycotoxin production by *Byssochlamys* and *Paecilomyces* species[a]

Species	Known mycotoxins
Byssochlamys fulva	Byssochlamic acid
Byssochlamys nivea	Patulin, mycophenolic acid, byssochlamic acid
Byssochlamys laguncularіae	Mycophenolic acid, byssochlamic acid
Byssochlamys spectabilis	Viriditoxin
Byssochlamys divaricatum	Emodin
Byssochlamys verrucosa	Byssochlamic acid
Byssochlamys zollerniae	No known mycotoxins detected
Paecilomyces maximus	No known mycotoxins detected
Paecilomyces dactylethromorphus	Patulin

[a] Many other metabolites are produced but not listed here

Table 4. Overview of isolates in CBS collection correlated with origin[a]

Species	No. isolates investigated	No. isolates from heat-treated products	Percentage from heat-treated products
Byssochlamys fulva	5	5	100
Byssochlamys nivea	5	3	60
Byssochlamys langucularíae	2	1	50
Byssochlamys spectabilis	17	6	35
Byssochlamys divaricatum	4	3	75
Byssochlamys verrucosa	2	0	0
Byssochlamys zollerniae	1	0	0
Paecilomyces maximus	6	0	0
Paecilomyces dactylethromorphus	5	0	0

[a] CBS is the culture collection of the Centraalbureau voor Schimmelcultures, Utrecht, Netherlands. Isolates of unknown origin are excluded

Two laboratory experiments were conducted on 45 day-old ascospores of *B. spectabilis* in ACES buffer. Conidia, chlamydospores and other fungal fragments were inactivated as described above (Section 2.7.). Figure 2 shows one of the two thermal death rate curves. Between 0 and 8 minutes of the heat-treatment, activation of the ascospores occurs, then a linear correlation exists between time and the logarithm of surviving ascospores. Regression analyses on the best fit resulted in two equations:

$$\log \text{CFU} = -0.0170^{*}\text{T (min)} + 5.9779 \ (r\ 0.972,\ p < 0.05\ );$$

and

$$\log \text{CFU} = -0.0179^{*}\text{T (min)} + 6.379 \ (r = 0.976,\ p < 0.05),$$

for experiments 1 and 2, respectively. The D_{85} value was calculated from these equations. Experiment 1 resulted in a D_{85} of 59 min, and the D_{85} value in experiment 2 appeared to be 56 min. Taking the 95% confidence level into account, the D_{85} value may vary between 49 and 75 min in experiment 1 and between 47 and 70 min in experiment 2.

4. DISCUSSION

This study has demonstrated that *Byssochlamys* and its associated anamorph species can be separated into at least nine taxa. *B. nivea* var.

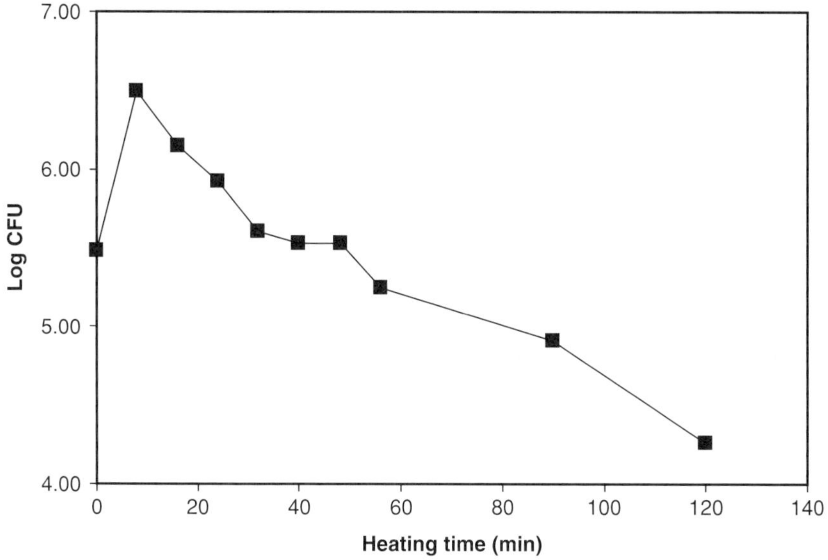

Figure 2. Thermal death curve of *Byssochlamys spectabilis* CBS 109073 at 85°C.

laguncularиae can be considered as a separate taxon, in addition to the known taxa, *B. fulva, B. nivea, B. verrucosa* and *B. zollerniae*. Within the species complex, *Byssochlamys* teleomorphs are observed in two other taxa, *B. spectabilis* and *B. divaricata*, which are clearly different from the other members of the genus *Byssochlamys*.

According to the literature, *Paec. variotii sensu lato* plays a role in mycotoxicoses described in several types of animals. In 1916, Turresson reported that rabbits died after ingestion of conidia and mycelium of *Pen. divaricatum* (= *B. divaricata*). *B. divaricata* produces the mycotoxin emodin, a genotoxic and diarrhoeagenic anthraquinone (Müller et al., 1996), which could be the cause of this described mycotoxicosis. Our investigation showed that byssochlamic acid is formed by *B. nivea*, *B. fulva*, *B. laguncularиae* and *B. verrucosa*. Byssochlamic acid was shown by Raistrick and Smith (1933) to be toxic to mice, and it is weakly hepatotoxic to guinea pigs (Gedek, 1971).

Patulin was reported to be produced by *B. nivea* (Karrow and Foster, 1944; Kis et al., 1969) and *B. fulva* (Rice et al., 1977). However, we could not detect patulin production by any of the investigated *B. fulva* isolates. Patulin production by strains of *B. nivea* was confirmed and its production by *Paec. dactylethromorphus* is described. Patulin was not produced by any of the strains of *B. laguncularιae* examined in this study. Mycophenolic acid was produced by strains of *B. nivea* and *B. laguncularιae*. This metabolite is an antibiotic, with anti-tumour, anti-psoriasis and immunosuppressive features (Bentley, 2000) and may be of relevance for secondary mycotoxicosis (bacterial infections caused by intake of an immunosuppressive mycotoxin).

Paec. variotii sensu lato is a rather common fungus in the air, in soil (subtropical and tropical climates), in compost (Knösel and Rész, 1973) and on wood (Ram, 1968). It is also common in foods such as rye bread, margarine, peanuts and peanut cake (Joffe, 1969; King et al., 1981), cereals (Pelhate, 1968) and heat treated fruit juices (S. Udagawa, A. D. Hocking , unpublished data).

From our study it can be concluded that the D_{85} value of *B. spectabilis* in ACES buffer was between 47 and 75 minutes. Comparing these results with other data, it seems that the ascospores of this species are one of the most heat resistant fungal ascospores. As this species is also capable of producing viriditoxin, it is an important spoilage fungus in pasteurized food and feed.

Paec. maximus commonly occurs in subtropical and tropical soils and *Paec. dactylethromorphus* is isolated from products such as acetic acid, leather and wood. Both species form chlamydospores, but we have never detected them from heat-treated samples. This indicates that ascospores, not thick walled chlamydospores, are the survival structures.

B. divaricata has also been isolated from heat-treated samples. This fungus does not form chlamydospores and therefore the mode of heat-survival is probably due to ascospores. *B. divaricata* and *B. spectabilis* make ascomata in culture only sparsely (and only after prolonged incubation at 30°C), nevertheless these structures should be present in nature. Soil (Udagawa et al., 1994) could be its natural habitat but also wood (Cartwright, 1937; Ram, 1968) should not be excluded.

5. ACKNOWLEDGEMENTS

The authors thank Joost Eleveld and Erik Dekker for their data on heat inactivation of various structures of *Byssochlamys spectabilis*. Jan Dijksterhuis kindly assisted with the statistical analyses of the thermal death curves. Jens C. Frisvad acknowledges the Danish Technical Research Council and Centre for Advanced Food Studies (LMC) for financial support.

6. REFERENCES

Bentley, R., 2000, Mycophenolic acid: A one hundred year odyssey from antibiotic to immunosuppressant, *Chem. Microbiol. Rev.* **16**:497-516.

Beuchat, L. R., and Rice, S. L., 1979, *Byssochlamys* spp. and processed fruits, *Adv. Food Res.* **25**:237-288.

Cartwright, K. T. S. G., 1937, A reinvestigation into the cause of 'Brown Oak', *Fistulina hepatica* (Hus.) Fr., *Trans. Br. Mycol. Soc.* 40:17-89.

Frisvad, J.C., and Thrane, U., 1993, Liquid column chromatography of mycotoxins, in: *Chromatography of Mycotoxins. Techniques and Applications*, V. Betina, ed., *J. Chromatogr. Libr.* **54**:253-372.

Gedek, B., 1971, Mycotoxine in Lebensmitteln und ihre Bedeutung für die Gesundheit des Menschen, *Materia Med. Normark* **23**:130-141.

Hull, R., 1939, Study of *Byssochlamys fulva* and control measures in processed fruits, *Ann. Appl. Biol.* **26**:800-822.

Joffe, A. Z., 1969, The mycoflora of fresh and stored groundnuts kernels in Israel, *Mycopathol. Mycol. Appl.* **39**:255-264.

Karrow, E. O., and Foster, J. W., 1944, An antibiotic substance from species of *Gymnoascus* and *Penicillium*, *Science, N. Y.* **99**:265-266.

King, A. D., Hocking, A. D. and Pitt, J. I., 1981, The mycoflora of some Australian foods, *Food Technol. Aust.* **33**:55-60.

Kis, Z., Furger, P., and Sigg, H. P., 1969, Isolation of pyrenophorol, *Experientia* **25**:123-124.

Knösel, D., and Rézs, A., 1973, Pilze ays Müllkompost. Enzymatischer Abbau von Pektin und Zellulose durch wärmeliebende Spezies, *Städtehygiene* **6**, 6 pp.

Kramer, R. K., Davis, N. D., and Diener, U. L., 1976, Byssotoxin A, a secondary metabolite of *Byssochlamys fulva, Appl. Environ. Microbiol.* **31**:249-253.

Mori, T., Shin-ya, K., Takatori, K., Aihara, M., and Hayakawa, Y., 2003, Byssochlamysol, a new antitumor steroid against IGF-1-dependent cells from *Byssochlamys nivea*. II. Physico-chemical properties and structure elucidation, *J Antibiot. (Tokyo)* **56**(1):6-8.

Müller, S. O., Eckert, I., Lutz, W. K., and Stopper, H., 1996, Genotoxicity of the laxative drug components emodin, aloe-emodin and danthron in mammalian cells: topoisomerase II mediated? *Mutation Res.* **371**:165-173.

Olliver, M., and Rendle, T., 1934, A new problem in food preservation. Studies on *Byssochlamys fulva* and its effect on the tissues of processed fruit, *J. Soc. Chem. Ind. London* **53**:66-172.

Pelhate, J., 1968, Inventaire de la mycoflore des blés de conservation, *Bull. Trimest. Soc. Mycol. Fr.* **84**:127-143.

Raistrick, H., and Smith, G., 1933, Studies on the biochemistry of micro-organisms XXXV. The metabolic products of *Byssochlamys fulva* Oliiver & Smith, *Biochem. J.* **27**:1814-1819,

Ram, C., 1968, Timber-attacking fungi from the state of Maranhao, Brazil. Some new species of *Paecilomyces* and its perfect state *Byssochlamys* Westl., VIII. *Nova Hedwigia* **16**:305-314

Rice S. L., Beuchat, L. R., and Worthington, R. E., 1977, Patulin production by *Byssochlamys* spp. in fruit juices, *Appl. Environ. Microbiol.* **34**:791-796.

Samson, R. A., 1974, *Paecilomyces* and some allied hyphomycetes, *Stud. Mycol. Baarn* **6**. Centraalbureau voor Schimmelcultures, Baarn, Netherlands

Samson, R. A., and Tansey M. R., 1975, *Byssochlamys verrucosa* sp. nov., *Trans. Br. Mycol. Soc.* **65**:512-514.

Samson, R. A., Hoekstra, E. S. and Frisvad, J. C., eds, 2004, *Introduction to Food- and Airborne Fungi,* 7th edition, Centraalbureau voor Schimmelcultures, Utrecht, Netherlands.

Samson, R. A., Houbraken, J., Kuijpers, A., and Frisvad, J. C., 2005, Revision of *Paecilomyces* with its teleomorph *Byssochlamys* and its mycotoxin production, *Mycol. Res.* (submitted).

Smedsgaard, J., 1997, Micro-scale extraction procedure for standardized screening of fungal metabolite production in cultures, *J. Chromatog. A,* **760**:264-270.

Splittstoesser, D. F., 1987, Fruits and fruit products, in: *Food and Beverage Mycology*, second edn, L.R. Beuchat, ed, AVI Van Nostrand Reinhold, New York, pp. 101-122.

Tanawaki, M. H., 1995, Growth and mycotoxin production by fungi under modified atmospheres, Ph.D. thesis, University of New South Wales, Kensington, N.S.W.

Thom, C., 1930, *The Penicillia,* Williams and Wilkins, Baltimore.

Turesson, G., 1916., The presence and significance of moulds in the alimentary canal of man and higher animals, *Svensk Bot. Tidskr.* **10**:1-27.

Udagawa, S., and Suzuki, S., 1994, *Talaromyces spectabilis*, a new species of food-borne ascomycetes, *Mycotaxon* **50**:91-88.

Vinokurova, N. G., Boichenko, D. M., Baskunov, B. P., Zelenkova, N. F., Vepritskaya, I. G., Arinbasarov, M. U. and Reshetilova T. A., 2001, Minor alkaloids of the fungus *Penicillium roquefortii* Thom 1906., *Appl. Biochem. Microbiol.* **37**:184-187.

EFFECT OF WATER ACTIVITY AND TEMPERATURE ON PRODUCTION OF AFLATOXIN AND CYCLOPIAZONIC ACID BY *ASPERGILLUS FLAVUS* IN PEANUTS

Graciela Vaamonde, Andrea Patriarca and Virginia E. Fernández Pinto[*]

1. INTRODUCTION

It is well known that some isolates of *Aspergillus flavus* are able to produce cyclopiazonic acid (CPA) in addition to aflatoxins (Luk et al., 1977; Gallagher et al., 1978). CPA producing strains of *A. flavus* are frequently isolated from substrates such as peanuts (Blaney et al., 1989; Vaamonde et al., 2003) and maize (Resnik et al., 1996), indicating that this toxin could be a common metabolite and thus is likely to be present in some aflatoxin contaminated foods. Natural co-occurrence of both toxins has been detected in peanuts (Urano et al., 1992; Fernández Pinto et al., 2001) and it has been hypothesized that the presence of both toxins in food and feeds may result in additive or synergistic effects.

CPA is toxic to poultry and may have contributed to the outbreak of the classic "Turkey X" disease which killed about 100,000 turkey poults in England in 1960 (Cole, 1986). Some disease outbreaks of unknown aetiology observed in chickens in Argentina could also be attributed to the presence of CPA in peanut meal used as a raw material in poultry feeds, as strains of *A. flavus* capable of producing

*Laboratorio de Microbiología de Alimentos, Departamento de Química Orgánica, Area Bromatología, Facultad de Ciencias Exactas y Naturales, Universidad de Buenos Aires, Ciudad Universitaria, Pabellón II, 3° Piso, 1428, Buenos Aires, Argentina. Correspondence to: vaamonde@qo.fcen.uba.ar

high levels of CPA are frequently isolated from peanuts grown in this country (Vaamonde et al., 2003). CPA accumulates in the muscle of chickens following oral dosing (Norred et al., 1988) so that a potential for contamination of meat and meat products exists.

In view of the health hazards for animals and humans caused by the co-occurrence of aflatoxins and CPA, the production of these toxins in agricultural commodities must be controlled. Water activity (a_w) and temperature are the most important environmental factors preventing fungal growth and mycotoxin biosynthesis. Conditions for mycotoxin production are generally more restrictive than those for growth and can differ between mycotoxins produced by the same fungal species, as well as between fungi producing the same mycotoxin (Frisvad and Samson, 1991). Interactions between factors such as a_w, temperature and time are also important and should be taken into account in the design of experiments to study their effects on mycotoxin production (Gqaleni et al., 1997).

The effects of temperature and a_w on the production of aflatoxins by *A. flavus* have been widely studied (ICMSF, 1996) but very little is known about the effect of these factors on CPA production. Besides, most published studies on mycotoxin formation have been concerned with single mycotoxins. Few studies have examined how environmental factors can affect the simultaneous production of these two toxins on agar culture media (Gqaleni et al., 1997) and natural substrates (Gqaleni et al.,1996b).

In the present work an experiment with a full factorial design was used to study the effects of, and interactions among, temperature, a_w and incubation period on co-production of AF and CPA on peanuts inoculated with a cocktail of *A. flavus* strains. Peanuts were sterilized by gamma irradiation to keep, as closely as possible, to natural conditions for mycotoxin production. In this way, it was hoped that a different fungal response to environmental conditions in relation to biosynthesis of both secondary metabolites could be observed.

2. MATERIALS AND METHODS

2.1. Substrate

Samples of peanuts (*Arachis hypogaea* cv Runner) from the peanut growing area in the province of Córdoba, Argentina, were used. Samples were analysed for aflatoxins and CPA and shown to be free of both toxins. Peanut kernels were distributed in plastic bags in portions of 2.5 kg of material and were treated by gamma irradiation (Cuero et al., 1986) to kill contaminant microorganisms. Preliminary experiments showed that a dosage rate of 6 kGy produced kernels free of viable microorganisms without adversely affecting seed germination. Irradiation was carried out at the Comisión Nacional de Energía Atómica, Buenos Aires, Argentina, using a ^{60}Co source.

Water activity was adjusted by adding sterile water according to data from the water sorption isotherms of peanuts (Karon and Hillery, 1949). The material was mixed and left to equilibrate at 7°C until measurements of a_w in three consecutives days showed variation <0.005. A Vaisala Humicap HMI31 hygrometer (Woburn, MA) was used to measure a_w at 25°C. Levels of a_w achieved were 0.86, 0.88, 0.92 and 0.94.

2.2. Microorganisms

Fungi used in this study were isolated from peanuts and identified as previously described (Vaamonde et al., 2003). Four *A. flavus* strains which were shown to produce both aflatoxin B_1 (AFB_1) and CPA were selected.

2.3. Inoculum and Incubation

A cocktail inoculum was prepared with the four selected strains. Spores of each strain from a 7 day old culture on Malt Extract Agar were harvested using glycerol solutions (2.5 ml) adjusted to the a_w appropriate for each sample of equilibrated peanuts. The four suspensions were mixed and diluted to obtain a solution of 10^6 spores/ml. The numbers of spores in the inoculating suspensions were counted in a haemocytometer. Samples of peanuts (25 g) equilibrated to the selected a_w were placed in Erlenmeyer flasks (250 ml) and inoculated with 1 ml of the spore suspension. The material was thoroughly mixed to obtain homogeneous distribution of the inoculum.

The Erlenmeyer flasks were incubated at 20°C, 25°C and 30°C for 7, 14, 21 and 28 d respectively under constant relative humidity. This was achieved by placing the flasks in polyethylene bags (gauge 0.04 mm) containing receptacles with glycerol solutions adjusted to the corresponding a_w. Each experiment was carried out with three replicates.

2.4. Mycotoxin Analysis

Three flasks were removed weekly from each set of conditions and analysed for AFB_1 and CPA. Analysis of AFB_1 was carried out according to the BF method (AOAC, 1995). Peanut kernels (25 g) were extracted with methanol-water (125 ml; 55+45), hexane (50 ml) and NaCl (1 g) in a high-speed blender (1 min). The extract was filtered through Whatman No 4 filter paper and centrifuged at 2000 rpm (5 min). Aliquots of the aqueous phase (25 ml) were transferred to a separatory funnel and extracted with chloroform (25 ml) by shaking for 1 min. The chloroform extract was filtered through anhydrous Na_2SO_4 and evaporated to dryness. AFB_1 was detected by TLC on silica gel G60 plates using chloroform:acetone (90:10) as a developing solvent. AFB_1 concentrations were determined by visual comparison of fluorescence under UV light (366 nm) with standard solutions (Sigma Chemical, St. Louis, MO, USA) dissolved in benzene:acetonitrile 98:2.

CPA was analysed by the method of Fernández Pinto et al. (2001). Peanuts (25 g) were mixed with methanol:2% sodium hydrogen carbonate mixture (7:3; 100 ml). The mixture was blended at high speed for 3 min, centrifuged (1500 rpm) for 15 min and then filtered. Filtrate (50 ml) was defatted twice with hexane (50 ml) by shaking in a wrist action shaker (3 min). KCl solution (10%; 25 ml) was added to the sample layer and acidified to pH 2 with 6N HCl. The solution was transferred quantitatively to a separatory funnel, extracted twice with chloroform (25 ml), and filtered over anhydrous Na_2SO_4. This extract was evaporated to dryness and redissolved in chloroform (200 μl). CPA was detected by TLC separation on silica gel G60 plates. TLC plates were immersed completely in oxalic acid solution (2% wt/wt) in methanol for 10 min, heated at 100°C for 1 h and cooled. Extracts (2 μl) were applied to the plates with standard solution every fourth track. The plates were developed in ethyl acetate:2-propanol: ammonium hydroxide (40:30:20). After development, the plates were dried 5 min at 50°C to drive off ammonia and then sprayed with Erlich's reagent [4-dimethylaminobenzaldehyde (1 g) in ethanol (75 ml) and concentrated HCl (25 ml)]. Blue spots were analysed after 10 min by visual comparison with a standard CPA solution (Sigma Chemical Co., St. Louis, MO, USA). The detection limit of the method was 50 μg/kg.

3. RESULTS AND DISCUSSION

Figure 1 shows the influence of water activity and temperature on CPA accumulation over a 28-day period. CPA production was inhibited at the lowest a_w studied (0.86) at all temperatures. At higher a_w levels, CPA was detected after a lag period of between 7 and 14 d. The amount of CPA produced was determined by the complex interaction of a_w, temperature and incubation time. Measurements performed only at a fixed time could lead to misinterpretation of the results. For example, from the curve at a_w 0.92 (Figure 1) different conclusions could be drawn regarding the influence of temperature on CPA production by considering the accumulation after 14 or 28 d. Taking into account the whole incubation period at the various a_w levels, CPA reached a maximal accumulation (4469.2 µg/kg) after 28 d at 0.94 a_w and 25°C, but also a considerable production (2690 ppb) was detected at the lowest temperature studied (20°C) after 21 d at relatively high

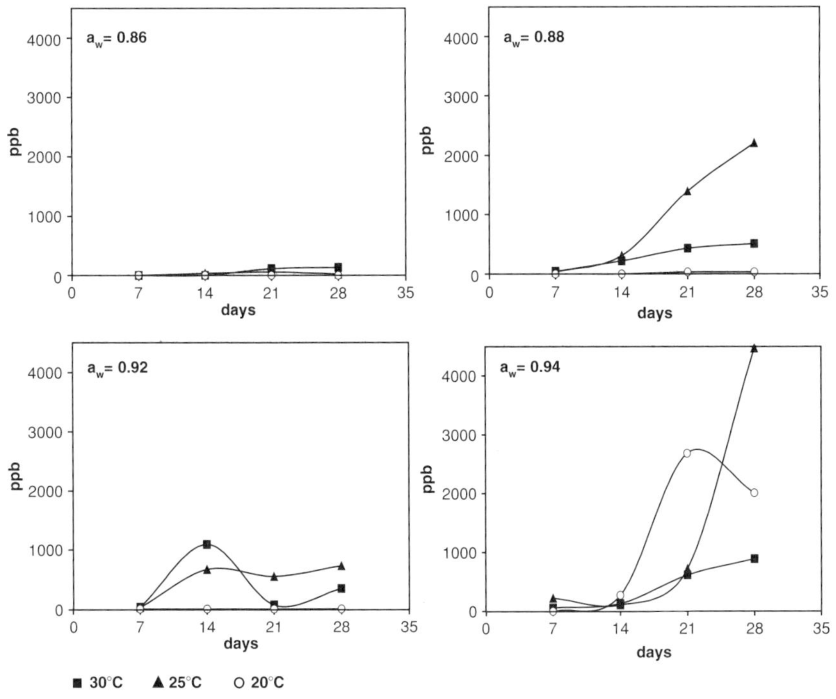

Figure 1. Interaction between a_w and temperature and their effect on cyclopiazonic acid production by *A. flavus* on peanuts over a 28 day period

a_w. These results are in concordance with those obtained by Gqaleni et al. (1996a) who reported that *A. flavus* produced higher levels of CPA at 20°C than at 30°C on maize grains held at constant a_w. The same effect was observed by Sosa et al. (2000) who found maximum CPA production by *Penicillium commune* at 20°C and 0.90 a_w in a medium based on meat extract. According to Gqaleni et al (1996b) CPA production proceeded in a similar pattern for *A. flavus* and *P. commune* with a combination of high a_w and low temperature favouring high CPA production and low a_w and high temperature supporting least CPA production.

AFB_1 was produced over the whole a_w range (Figure 2). AFB_1 concentrations were low after 7 d at all temperatures, but as time progressed, the maximal accumulation was observed at the lowest a_w (0.86) with temperatures of 25°C and 30°C, temperatures reported as favourable for aflatoxin production by other authors (ICMSF, 1996). A combined effect of a_w and temperature was observed as AFB_1 production was inhibited at low a_w (0.86) and low temperature (20°C). At

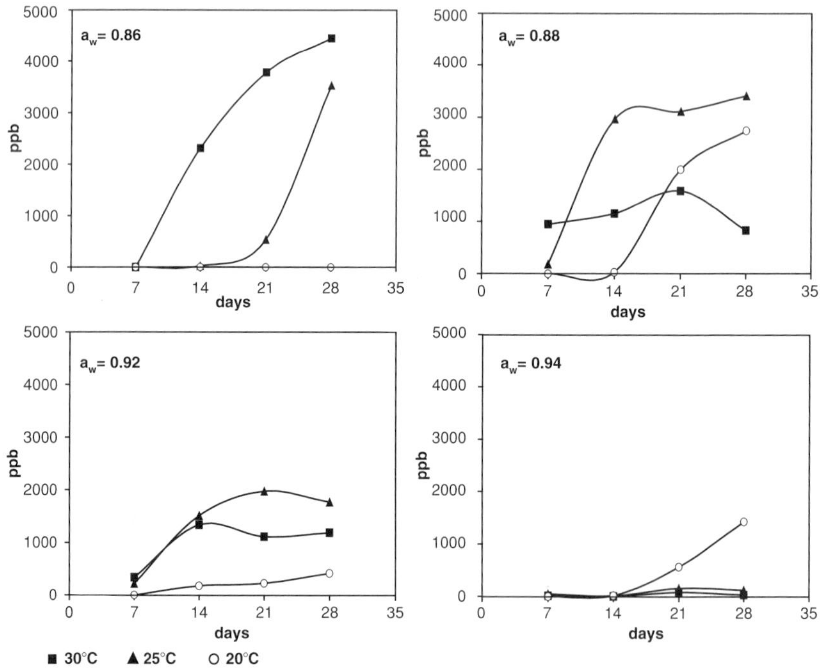

Figure 2. Interaction between a_w and temperature and their effect on aflatoxin B_1 production by *A. flavus* on peanuts over a 28 day period

higher a_w, AFB_1 production taken over the whole incubation period was lower than at low a_w. The lowest accumulation was detected at 0.94 a_w. At this a_w level the fungus grew vigorously at 25°C and 30°C but very low quantities of AFB_1 were detected in such conditions, while at 20°C slower growth was observed but a considerable amount of AFB_1 (1375 ppb) was produced after 28 d.

As mycotoxins are fungal secondary metabolites, production need not to be correlated with the growth of the fungus and factors such as induction, end product inhibition, catabolite repression and phosphate regulation can influence production (Tuomi et al., 2000). However, the results obtained for AFB_1 accumulation were unexpected and contradict data from previous studies which reported that aflatoxin production increases with increasing a_w (Diener and Davis, 1970; Montani et al., 1988; Gqaleni et al., 1996b).

Differences could be due to several factors. Gqaleni et al. (1996a), using several *A. flavus* strains known as co-producers of aflatoxins and CPA, reported that the pattern of co-production of these toxins by a particular isolate can vary widely depending on the type of cultural system. When they used yeast extract sucrose agar or a liquid medium, aflatoxins were produced in very low quantities over a period of 21 d of incubation at 30°C despite the high a_w of these substrates. When they used autoclaved maize grains (a_w 0.992) higher concentrations of both toxins were detected and the ratios between concentrations of aflatoxins and CPA were 16.6:1, 5.7:1 and 1.6:1 after 7, 15 and 21 d respectively (Gqaleni et al., 1996a). The heat sterilised maize grains used by Gqaleni et al. (1996a; 1996b) differed from the cultural system used in the present work, where living seeds were used. In this case, the substrate conditioned at higher a_w levels might perhaps have had some defence mechanism, e.g. synthesis of inhibitors of aflatoxin biosynthesis. The fungus might respond by increasing production of CPA since the biosynthetic pathways of the two toxins are different. CPA is derived from tryptophan (Holzapfel, 1971) whereas aflatoxins are decaketides (Smith and Moss, 1985). On the other hand, reduced metabolic activity of viable kernels associated with decreased a_w could increase susceptibility to *A. flavus* growth and aflatoxin production, an effect observed in the field with decreased pod moisture content under drought stress (Mehan et al., 1991).

Other factors such as the peanut variety, the size of inoculum and oxygen availability could also be responsible for the unexpected results obtained in the present work. The possibility of competitive growth of microorganisms that could have survived the disinfection treatment should be also considered since the radiation dose employed (6 kGy) was relatively low when compared with that applied in other studies (Marín et al., 1999), although microbial contamination was not visually detected. Further experiments are currently in progress in order to clarify the possible influence of some of the above mentioned factors, e.g. to perform mycotoxin accumulation curves using *A. flavus* and *A. parasiticus* strains capable of producing only aflatoxins with other peanut varieties and different inoculum levels.

It is evident that accumulation of two (or more) toxins produced simultaneously by a fungus is affected by numerous factors, as well as their interactions, as the result of a very complex relationship between the microorganism, the substrate and the environment. The influence of each of these factors on the production of each toxin may be different. In fact, variation of the relative amounts of the toxins produced with changing a_w and temperature has been observed for other fungal species producing several toxins (Magan et al., 1984; Wagener et al., 1980). Biosynthesis of different mycotoxins produced by *Alternaria* spp. is favoured by different temperatures. Production of tenuazonic acid, a tetramic acid like CPA, occurred at 20°C (Young et al., 1980), and was affected in a very different way by environmental factors compared with other *Alternaria* toxins such as alternariol and alternariol monomethyl ether (Magan et al.,1984).

The conditions for maximum production of aflatoxins and CPA by *A. flavus* are different, according to results of the present work and those obtained by Gqaleni et al. (1996a; 1996b). Table 1 illustrates the relative concentrations of both toxins in some of the conditions used in our study. It can be pointed out that these conditions are different from those in most published studies on mycotoxin formation because a) a cocktail of native strains that co-produce aflatoxins and CPA was used; b) the substrate was not autoclaved and; c) the study covered sufficient time to observe the evolution of the interaction between the

Table 1. Mean levels of cyclopiazonic acid (CPA) and aflatoxin B_1 (AFB_1) produced by *Aspergillus flavus* in peanuts under conditions of controlled temperature, a_w and time

Water activity	Temperature (°C)	Time (days)	CPA (μg/kg)	AFB_1 (μg/kg)
0.94	25	28	4469.2[a]	118.7
0.94	25	14	109.8	14.8
0.92	25	21	550.2	1972.8
0.88	25	28	2205.7	3411.7
0.86	30	28	134.5	4450.0
0.86	20	28	ND[b]	0.4

[a] Optimal conditions for CPA production; [b]Limiting conditions for CPA production

fungus and the substrate under different environmental conditions. It can be seen that at the points at which one of the toxins reaches the highest concentration the other is accumulated in minimal quantities. While concentrations of both toxins in extreme conditions are quite different, at intermediate points the relative concentrations are more similar as can be observed in Table 1 for 0.88 a_w, 25°C and 28 d.

The ability to respond in a different way to environmental factors in relation to the biosynthesis of toxic secondary metabolites may aid survival of the fungus in a particular ecological niche allowing it to colonize the substrate more efficiently than other competitors.

4. REFERENCES

AOAC (Association of Official Analytical Chemists), 1995, *Official Methods of Analysis of the Association of Official Analytical Chemists*, 16th edition, Arlington, USA, p. 970.45.

Blaney, B. J., Kelly, M. A., Tyler, A. L., and Connole, M. D., 1989, Aflatoxin and cyclopiazonic acid production by Queensland isolates of *Aspergillus flavus* and *Aspergillus parasiticus, Aust. J. Agric. Res.* **40**:395-400.

Cole, R. J., 1986, Etiology of turkey "X" disease in retrospect: a case for the involvement of cyclopiazonic acid, *Mycotoxin Res.* **2**:3-7.

Cuero, R. G., Smith, J. E., and Lacey, J., 1986, The influence of gamma irradiation and sodium hypochlorite sterilization on maize seed microflora and germination, *Food Microbiol.* **3**:107-113.

Diener, U. L., and Davis, N. D., 1970, Limiting temperature and relative humidity for aflatoxin production by *Aspergillus flavus* in stored peanuts, *J. Am. Oil Chem. Soc.* **47**:347-351.

Fernández Pinto, V., Patriarca, A., Locani, O., and Vaamonde, G., 2001, Natural co-occurrence of aflatoxin and cyclopiazonic acid in peanuts grown in Argentina, *Food Addit. Contam.* **18**:1017-1020.

Frisvad, J. C., and Samson, R. A., 1991, Filamentous fungi in foods and feeds: ecology, spoilage and mycotoxin production, in: *Handbook of Applied Mycology: Foods and Feeds*, D. K. Arora, K. G. Mukerji and E. H. Marth, eds, Marcel Dekker, Inc., New York, N.Y., pp. 31-68.

Gallagher, R. T., Richard, J. L., Stahr, H. M., and Cole, R. J., 1978, Cyclopiazonic acid production by aflatoxigenic and non-aflatoxigenic strains of *Aspergillus flavus*, *Mycopathologia*, **66**:31-36.

Gqaleni, N., Smith, J. E., and Lacey, J., 1996a, Co-production of aflatoxins and cyclopiazonic acid in isolates of *Aspergillus flavus*, *Food Addit. Contam.* **13**: 677-685

Gqaleni, N., Smith, J. E., Lacey, J., and Gettinby, G., 1996b, The production of cyclopiazonic acid by *Penicillium commune* and cyclopiazonic acid and aflatoxins by *Aspergillus flavus* as affected by water activity and temperature on maize grains, *Mycopathologia*, **136**:103-108.

Gqaleni, N., Smith, J. E., Lacey, J., and Gettinby, G., 1997, Effects of temperature, water activity and incubation time on production of aflatoxins and cyclopiazonic acid by an isolate of *Aspergillus flavus* in surface agar culture, *Appl. Environ. Microbiol.* **63**:1048-1053.

Holzapfel, C. W., 1971, On the biosynthesis of cyclopiazonic acid, *Phytochemistry*, **10**:351-358.

ICMSF (International Commission on Microbiological Specifications for Foods), 1996, *Microorganisms in Foods 5. Characteristics of Microbial Pathogens*, Blackie Academic and Professional, London, pp. 347-381.

Karon, M. L., and Hillery, B. E., 1949, Hygroscopic equilibrium of peanuts. *J. Am. Oil Chem. Soc.* **26**:16. [89]

Luk, K. C., Kobbe, B., and Townsend, J. M., 1977, Production of cyclopiazonic acid by *Aspergillus flavus* Link, *Appl. Environ. Microbiol.* **33**:211-212.

Magan, N., Cayley, G. R., and Lacey, J., 1984, Effect of water activity and temperature on mycotoxin production by *Alternaria alternata* in culture and on wheat grain, *Appl. Environ. Microbiol.* **47**:1113-1117.

Marín, S., Sanchis, V., Sanz, D., Castel, I., Ramos, A. J., Canela, R., and Magan, N., 1999, Control of growth and fumonisin B_1 production by *Fusarium verticillioides* and *Fusarium proliferatum* isolates in moist maize with propionate preservatives, *Food Addit.Contam.* **16**: 555-563.

Mehan, V. K., Mc Donald, D., Haravu, L. J., and Jayanthi, S., 1991, The groundnut aflatoxin problem. Review and Literature Database, ICRISAT, International Crops Research Institute for the Semi-Arid Tropics, Patancheru, Andra Pradesh 502 324, India, p. 59.

Montani, M. L., Vaamonde, G., Resnik, S. L., and Buera, P., 1988, Water activity influence on aflatoxin accumulation in corn, *Int. J. Food Microbiol.* **6**:349-353.

Norred, W. P., Porter, J. K., Dorner, J.W., and Cole, R. J., 1988, Occurrence of the mycotoxin, cyclopiazonic acid, in meat after oral administration to chickens, *J. Agric. Food Chem.* **36**:113-116.

Resnik, S. L., González, H. H. L., Pacin, A. M., Viora, M., Caballero, G. M., and Gros, E. G., 1996, Cyclopiazonic acid and aflatoxins production by *Aspergillus flavus* isolated from Argentinian corn, *Mycotoxin Res.* **12**:61-66.

Sosa, M. J., Córdoba, J. J., Díaz, C., Rodríguez, M., Bermúdez, E., Asensio, M. A., and Núñez, F., 2002, Production of cyclopiazonic acid by *Penicillium commune* isolated from dry-cured ham on a meat extract-based substrate, *J. Food Prot.* **65**:988-992.

Tuomi, T., Reijula, K., Johnsson, T., Hemminki, K., Hintikka, E., Lindroos, O., Kalso, S., Koukila-Kähkölä, P., Mussalo-Rauhamaa, H., and Haahtela, T., 2000, Mycotoxins in crude building materials from water-damaged buildings, *Appl. Environ. Microbiol.* **66**:1899-1904.

Urano, T., Trucksess, M. W., Beaver, R. W., Wilson, D. M., Dorner, J. W., and Dowell, F. E., 1992, Co-occurrence of cyclopiazonic acid and aflatoxins in corn and peanuts, *J. AOAC Int.* **75**:838-841.

Vaamonde, G., Patriarca, A., Fernández Pinto, V., Comerio, R., and Degrossi, C., 2003, Variability of aflatoxin and cyclopiazonic acid production by *Aspergillus* section *flavi* from different substrates in Argentina, *Int. J. Food Microbiol.* **88**:79-84.

Wagener, R. E., Davies, N. D. and Diener, U. L., 1980, Penitrem A and roquefortine production by *Penicillium commune*, *Appl. Environ. Microbiol.* **39**:882-887.

Young, A. B. , Davis, N. D., and Diener, U. L., 1980, The effect of temperature and moisture on tenuazonic acid production by *Alternaria tenuissima*, *Phytopathology* **70**:607-609.

Editors' note

As the authors point out, the results of this work are unexpected, i.e. higher mycotoxin production at lower water activities. Of the explanations suggested by the authors, the most likely reason is limitation of oxygen during growth of the fungi, due to the use of Erlenmeyer flasks as experimental vessels. Oxygen limitation during growth of fungi in narrow necked flasks such as Erlenmeyers has long been known. Fungal growth under optimal conditions is characterised by very high oxygen consumption. The wide mouthed Fernbach flask (illustrated in Raper and Thom, *Manual of the Penicillia*, Williams and Wilkins, Baltimore, 1949, p. 91) was found to be superior for growth of fungi. Even with the use of wide mouthed flasks, care needs to be taken to use thin cotton wool closures, to permit the maximum possible diffusion of oxygen.

The view point that oxygen limitation produced these unusual results is supported by the data: at lower water activities where growth is slower, oxygen consumption is much lower, and sufficient for normal growth and mycotoxin production.

Section 4. Control of fungi and mycotoxins in foods

INACTIVATION OF FRUIT SPOILAGE YEASTS AND MOULDS USING HIGH PRESSURE PROCESSING

Ailsa D. Hocking, Mariam Begum and Cindy M. Stewart†*

1. INTRODUCTION

Processing of foods using ultra high pressures offers an alternative non-thermal treatment for the production of high quality processed food products which maintain many of the organoleptic qualities of fresh foods (Stewart and Cole, 2001). High pressure processing (HPP) is particularly useful for acid foods such as fruit pieces, purees and juices as although it does not inactivate bacterial spores, it provides a pasteurisation process. Many studies on the application of high pressure treatments have focused on the destruction of pathogenic bacteria, with less attention being paid to spoilage microorganisms, particularly yeasts and moulds.

This work described here was undertaken as part of a larger project to develop a high pressure processed shelf-stable pear product (Gamage et al., 2004). The project encompassed enzyme (polyphenyloxidase) inactivation and shelf life studies, as well as microbiological stability. As yeasts and moulds, including heat resistant moulds, are the most common spoilage microorganisms in processed fruit products, representative species of these fungi were chosen for the microbiological investigations.

* A. D. Hocking and M. Begum, Food Science Australia, CSIRO, P.O. Box 52, North Ryde, NSW Australia 1670. Correspondence to ailsa.hocking@csiro.au
†C. Stewart, National Center for Food Safety & Technology, 6502 S. Archer Rd, Summit-Argo, IL 60501, USA

2. MATERIALS AND METHODS

2.1. Yeast and mould cultures

A single strain of each of two yeasts and four filamentous fungi was chosen for inclusion in this study. The species/strains examined were *Saccharomyces cerevisiae* FRR 1813, beer fermentation strain; *Pichia anomala* FRR 5220 isolated from fermenting vanilla-blueberry yoghurt; *Penicillium expansum* FRR 1536 isolated from mouldy pears; *Fusarium oxysporum* FRR 5610, isolated from spoiled UHT treated fruit juice; *Byssochlamys fulva* FRR 3792 from heat treated strawberry puree; and *Neosartorya fischeri* FRR 4595 from heat treated strawberries. FRR is the acronym of the fungal culture collection of Food Science Australia, CSIRO, North Ryde, NSW. The two yeast species were selected because of their propensity for spoilage of fruit products and their ability to form ascospores, which may be more resistant to high pressure treatment than vegetative cells. *P. expansum* was included because of its significance in post harvest spoilage of apple and pears and the consequent possibility of high numbers of spores on pears before processing. *F. oxysporum* has recently been causing spoilage problems in UHT processed juice products, indicating that it may be heat resistant and, therefore, possibly pressure resistant also. The two heat resistant moulds were included because it was considered important to be able to control these species if a shelf stable fruit product were to be developed.

2.2. Preparation of cell suspensions

The two yeasts were grown on Malt Extract agar (MEA) at 25°C for 10 days. *P. expansum* was grown on Czapek Yeast Extract agar (CYA) at 25°C for 7 days. *F. oxysporum* was grown on Tap Water Agar with carnation leaf pieces at 25°C for 14 days under a light bank providing a 12 hour photoperiod to induce formation of chlamydoconidia. The formulae for these media are from Pitt and Hocking (1997). The heat resistant moulds *B. fulva* and *N. fischeri* were grown on Malt Extract Agar (MEA) at 30°C for 14 days. Ascospore production in the yeasts and the heat resistant fungi was confirmed by microscopy. Cells from culture plates were suspended in sucrose solution (20° Brix) adjusted to pH 4.2 with citric acid, to yield *ca* 10^4 ascospores/ml for heat resistant moulds, and 10^7 cfu/ml for other microorganisms. Suspensions of cells from the two yeast species contained 20-25% ascospores. Duplicate suspensions were prepared.

2.3. High pressure treatment

Cell suspensions (1.2 ml) were transferred to sterile high pressure processing vials (1.5 ml). High pressure treatments were applied using a U-111 High Pressure Multi-vessel Apparatus (High Pressure Research Centre, Polish Academy of Sciences, Warsaw, Poland), comprising five separate vessels which can be pressurised simultaneously, but depressurised individually. The equipment also has the capacity for temperature control, but these experiments were performed at ambient temperature (20°C).

For the two yeasts and *P. expansum* and *F. oxysporum,* the unit was pressurised to 400 MPa, with pressure applied for 15, 30, 45, 60 and 120 sec. This pressure treatment regimen was selected to allow the acquisition of data that would provide an inactivation response curve. Treatment at the pressure intended for the pear product (600 MPa) would have resulted in inactivation times that were too short to measure with the equipment available.

To assess the effect of a proposed blanching process for the fruit product, cell suspensions of the two heat resistant moulds, *B. fulva* and *N. fischeri,* were heated at 95°C for five minutes before they were subjected to high pressure treatment. Cell suspensions (approx. 10 ml) were filled into sterile plastic vials which were then immersed in a water bath at 95°C. A thermocouple attached to a data logger was placed in a vial containing a similar volume of suspending fluid (20 °Brix sucrose, pH 4.2). The 5 min blanching period was taken from the time the control vial reached 95°C.

For the two heat resistant moulds, the pressure treatment was 600 MPa applied at the same time intervals as used for the yeasts and heat sensitive moulds, using a 2 l high pressure processing unit (Flow International Corporation, USA). Spore suspensions (approximately 4 ml) were aseptically filled into the bulb of sterile disposable plastic Pasteur pipettes (5 ml) (Copan Italia S.P.A., Italy), and the end heat sealed. To protect the high pressure unit from contamination in the event of sample tube leakage, the sealed tubes were immersed in peroxyacetic acid (250 ppm) (Proxitane Sanitiser, Solvay Interox, Australia) within high barrier vacuum bags (Cryovac Australia Pty Ltd, Australia), which were heat sealed. The pressure fluid was water and pressurization was carried out at ambient temperature (18-20°C). Come up times to the designated pressure (600 MPa) were less than 10 sec, and depressurization required less than 5 sec. The pressure run was repeated using separate cell suspensions of each species to provide duplicate data.

2.4. Assessment of cell survival

Initial and surviving populations were enumerated in duplicate using the spread plating technique on appropriate growth media, being MEA for the yeasts and heat resistant moulds, and CYA for *P. expansum* and *F. oxysporum*, with incubation at the same temperatures used to grow the cultures initially. The limit of detection for the method was 10 cfu/ml.

3. RESULTS

3.1. Inactivation of yeasts

After 60 sec pressurisation at 400 MPa, a 3 to 4 $\log_{10}$ reduction was achieved for both *S. cerevisiae* and *P. anomala,* with a 4-5 $\log_{10}$ reduction after 120 sec at 400 MPa (Figure 1). The inactivation curve for *S. cerevisiae* showed a quite linear response between zero and 60 sec at 400 MPa for the first inactivation run, dropping from an initial level of 9×10^7 cfu/ml to 3.3×10^3 cfu/ml after 60 sec. The duplicate run yielded a curve more similar to that of *P. anomala,* dropping rapidly from an initial level of 1.4×10^7 to 4.9×10^4 after 30 sec, then tailing off to 2.1×10^4 after 60 sec, and 1.9×10^3 cfu/ml after 120 sec (Figure 1a).

P. anomala exhibited a relatively sharp drop in viable cells after 15 sec HPP treatment, from 1-2 $\times 10^8$ to 1-2 $\times 10^5$ cfu/ml, with a steady,

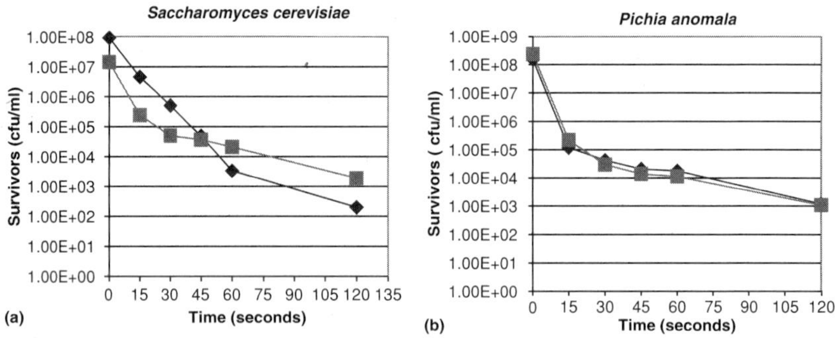

Figure 1. Effect of pressure treatment of 400 MPa at 25°C on (a) *Saccharomyces cerevisiae* and (b) *Pichia anomala* suspensions of mixed vegetative cells and ascospores. Cells were suspended in 20 °Brix sucrose solution, pH 4.2 (acidified with citric acid). Graphs show replicates 1 (◆) and 2 (■) for each pressure run.

gradual decline with increasing treatment times to 1-2 × 10^4 cfu/ml after 60 sec, and 10^3 after 120 sec (Figure 1b).

3.2. Inactivation of heat sensitive moulds

Both *P. expansum* and *F. oxysporum* were pressure treated at 400 MPa. *P. expansum* was quite sensitive to pressure treatment. A 3-4 $\log_{10}$ reduction was achieved after 15-30 sec (Figure 2). After 60 sec, the reduction was >5 $\log_{10}$ and by 120 sec, no survivors were detected (limit of detection 10 cfu/ml). *F. oxysporum* conidia and chlamydoconidia were even more sensitive to pressure treatment. A suspension of 2.5 × 10^7 mixed microconidia and chlamydoconidia was reduced to 1.4 × 10^2 cfu/ml after 15 sec at 400 MPa, and after 30 sec, no survivors were detected (results not shown).

3.3. Inactivation of heat resistant moulds

As the ascospores of heat resistant moulds are known also to be pressure resistant (Butz et al., 1996; Reyns et al., 2003; Voldrich et al., 2004), both *B. fulva* and *N. fischeri* spore suspensions were pressure treated at 600 MPa. Spore suspensions were divided into two portions, and one of each was blanched at 95°C for 5 min. Both blanched and unblanched spore suspensions were then pressure treated simultaneously to ensure that they were exposed to exactly the same conditions.

B. fulva ascospores were more pressure sensitive than those of *N. fischeri* (Figure 3). Unblanched ascospore suspensions of *B. fulva,*

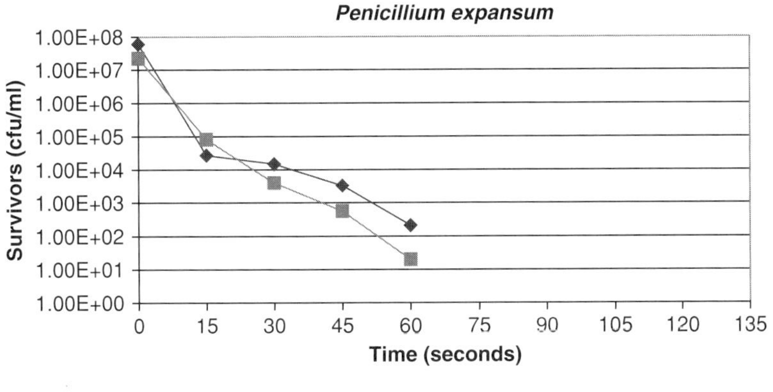

Figure 2. Effect of pressure treatment of 400 MPa at 25°C on *Penicillium expansum* conidia. Cells were suspended in 20 °Brix sucrose solution, pH 4.2 (acidified with citric acid). Graphs show replicates 1 (◆) and 2 (■) for each pressure run.

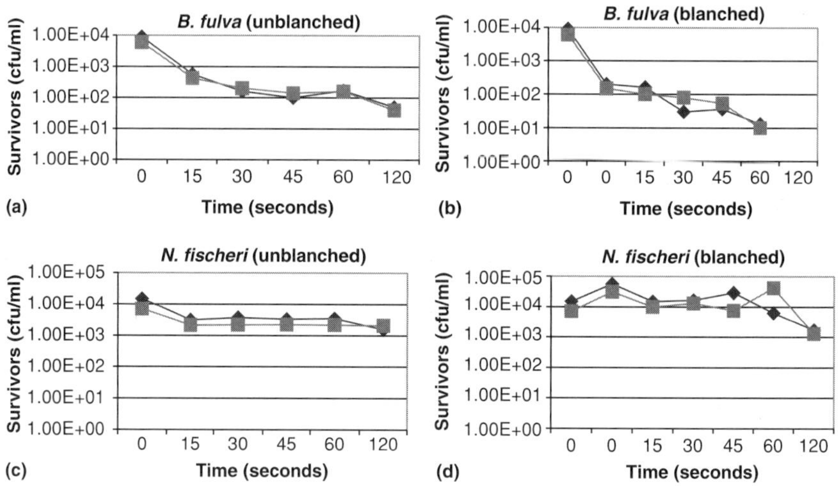

Figure 3. Effect of pressure treatment of 600 MPa at 25°C on *Byssochlamys fulva* (a, b) and *Neosartorya fischeri* (c, d) ascospores. Cells were suspended in 20 °Brix sucrose solution, pH 4.2 (acidified with citric acid). Blanched samples (b, d) were heated to 95°C for 5 min before pressure treatment. Graphs show replicates 1 (◆) and 2 (■) for each pressure run.

showed a 1.5 $\log_{10}$ reduction after 15 sec at 600 MPa (Figure 3a). The decrease with longer pressure treatments was more gradual, with only 2 $\log_{10}$ reduction after 60 sec and 2.5 $\log_{10}$ reduction after 120 sec. Blanching at 95°C for 5 min resulted in a reduction of 1-2 $\log_{10}$ (Figure 3b). Pressure treatment of the blanched ascospores resulted in a 2 $\log_{10}$ reduction after 60 sec, and after 120 sec at 600 MPa, there were less than 10^1 ascospores/ml.

Unblanched ascospores of *N. fischeri* showed a slight reduction (less than 1 $\log_{10}$) after 15 sec at 600 MPa (Figure 3c). Longer treatments (up to 120 sec) appeared to have no further effect on ascospore viability, with barely 1 $\log_{10}$ reduction after 120 sec treatment. In the first experiment, numbers were reduced from 1.50×10^4 to 1.53×10^3 after 120 sec at 600 MPa, while in the second experiment, initial numbers of 7.0×10^3 were reduced only to 2.0×10^3.

Blanching at 95°C caused activation of *N. fischeri* ascospores (Figure 3d) with viable numbers rising by 0.5 $\log_{10}$ or slightly more. Subsequent high pressure treatment at 600 MPa resulted in a slight decrease (less than 0.5 $\log_{10}$), but longer treatment times to 60 sec had little further effect. After 120 sec at 600 MPa, final numbers were reduced by slightly less than one $\log_{10}$ from the initial, unblanched levels.

4. DISCUSSION

Inactivation of yeasts by ultra high pressure has been reported previously (Ogawa et al., 1990; Palou et al., 1997; Parish, 1998; Zook et al., 1999; Basak et al., 2002). Most studies have investigated the effects of HPP on the common spoilage yeasts *Saccharomyces cerevisiae* and *Zygosaccharomyces bailii*. The responses of *Pichia anomala* to HPP treatment have not previously been reported. As observed in this study, other workers have shown that treatment at 400 MPa or higher will result in inactivation of yeasts provided that the pressure is applied for several minutes. However, the time required for inactivation also depends on the composition of the menstruum in which the yeasts are suspended. Yeasts are less susceptible to pressure treatment when suspended in juice concentrate than in single strength juice (Ogawa et al., 1990; Basak et al., 2002). Although ascospores of yeasts are more resistant to pressure treatment than vegetative cells, they are relatively quickly inactivated at 500 MPa, with D values of less than 0.2 min in single strength juice (Zook et al., 1999).

The conidia of *Penicillium expansum* were more sensitive to pressure treatment than yeast cells. Our study showed a 6 $\log_{10}$ reduction in viability of *P. expansum* conidia pressure treated at 400 MPa for 60 sec. A suspension of microconidia and chlamydoconidia of *F. oxysporum* was even more sensitive to pressure treatment than *P. expansum*. These results are in agreement with results reported by Ogawa et al. (1990) who found that treatment at 400 MPa at room temperature resulted in at least 4 $\log_{10}$ reduction in viability of conidia of *Aspergillus awamori* and sporangiospores of *Mucor plumbeus*. Voldrich et al. (2004) also reported that conidia of *Talaromyces avellaneus* exhibited comparable pressure sensitivity to vegetative cells of other yeast and mould species.

Ascospores of the two heat resistant mould species were relatively resistant to the pressure treatments received, as observed by other workers (Butz et al., 1996; Reyns et al., 2003; Voldrich et al., 2004). Blanching at 95°C for 5 min did not affect the pressure resistance, and in fact, ascospores of *N. fischeri* became more resistant to pressure after this mild heat treatment. The ascospores used in these studies were relatively young (from 14 day old cultures), and it could be expected that older ascospores would be more pressure resistant, as heat resistance in *N. fischeri* ascospores increases with age (Conner at al., 1987).

The results of this study indicate that high pressure processing (600 MPa for several minutes) should be sufficient to inactivate vegetative cells of yeasts and moulds and also the ascospores of yeasts, providing a process that is equivalent to pasteurisation. However, treatment with high pressure alone is insufficient to inactivate even relatively young ascospores of heat resistant moulds, indicating that combined treatments (for example heat and pressure) will be needed to control the outgrowth of these fungi during storage of HPP fruit products at ambient temperature.

5. REFERENCES

Basak, S., Ramaswamy, H. S., and Piette, J. P. G., 2002, High pressure destruction kinetics of *Leuconostoc mesenteroides* and *Saccharomyces cerevisiae* in single strength and concentrated orange juice, *Innov. Food Sci. Emerging Technol.* **3**:223-231.

Butz., P., Funtenberger, S., Haberditzl, T., and Tauscher, B., 1996, High pressure inactivation of *Byssochlamys nivea* ascospores and other heat resistant moulds, *Lebensm.-Wiss. u.-Technol.* **29**:404-410.

Conner, D. E., Beuchat, L. R., and Chang, C. J., 1987, Age-related changes in ultrastructure and chemical composition associated with changes in heat resistance of *Neosartorya fischeri* ascospores, *Trans. Br. Mycol. Soc.* **89**:539-550.

Gamage, T.V., Hocking, A., Begum, M., Stewart, C.M., Vu, T., Ng, S., Sellahewa, J., and Versteeg, C., 2004, Quality attributes of high pressure processed pears, Proceedings of the 9th International Congress on Engineering and Food, Montpellier, France, March 2004, pp. 325-330.

Ogawa, H., Fukuhisa, K., Kubo, Y., and Fukomoto, H., 1990, Pressure inactivation of yeasts, molds, and pectinesterase in Satsuma mandarin juice: effects of juice concentration, pH, and organic acids, and comparison with heat sanitation, *Agric. Biol. Chem.* **54**:1219-1225.

Palou, E., López-Malo, A., Barbosa-Cánovas, G. V., Welti-Chanes, J., and Swanson, B. G., 1997, Kinetic analysis of *Zygosaccharomyces bailii* inactivation by high hydrostatic pressure, *Lebensm.-Wiss. u.-Technol.* **30**:703-708.

Parish, M. E., 1998, High pressure inactivation of *Saccharomyces cerevisiae*, endogenous microflora and pectinmethylesterase in orange juice, *J. Food Safety* **18**:57-65.

Pitt, J. I. and Hocking, A. D., 1997, *Fungi and Food Spoilage*, 2nd edition, Blackie Academic & Professional, London.

Reyns, K. M. F. A., Veberke, E. A. V., and Michiels, C. W., 2003, Activation and inactivation of *Talaromyces macrosporus* ascospores by high hydrostatic pressure, *J. Food Prot.* **66**:1035-1042.

Stewart, C. M., and Cole, M. B., 2001, Preservation by the application of nonthermal processing, in: *Spoilage of Processed Foods: Causes and Diagnosis*, C. J. Moir, C. Andrew-Kabilafkas, G. Arnold, B. M. Cox, A. D. Hocking and I. Jenson, eds, AIFST (Inc.), Sydney, NSW Australia, pp. 53-61.

Voldrich, M., Dobiás, J., Tichá, L., Cerovsky, M., and Krátká, J., 2004, Resistance of vegetative cells and ascospores of heat resistant mould *Talaromyces avellaneus* to the high pressure treatment in apple juice, *J. Food Eng.* **61**:541-543.

Zook, C. D., Parish, M.E., Braddock, R. J., and Balaban, M. O., 1999, High pressure inactivation kinetics of *Saccharomyces cerevisiae* ascospores in orange and apple juice, *J. Food Sci.* **64**:533-535.

ACTIVATION OF ASCOSPORES BY NOVEL FOOD PRESERVATION TECHNIQUES

*Jan Dijksterhuis and Robert A. Samson**

1. INTRODUCTION

Most fungal survival structures can be regarded as heat resistant to some extent: sclerotia, conidia and ascospores can survive temperatures between 55 and 95°C. *Byssochlamys*, *Neosartorya* and *Talaromyces* are the most well known heat-resistant fungal genera. Ascospores of these fungi are the most resilient eukaryotic structures currently known. A decimal reduction time of 1.5-11 min at 90°C has been reported for some species (Scholte et al., 2004). Recently, Panagou et al. (2002) reported moderate heat resistance (D_{75} 4.9-7.8 min) in ascospores of *Monascus ruber* isolated from brine of a commercial thermally processed can of green olives. There appeared to be a complex interaction between pH and salt content of the heating menstruum and decimal reduction time for this fungus.

During recent work undertaken in our laboratory, moderate heat resistance has been observed for *Talaromyces stipitatus* and *T. helicus* (J. Dijksterhuis, unpublished results), and studies involving ascospores of *Byssochlamys spectabilis* have resulted in a calculated decimal reduction time at 85°C of 47-75 min (J. Houbraken, unpublished results). Table 1 shows a compilation of heat resistance data for many known heat resistant species and their D values in various heating menstrua (modified from Scholte et al., 2004).

*Jan Dijksterhuis and Robert Samson, Department of Applied Research and Services, Centraalbureau voor Schimmelcultures, Fungal Biodiversity Centre, Uppsalalaan 8, 3584 CT, Utrecht, The Netherlands. Correspondence to dijksterhuis @cbs.knaw.nl

Table 1. Heat-resistance of ascospores at different temperatures and medium composition

Fungal species	D-value (min)	Medium	Reference
Byssochlamys fulva	86°C, 13-14	Grape Juice	Michener and King (1974)
	90°C, 4-36 (3 $\log_{10}$ reduction)	Buffer pH 3.6 and 5.0, 16 °Brix	Bayne and Michener (1979)
	90°C, 8.1	Tomato juice	Kotzekidou (1997)
Byssochlamys nivea	85°C, 1.3-4.5	Buffer pH 3.5	Casella et al. (1990)
	88°C, 8-9 sec	Ringer solution	Engel and Teuber (1991)
	90°C, 1.5	Tomato juice	Kotzekidou (1997)
Byssochlamys spectabilis	85°C, 47-75	Buffer, pH 6.8	Authors' unpublished data
Eurotium herbariorum	70°C, 1.1-4.6	Grape Juice, 65 °Brix	Splittstoesser et al. (1989)
Eurotium chevalieri	70°C, 17.2	Plum extract (pH 3.8,	Pitt and Christian (1970)
	80°C, 3.3	20 °Brix, a_w 0.98)	
Monascus ruber	80°C, 1.7-2.0	Buffers (pH 3,0 ; pH 7,0)	Panagou et al. (2002)
	80°C, 0.9-1.0	Brine	
Neosartorya fischeri	85°C, 13.2	Apple Juice	Conner and Beuchat (1987b)
	85°C, 10.1	Grape Juice	Conner and Beuchat (1987b)
	85°C, 10-60	In ACES-buffer, 10 mM, pH 6.8	Authors' unpublished data CBS 133.64
	85°C, 10.4	Buffer pH 7.0	Conner and Beuchat (1987b)
	85°C, 35.3	Buffer pH 7.0	Rajashekhara et el. (1996)
	88°C, 1.4	Apple Juice	Scott and Bernard (1987)
	88°C, 4.2-16.2	Heated fruit fillings	Beuchat (1986)
	90°C, 4.4-6.6	Tomato Juice	Kotzekidou (1997)
	91°C, < 2	Heated fruit fillings	Beuchat (1986)
Neosartorya pseudofischeri	95°C, 20 sec		Authors' unpublished data
Talaromyces flavus (macrosporus)	85°C, 39	Buffer pH 5.0, glucose, 16 °Brix	King (1997)
	85°C, 20-26	Buffer pH 5.0, glucose	King and Halbrook (1987)

Table 1. Heat-resistance of ascospores at different temperatures and medium composition

Fungal species	D-value (min)	Medium	Reference
	85°C, 30-100	ACES-buffer, 10 mM, pH 6.8	Dijksterhuis and Teunissen (2004)
	88°C, 7.8	Apple Juice	Scott and Bernard (1987)
	88°C, 7.1-22.3	Heated fruit fillings	Beuchat (1986)
	90°C, 2-8	Buffer pH 5.0, glucose	King and Halbrook (1987)
	90°C, 6.2	Buffer pH 5.0, glucose	King (1997)
	90°C, 6.0	Buffer pH 5.0, glucose Slug flow heat exchanger	King (1997)
	90°C, 2.7-4.1	Organic acids	King and Whitehand (1990)
Talaromyces flavus	90°C, 2.5-11.1	Sugar 0-60 °Brix)	King and Whitehand (1990)
	90°C, 5.2-7.1	pH 3.6-6.6	King and Whitehand (1990)
	91°C, 2.1-11.7	Heated fruit fillings	Beuchat (1986)
T. trachyspermus	85°C, 45 sec		Authors' unpublished data
T. helicus	70°C, approx. 20		"
T. stipitatus	72°C, approx. 85		
Xeromyces bisporus	82°C, 2.3		Pitt and Hocking (1982)

2. HEAT-RESISTANT ASCOSPORES OF *TALAROMYCES MACROSPORUS*

Germination of heat-resistant ascospores of different species is activated and also synchronised by a heat treatment (see for a survey, Dijksterhuis and Samson, 2002). Germination of these spores has been studied in detail by Dijksterhuis et al. (2002) in *Talaromyces macrosporus*. This fungus is a candidate to become a model system for research on heat-resistant ascospores. On oatmeal agar at 30°C *T. macrosporus* forms numerous yellow ascoma (Figure 1) within a few weeks and a dense homogenous suspension of ascospores can be harvested by a simple procedure. These cells do not germinate when left in malt extract broth for prolonged times. A 7-10 min treatment at

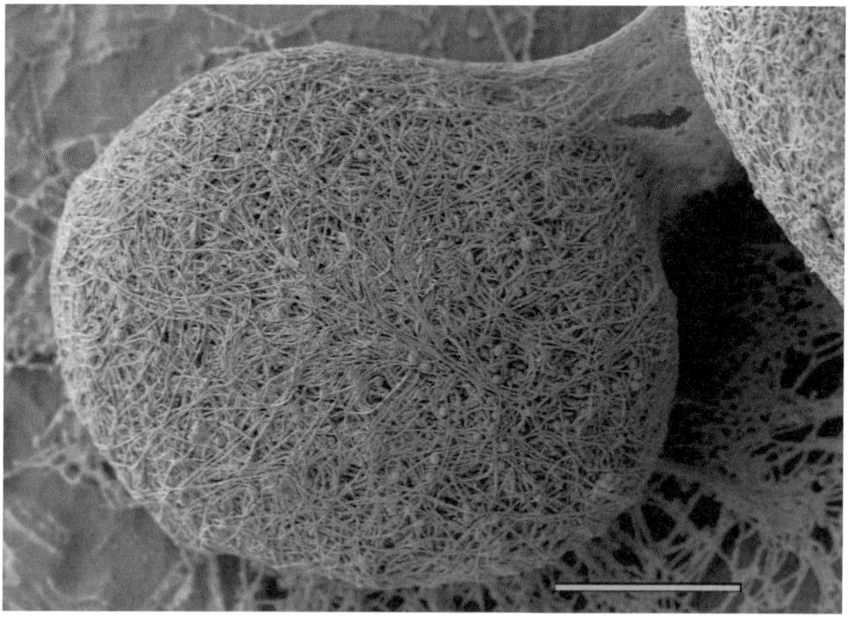

Figure 1. Ascoma of *Talaromyces macrosporus*. Note the intricate network on the outside of the structure. The ascospores visible on the outside of the fruit body have been released from other broken ascomata. Bar = 100 μm.

85°C, however, results in germination of the majority of the cells. The ascospores contain high levels of trehalose, low amounts of water and are bound by a very thick multi-layered cell wall. Upon heat treatment the trehalose is broken down by a very active trehalase and the product of hydrolysis, glucose, is accumulated inside the spore. Figure 2 shows

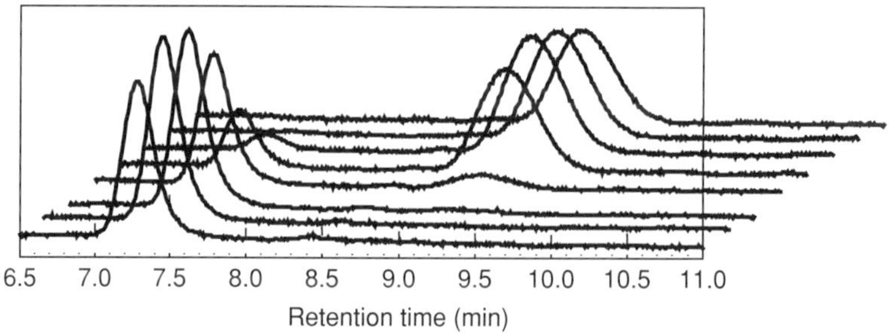

Figure 2. HPLC profiles of cell free extracts of broken spores during early germination of ascospores. The left peak shows trehalose, the right peak co-elutes with a glucose standard. The different profiles are at 0, 24, 39, 56, 73, 90, 106 and124 min after the start of the heat treatment (from front to rear).

the degradation of trehalose and the simultaneous formation of glucose early in the germination process. The glucose is present in the cells for only a short time. After that it occurs in measurable quantities in the substrate, indicating a massive release of glucose from the germinating cell.

After 150 min or more the inner cell emerges rapidly from within the outer cell wall, which is ruptured. The emptied outer cell wall remains attached to the protoplast, which is encompassed by the inner layer of the ascospore cell wall. This process is very sudden, it only takes a second or less, and is termed prosilition (Lat: *prosilire*, to jump out). After this remarkable phenomenon, the respiration of the cells increases quickly and the cells swell and form a germ tube 6 hours after the heat treatment. Figure 3 shows snap frozen cells that are in various stages of the process, as observed by Cryo-Scanning Electron Microscopy using a JEOL JSM 840 scanning electron microscope (Dijksterhuis et al., 2002).

Recently, prosilition was confirmed in other species of *Talaromyces* namely *T. stipitatus, T. helicus* and *T. bacillosporus* (our unpublished observations). However, ascospores of *Neosartorya* species seem to germinate by a slow separation of the two shell-like ornamented halves and subsequent formation of a germ tube. Apparently, these two genera show different modes of ascospore germination.

3. FUNGI IN FOOD AND HIGH PRESSURE TREATMENT

Ultra high pressure is a suitable candidate for non-thermal treatment of food products. These treatments have the benefit that the organoleptic properties of the food are less affected compared to pasteurisation or more severe heat treatment. In addition, vitamins are better preserved after the application of this alternative preservation technique. While vegetative microbial cells are inactivated at relatively low pressures (200-300 MPa), spores are more resistant to these treatments. Bacterial spores even are activated to germinate at a treatment of 200 MPa, but bacterial spores germinate so quickly that they are killed during prolonged treatment. When higher pressures are applied the germination sequence of the bacterial spores is blocked rendering the spores less vulnerable to ultra high pressures

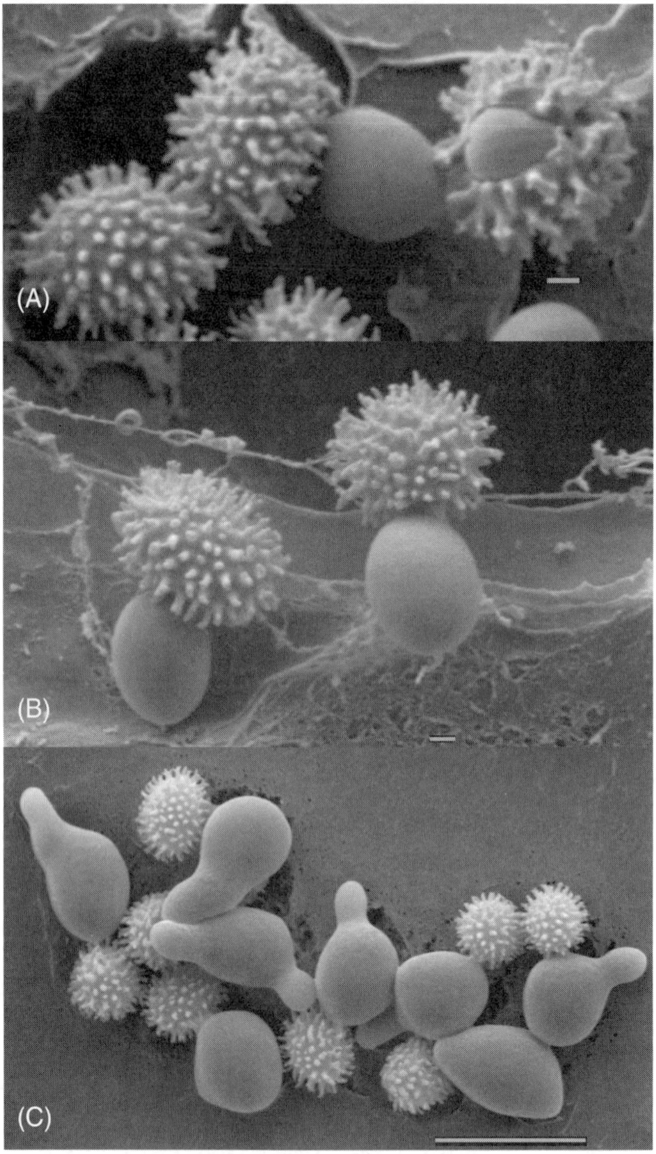

Figure 3. Germinating ascospores of *T. macrosporus*. In the top panel (A) an unprosilited cell (left) and a fully prosilited spore (middle) are shown. At the right a spore in the process of prosilition is captured. The outer cell wall has opened and the smooth inner cell wall is visible. In the middle panel (B) two fully prosilited spores are shown. Note the connection between the released cell and the empty outer cell wall. In the bottom panel (C), spores 6 h after heat treatment are shown, most of the swollen cells exhibit a germ tube, in one case two germ tubes are formed. Bars are 1 μm (A and B) and 10 μm (C).

(above 600 MPa), but also with a higher sensitivity for heat treatments (Wuytack et al., 1998).

Do resistant fungal spores such as those produced by heat-resistant fungi show extended endurance against this novel food treatment? Ascospores of *Byssochlames nivea* survive pressures at or above 600 MPa for many minutes (Butz et al., 1996), but are killed after repetitive treatments at these pressures, which are designated as "oscillative treatments" (Palou et al., 1998). These authors also describe that the application of elevated temperatures in the presence of high pressures effectively kills the ascospores. However, care is necessary with such applications to ensure that the organoleptic properties of the food are changed as minimally as possible. Probably a "happy medium" approach is important with such treatments.

4. BREAKING ASCOSPORE DORMANCY BY ULTRA-HIGH PRESSURE

Is heat the only factor that can break the dormancy of these ascospores? Recently activation of ascospores of *T. macrosporus* by high-pressure treatments was reported by Reyns et al. (2003) and Dijksterhuis and Teunissen (2004). Both studies used the same strain of fungus and a similar pressurisation equipment. The most important shared observation was that activation of ascospores occurred after a pressure treatment and that even a very short treatment at high pressure caused maximal activation. This is relevant for the food industry, because short treatments are important for economic reasons.

Dijksterhuis and Teunissen (2004) observed no activation at 200 MPa and activation of only part of the spores (up to 7% of cells) between 400 and 800 MPa. This could indicate that treatments at approx. 300 MPa would be of interest for the food industry to prevent contamination with microorganisms without activating these fungi. Reyns et al. (2003), however, observed partial activation of *T. macrosporus* at 200 MPa and activation of all spores at 600 MPa after 15 sec treatment.

A number of factors could have caused of these differences between the studies. Firstly, the growth conditions of the fungal cultures were different. As well as age of the culture, growth temperature (Conner and Beuchat, 1987a) and the growth medium (Beuchat, 1988a) all influence the heat resistance of spores. Beuchat (1988a) reported that ascospores harvested from malt extract agar (which was used by

Reyns et al., 2003) exhibited a somewhat lower heat resistance than spores formed on oatmeal agar (used by Dijksterhuis and Teunissen, 2004). These factors also may have a bearing on the extent of activation of the spores. The spores used by Reyns et al. (2003) were younger and grown at a lower temperature. Dijksterhuis and Teunissen (2004) report that the age of the fungal culture from which ascospores are harvested correlates with the acquisition of heat resistance, and that the major increase in heat resistance occurs between 20 and 40 days incubation. The combined results of the two papers point to the importance of ascospore maturity in influencing the ability of the cells to remain dormant.

The second parameter that may affect activation is the treatment of the spores before and during pressurisation. Reyns et al. (2003) pretreated their ascospore suspension for 20 min at 65°C to kill vegetative cells. In a buffer at a pH 6.8, this treatment should not activate ascospores, although at 70°C a significant increase in activation is observed in our laboratory (J. Dijksterhuis, unpublished results). At lower pH (3.0) Reyns et al. (2003) observed that nearly complete activation occurred after heat treatment at 65°C. This indicates a clear lowering of the heat activation temperature at low pH. Recently, we observed activation in a low pH medium at room temperature or only slightly elevated temperatures (J. Dijksterhuis, unpublished results). Heat resistant ascospores cause problems in fruit juices, and this lowering in heat activation could also occur in these food products. The lower pH of the environment in fruit juices may reduce the heat resistance of the ascospores somewhat, but the protective effect of the increasing the sugar content is much greater (King and Whitehand, 1990). Organic acids, particularly fumaric acid and to a lesser extent sorbic, benzoic and acetic acids have clear effect on heat resistance below pH 4.0 (Beuchat, 1988b). We have observed that the spores also become more resistant to high pressure under these conditions (J. Dijksterhuis, unpublished results).

It is also possible that menstruum in which the spores are suspended during the high pressure treatments has an effect. Dijksterhuis and Teunissen (2004) used buffer (10 mM ACES, pH 6.8) whereas Reyns et al. (2003) suspended ascospores in distilled water. During high pressure treatment the acidity of the medium decreases, but this drop would be less extensive in a buffered system. Spores suspended in distilled water may be confronted with a temporary drop in pH, resulting in more extensive activation.

The authors of both papers conclude that activation by high pressure might be related to the barrier function of the ascospore cell wall.

Dijksterhuis and Teunissen (2004) performed cryo-electron microscopy on the spores and showed alterations of the cell after very short treatments at high temperature. Reyns et al. (2003) illustrated that the spores collapsed when air dried after a high pressure treatments, whereas untreated spores maintained their shape. These observations indicate that structural changes occur in the cell wall and that these have a direct influence of the process of activation.

5. THE CONNECTION BETWEEN THE DIFFERENT TREATMENTS

Ascospores of *Eurotium herbariorum* are recognised as heat resistant, albeit less so than *Byssochlamys, Neosartorya* and *Talaromyces* species (Splittstoesser et al., 1989). Eicher and Ludwig (2002) showed that a proportion of the spores of *Eurotium repens* (8%) is activated from dormancy by a treatment of 200 MPa for 60 min. The ascospores were also activated by a heat treatment: after 8 min at 60°C approximately 50% of the spores germinated (on Sabouraud agar) after 5 days. At 50°C, 60 min were needed to activate this population of cells. At room temperature in an isotonic salt solution ascospores did germinate, though after a delay: after 18 hours approximately 15% of the cells showed signs of germination. Ascospores that were heat activated (15 min, 60°C) were more sensitive to a subsequent high pressure treatment at 500 MPa. A 30 min treatment at 500 MPa reduced the number of colony forming units even compared with the unactivated spore suspensions. When the pressure treatment was applied immediately after a heat treatment at 60°C for 15 min the number of colony forming cells reduced by a factor of 40. When a pause was introduced between the treatments during which the spores were stored at 20°C in an isotonic salt solution, the number of colony forming units restored 10-fold. This phenomenon was designated "re-stabilisation" (Eicher and Ludwig, 2002). After treatments with high pressure (500 MPA, 30 min), heat activation did not result in any enhancement of germinating cells of *E. repens*. In fact, heat treatment resulted in some further reduction of colony forming units, viability counts only reduced further when a pause was present between the treatments (Eicher and Ludwig, 2002).

In the case of *T. macrosporus,* Dijksterhuis and Teunissen (2004) described near total activation of germination by heat (7 min, 85°C) after 5 min pressure treatments up to 800 MPa. Short pressure treatments at least did not lead to heat sensitisation in their experiments. However, Reyns et al. (2003) show clear sensitisation of ascospores for the activation heat treatment (30 min, 80°C) after all pressure treatments at 600 MPa and 700 MPa.

6. DORMANCY REVISITED

The observed re-stabilisation phenomenon of *E. repens* ascospores poses the question of whether ascospores can return to their dormant state once activated. We addressed this question in a number of experiments where heat activated ascospores of *T. macrosporus* were confronted with a sudden lowering of the temperature or with drying conditions. The high amount of trehalose (10-20% wet weight) and the low water content of the spores (38%) may introduce a very high viscosity inside the spores. This has recently been confirmed by electron-(para)magnetic-resonance studies (J. Dijksterhuis, unpublished results). A sudden lowering of the temperature or a reduction of the water content will most likely introduce a glass transition situation inside the cell. The glassy state is an amorphous phase characterized by very low movement speeds of the cell components. Reduction of the water content or lowering of the temperature are two factors that favour the glass transition of the ascospores. Plunging of the cells into liquid nitrogen or controlled drying below 3% water-content especially will introduce a glassy state in these cells. This transition also may re-establish the dormancy of these cells. "Biological glasses", a term introduced by Buitink (2000) are characterised by a melting temperature that can be high (Wolkers et al., 1996) and cells may need to be exposed to high temperatures again in order to re-activate them.

A number of experiments were done by storing heat-activated ascospores (7 min, 85°C) on ice for 15 min or plunging the cells into liquid nitrogen and keeping them there for 15 min. The samples were allowed to warm to room temperature and subsequently plated out. In addition, activated ascospores were incubated at 30°C for 1 h before the cold treatment. Table 2 summarises these experiments and shows that dormant ascospores showed only low levels of germination whether they are cooled or not, while heat activated cells under all cases showed very high percentages of germinating cells.

Ice-treatment directly after or 1 h following heat activation did not show any effect. These experiments indicate that a sudden lowering of

Table 2. Influence of a sudden cooling treatment on extent of germination of activated ascospores of *T. macrosporus*. Numbers represent the ratio between the treated samples and the untreated controls.

	Cooled on ice		Cooled in liquid nitrogen	
	Experiment 1	*Experiment 2*	*Experiment 1*	*Experiment 2*
Controls	1	1	1	1
Controls, cooled	1.2	0.3	0.5	0.2
Activated	150	9.7	11	13
Activated, cooled	248	11.2	7.4	14.1
Activated, cooled after 1 h	198	31	12.9	10.2

Heat activation of ascospores was in 3-5 ml suspensions at 85°C for 7 min and shaken at 140 rpm in a waterbath. When spores were incubated for 1 h this was at 30° C (140 rpm). Spores were placed in Eppendorf tubes and put on ice or into liquid nitrogen. Samples were plated out in duplicate and colonies counted. The ratio between treated samples and untreated controls is given in this table.

the temperature, irrespective of how fast and large it might be, does not "reset" the ascospores to the dormant mode.

In a further experiment, ascospores were dried according to the procedures used at the CBS. The latter includes controlled freezing to –40°C (1°C/min), storage at –80°C and drying under vacuum. Dried dormant ascospores remained dormant after drying and could be effectively activated by a heat treatment after resuspending them in buffer. Dried freshly activated ascospores produced the similar numbers of activated spores and the cells showed similar tolerance to the drying treatment as dormant cells. When these cells were heat treated (in case dormancy had re-established) no additional increase or decrease of cell numbers occurred. This indicated that these cells had retained their activated state, but still showed heat resistance. Both drying tolerance and heat resistance had decreased markedly after incubating the activated cells for 2 h at 30°C. These combined observations show that important phase transitions of the inner cell do not change the status of activation or dormancy in *T. macrosporus*.

7. THE SPEED OF HEAT ACTIVATION

According to Sussman (1966) dormancy is defined as a hypometabolic state; i.e. a rest period or reversible interruption of (phenotypic) development. He discerns exogenous dormany (quiescence) which include delayed development due to physical or chemical cues.

Constitutive dormancy is a condition in which development is delayed as an innate property such as a barrier to the penetration of nutrients, a metabolic block, or a self-inhibitory compound. In case of *T. macrosporus,* ascospores can be activated and also synchronised to germinate by a robust physical signal such as heat and/or high pressure. Careful examination of the data provided by Beuchat (1986) suggests that the speed of activation is increased at higher temperatures. Ascospores of *Neosartorya fischeri* exhibited constant rates of heat activation between 70° and 85°C (King and Halbrook, 1987, Figure 1). In our laboratory we observed that ascospores added to preheated buffer were fully activated within 2 min.

Kikoku (2003) reported full activation of *T. macrosporus* ascospores in a citrate-phosphate buffer (pH 6.5) within 100 s at 81° and 82.5°C. Above this temperature this time became even shorter, namely 60 s at 86.5 and 87°C, and 35 s at 91°C. From these data the author extracted rate constants of heat activation (expressed as k), which range from 1.2 to 4.1/min between 81° and 91°C. At 84°C no difference in k was observed between pH 3.5 or 6.5 (2.9 and 2.8/min respectively) and also in phosphate buffer (pH 6.6) the k value was 2.8/min. However in grape juice (5 °Brix) a very high k value was observed (7.7/min). Thus, the presence of the sugars or organic acids or some other compound in the fruit juice resulted in a very rapid activation at this temperature (100% in 20 s).

Activation energy (Ea) can be calculated using an Arrhenius plot where ln (0.23303.k) is plotted against 1/T. When activation is the result of the conformation or chemical change of one defined compound in the ascospore, for instance a compound of the (plasma) membrane or a receptor protein, the Ea reflects the energy needed to convert 1 mole of such compound. Changes in proteinaceous compounds do need a different energy absorption than lipid compounds and the Ea could give clues about the nature of activation. However, when more systemic changes in the ascospore absorb the energy delivered by the heat, the Ea calculated does not give much information. Recent findings at our laboratory show changes in the ascospore during heat activation involving both proteins and cell wall components, and this multitude of changes may make a interpretation of the Ea value difficult.

8. ACKNOWLEDGEMENT

The authors are indebted to Kenneth van Driel for assistance with Low-Temperature Scanning Electron Microscopy experiments, and

Yasmin Stoop for her assistance with the dormancy experiments. Joost Eleveld and Mark Kweens are thanked for their data on heat inactivation of *Talaromyces* ascospores.

9. REFERENCES

Bayne, H. G., and Michener, H. D., 1979, Heat resistance of *Byssochlamys* ascospores, *Appl. Environ. Microbiol.* **37**:449-453.

Beuchat, L. R., 1986, Extraordinary heat resistance of *Talaromyces flavus* and *Neosartorya fischeri* ascospores in fruit products, *J. Food Sci.* **51**:1506-1510.

Beuchat, L. R., 1988a, Thermal tolerance of *Talaromyces flavus* ascospores as affected by growth medium and temperature, age and sugar content in the inactivation medium, *Trans Br. Mycol. Soc.* **90**:359-374.

Beuchat, L. R., 1988b, Influence of organic acids on heat resistance characteristics of *Talaromyces flavus* ascospore, *Int. J. Food Microbiol.* **6**:97-105.

Buitink, J., 2000, Biological Glasses: Nature's way to preserve life, Ph.D. Thesis, University of Wageningen, The Netherlands, pp 5-17.

Butz, P., Funtenberger, S., Haberditzl, T., and Tausher, B., 1996, High pressure inactivation of *Byssochlamys nivea* ascospores and other heat resistant moulds, *Lebensm-Wiss. U-Technol.* **29**:404-410.

Casella, M. L. A., Matasci, F., and Schmidt-Lorenz, W., 1990, Influence of age, growth medium, and temperature on heat resistance of *Byssochlamys nivea* ascospores, *Lebensm-Wiss. U-Technol.* **23**:404-411.

Conner, D. R., and Beuchat, L. R., 1987a, Efficacy of media for promoting ascospore formation by *Neosartorya fischeri*, and the influence of age and culture temperature on heat resistance of ascospores, *Food Microbiol.* **4**:229-238.

Conner, D. R., Beuchat, L. R., and Chang, C. J., 1987b, Age-related changes in ultrastructure and chemical composition associated with changes in heat resistance of *Neosartorya fischeri* ascospores, *Trans. Br. Mycol. Soc.* **89**:539-550.

Dijksterhuis, J., and Samson, R. A., 2002, Food and crop spoilage on storage, in: *The Mycota XI Agricultural Applications*, F. Kempken, ed., Springer-Verlag, Berlin Heidelberg, Germany, pp. 39-52.

Dijksterhuis, J., van Driel, K. G. A., Sanders, M. G., Molenaar, D., Houbraken, J. A. M. P., Samson, R. A., and Kets, E. P. W., 2002, Trehalose degradation and glucose efflux precede cell ejection during germination of heat-resistant ascospores of *Talaromyces macrosporus, Arch. Microbiol.* **178**:1-7.

Dijksterhuis, J., and Teunissen, P. G. M., 2004, Dormant ascospores of *Talaromyces macrosporus* are activated to germinate after treatment with ultra high pressure, *J. Appl. Microbiol.* **96**:162-169.

Engel, G., and Teuber, M., 1991, Heat resistance of ascospores of *Byssochlamys nivea* in milk and cream, *Int. J. Food Microbiol.* **12**:225-234.

Eicher, R., and Ludwig, H., 2002, Influence of activation and germination on high pressure inactivation of ascospores of the mould *Eurotium repens, Comp. Biochem Physiol Part A* **131**:595-604.

Kikoku, Y., 2003, Heat activation characteristics of *Talaromyces* ascospores, *J. Food Sci.* **68**:2331-2335.

King, A. D., 1997, Heat resistance of *Talaromyces flavus* ascospores as determined by a two phase slug flow heat exchanger, *Int. J. Food Microbiol.* **35**:147-151.

King, A. D., and Halbrook, U., 1987, Ascospore heat resistance and control measures for *Talaromyces flavus* isolated from fruit juice concentrate, *J. Food Sci.* **52**: 1252-1254, 1266.

King, A. D., and Whitehand, L. C., 1990, Alteration of *Talaromyces flavus* heat resistance by growth conditions and heating medium composition, *J. Food Sci.* **55**: 830-832.

Kotzekidou, P., 1997, Heat resistance of *Byssochlamys nivea, Byssochlamys fulva* and *Neosartorya fischeri* isolated from canned tomato paste, *J. Food Sci.* **62**:410-412.

Michener, H. D., and King, A. D., 1974, Preparation of free heat-resistant ascospores from *Byssochlamys* asci, *Appl. Microbiol.* **27**: 671-673.

Palou, E., Lopez-Malo, A., Barbosa-Cánovas, G. V., Welti-Chanes, J., Davidson, P. M., and Swanson, B. G., 1998, Effect of oscillatory high hydrostatic pressure treatments on *Byssochlamys nivea* ascospores suspended in fruit juice concentrates, *Lett. Appl. Microbiol.* **27**:375-378.

Panagou, E .Z., Katsaboxalis, C. Z., and Nychas, G-J. E., 2002, Heat resistance of *Monascus ruber* ascospores isolated from thermally processed green olives of the Conservolea variety, *Int. J. Food Microbiol.* **76**:11-18.

Pitt, J. I., and Christian, J. H. B., 1970, Heat resistance of xerophilic fungi based on microscopic assessment of spore survival, *Appl. Microbiol.* **20**:682-686.

Pitt, J. I., and Hocking, A. D., 1982, Food spoilage fungi. I. *Xeromyces bisporus* Fraser, *CSIRO Food Res.Q.* **42**:1-6.

Rajashekhara E., Suresh, E. R., and Ethiraj, S., 1996, Influence of different heating media on thermal resistance of *Neosartorya fischeri* isolated from papaya fruit, *J. Appl. Bacteriol.* **81**:337-340.

Reyns, K. M. F. A., Verbeke, E. A. V., and Michiels, C. E. W., 2003. Activation and inactivation of *Talaromyces macrosporus* ascospores by high hydrostatic pressure, *J. Food Prot.* **66**:1035-1042.

Scholte, R. P. M., Samson, R. A. and Dijksterhuis, J., 2004, Spoilage fungi in the industrial processing of food, in: *Introduction to Food- and Airborne Fungi,* R. A. Samson, E. S. Hoekstra, J. C. Frisvad, eds, 7th Edition, Centraalbureau voor Schimmelcultures, Utrecht, The Netherlands, pp 339-356.

Scott, V. N., and Bernard, D. T., 1987, Heat resistance of *Talaromyces flavus* and *Neosartorya fischeri* from commercial fruit juices, *J. Food Prot.* **50**:18-20.

Splittstoesser, D. F., Lammers, J. M., Downing, D. L., and Churey, J. J. 1989, Heat resistance of *Eurotium herbariorum*, a xerophilic mold, *J. Food Sci.* **54**:683-685.

Sussman, S., and Halvorson, H. O., 1966, *Spores, their Dormancy and Germination,* Harper & Row, New York, USA.

Wuytack, E. Y., Boven, S., and Michiels, C. W., 1998, Comparative study of pressure-induced germination of *Bacillus subtilis* spores at low and high pressures, *Appl. Environ. Microbiol.* **64**:3220-3224.

MIXTURES OF NATURAL AND SYNTHETIC ANTIFUNGAL AGENTS

*Aurelio López-Malo, Enrique Palou, Reyna León-Cruz and Stella M. Alzamora**

1. INTRODUCTION

Antimicrobial agents are chemical compounds present in, or added to, foods that retard microbial growth or cause microbial death (López-Malo et al., 2000). The use of such agents is one of the oldest and most traditional food preservation techniques (López-Malo et al., 2005a).

Antimicrobial agents are somewhat arbitrarily classified as traditional and naturally occurring (Davidson, 2001). Traditional antimicrobials are those that have been used for long time, approved by many countries as antimicrobials in foods or are produced by synthetic means, or are inorganic. Antimicrobial agents may be either synthetic compounds intentionally added to foods or naturally occurring, biologically derived substances (the so called *naturally occurring antimicrobials*), which may be used commercially as additives for food preservation as well as exhibiting antimicrobial properties in the biological systems from which they originate (Sofos et al., 1998). However, as Davidson (2001) stated, antimicrobial agents now produced synthetically are also found in nature, including acetic, benzoic or sorbic acids.

* A. López-Malo, E. Palou, R. León-Cruz, Ingeniería Química y Alimentos, Universidad de las Américas, Puebla, Cholula 72820, Mexico. S. Alzamora, Dept Industrias, Fac. de Ciencias Exactas y Naturales, Univ. de Buenos Aires, Ciudad Universitaria 1428, Buenos Aires, Argentina. Correspondence to aurelio.lopezm@udlap.mx

Concerns about the use of antimicrobial agents in food products have been discussed for decades (Parish and Carroll, 1988). The increasing demand for more "natural" foods with reduced additives (including antimicrobial agents) and the increasing demand for greater convenience, have promoted the search for alternative antimicrobial agents or combinations to be used by the food industry (Alzamora et al., 2003; López-Malo et al., 2000, 2005a). In this search, a wide range of natural systems from animals, plants and microorganisms have been studied (Beuchat and Golden, 1989; Board, 1995; Hill, 1995; Nychas, 1995; Sofos et al., 1998; Davidson, 2001; Chikindas and Montville, 2002; Gould, 2002; Alzamora et al., 2003; López-Malo et al., 2000, 2005a), and the studies are continuing. However, the strict requirements to obtain approval and the economic cost of getting the product onto the market restrict the spectrum of new chemical compounds that can be used in the preservation of foods. These obstacles have prompted the renewed search for preservatives by examining compounds already used in the food industry, perhaps for other purposes, but with potential antimicrobial activity. Such compounds are already approved and not toxic in the levels used; many of them classified in the USA as generally recognized as safe (GRAS). Included in these compounds are, for example, the so-called "green chemicals" present in plants that are utilized as flavour ingredients (Davidson, 2001; Alzamora et al., 2003; López-Malo et al., 2005a).

The antimicrobial activities of several plant derivatives used today as seasoning agents in foods and beverages have been recognized for centuries. Although ancient civilizations recognized the antiseptic or antimicrobial potential of many plant extracts, it was not until the eighteenth century that scientific information underpinned the observed effects. Naturally occurring antimicrobial compounds are abundant in the environment. Some of these natural antimicrobial systems are already employed for food preservation, while many others are still being studied (Gould, 1995). The exploration and use of naturally occurring antimicrobials in foods, as well as their chemistry, food safety and toxicity aspects, antimicrobial activity and mechanisms of action, are covered in excellent reviews by Branen et al. (1980), Shelef (1983), Zaika (1988), Beuchat and Golden (1989), Wilkins and Board (1989), Conner (1993), Board (1995), Nychas (1995), Sofos et al. (1998), Smid and Gorris (1999), Davidson (2001), Gould (2002), and López-Malo et al. (2000, 2005a).

1.1. Sources of Natural Antimicrobials From Plants

Plants, herbs and spices as well as their derived essential oils and isolated compounds contain a large number of substances that are known to inhibit various metabolic activities of bacteria, and fungi, although many have not been fully exploited. Hundreds of plants are known to be potential sources of antimicrobial compounds (Wilkins and Board, 1989; Beuchat, 1994; Nychas, 1995). Antimicrobial compounds in plant materials are commonly contained in the essential oil fraction of leaves, flowers and flower buds, bulbs, rhizomes, fruit, or other parts of the plant (Shelef, 1983; Nychas, 1995). Antimicrobial compounds may be lethal to microorganisms or they may simply inhibit the production of metabolites such as mycotoxins (Conner and Beuchat, 1984a, b; Zaika, 1988; Beuchat, 1994; Davidson, 2001).

Major components with antimicrobial activity found in plants, herbs and spices are phenolic compounds, terpenes, aliphatic alcohols, aldehydes, ketones, acids and isoflavonoids. As a rule, the antimicrobial activity of essential oils depends on the chemical structure of their components and on their concentration (Davidson, 2001; López-Malo et al., 2005a). Simple and complex derivatives of phenol are the main antimicrobial compounds in essential oils from spices (Shelef, 1983). The antimicrobial activity of cinnamon, allspice, and cloves is attributed to eugenol (2-methoxy-4-allyl phenol) and cinnamic aldehyde, which are major constituents of the volatile oils of these spices (Bullerman et al., 1977; Farrell, 1990). Eugenol, carvacrol, thymol and vanillin have been recognized as active antimicrobial compounds from plant essential oils (López-Malo et al., 2005a), and aliphatic alcohols and phenolics have also been reported as fungal growth inhibitors (Katayama and Nagai, 1960; Farag et al., 1989).

A wide antimicrobial spectrum has been found in phenolic compounds such as thymol extracted from thyme and oregano, cinnamic aldehyde extracted from cinnamon, and eugenol extracted from cloves (Beuchat and Golden, 1989; Wilkins and Board, 1989; Davidson and Branen, 1993). Vanillin, a phenolic compound present in vanilla pods also has antifungal activity (Beuchat, 1976; Zaika, 1988; Beuchat and Golden, 1989; Farag et al., 1989; Cerrutti and Alzamora, 1996; Cerrutti et al., 1997; López-Malo et al., 1995, 1997, 1998). Cinnamon bark is highly inhibitory to the moulds *Aspergillus flavus*, *A. parasiticus*, *A. versicolor*, and *A. ochraceus* (Hitokoto et al., 1978). Ground cinnamon at a concentration of 1-2% in broth allowed some growth of *A. parasiticus*, but eliminated approximately 99% of the production of aflatoxin (Bullerman, 1974). The active compounds were reported to be cinnamic aldehyde and eugenol, which are the major constituents of essential oils from cinnamon and clove (Bullerman et al., 1977). Eugenol has been reported as one of the most effective natural antimicrobials from plant origin acting as a sporostatic agent (Al-Khayat and Blank, 1985). Thymol at 100 ppm inhibited *A. parasiticus* growth for 7 days at 28°C (Buchanan and Shepherd, 1981). *A. flavus* was shown to be more sensitive than two other *Aspergillus* species when exposed to essential oils of oregano and cloves (Paster et al., 1990).

1.2. Mechanisms of Action

Some essential oils, plant extracts, and oleoresins influence certain biochemical and/or metabolic functions, such as respiration or production of toxins or acids, indicating that the active components in various oils and oleoresins may have different specificities for target sites on or in microbial cells. Much of the research on spices has included speculation on the contribution of the terpene fraction to their antimicrobial activity. Few of the studies, however, have attempted to isolate and identify the antimicrobial fraction, and no references are found which relate to the mechanism by which spices inhibit microorganisms. It seems reasonable that since many of the components of the essential oils, such as eugenol and thymol, are similar in structure to active phenolic antimicrobials, their modes of action could be assumed to be similar (Davidson, 2001; López-Malo et al., 2005a).

The possible modes of action of phenolic compounds have been reported in several reviews (Wilkins and Board; 1989; Beuchat, 1994; Nychas, 1995; Sofos et al., 1998; Davidson, 2001; López-Malo et al., 2000, 2005a). These mechanisms have not been completely elucidated, however, the effect of phenolic compounds is concentration dependent (Prindle and Wright, 1977). At low concentration, phenols affected enzyme activity, especially of those enzymes associated with energy production, whereas at higher concentrations, protein denaturation occurred. The effect of phenolic antioxidants on microbial growth and toxin production could be the result of the ability of phenolic compounds to alter microbial cell permeability, permitting the loss of macromolecules from the interior. They could also interact with membrane proteins causing a deformation in structure and functionality (Fung et al., 1977). Conner and Beuchat (1984a, b) suggested that antimicrobial activity of essential oils on yeasts could be the result of disturbance in several enzymatic systems involved in energy production and synthesis of structural components. Once phenolic compounds cross the cell membrane, interactions with membrane enzymes and proteins would cause an opposite flow of protons, affecting cellular activity.

Increasing the concentration of thyme essential oil, thymol or carvacrol was not reflected in a direct relationship with antimicrobial effects. However, after exceeding a certain critical concentration, a rapid and drastic reduction in viable cells was observed (Juven et al., 1994). Phenolic compounds could sensitize cellular membranes and when sites are saturated, serious damage and a rapid collapse of cytoplasmatic membrane integrity could be present, with the consequent loss of cytoplasmatic constituents. It has been suggested that the effects of phenolic compound could be at two levels, on cellular wall and membrane integrity as well as on microbial physiological responses (Kabara and Eklund, 1991). Phenolic compounds could also denature enzymes responsible for spore germination or interfere with amino acids that are necessary in germination processes (Nychas, 1995).

1.3. Factors Affecting Activity

The antimicrobial activities of extracts from several types of plants and plant parts used as flavouring agents in foods have been recognized for many years. However, few studies have reported on the effect of extracts in combination with other factors on microbial growth. The potential of a compound as a total or partial substitute for common preservatives to inhibit growth of spoilage and pathogenic microorganisms needs to be evaluated alone and in combination with traditional preservation factors, such as storage temperature, pH, water activity (a_w), other antimicrobials, and modified atmospheres. Results from such studies could be very useful, permitting researchers involved in the development of multifactorial preservation of foods to assess quickly the impact of altering any combination of the studied variables.

1.4. Combined Antimicrobial Agents

Traditionally, only one chemical antimicrobial agent was used to preserve a food (Busta and Foegeding, 1983). However, more recently, the use of combined agents in a single food system has become more common. The use of combined antimicrobial agents theoretically provides a greater spectrum of activity, with enhanced action against pathogenic and/or spoilage microorganisms. It is thought that combined agents will act on different species in a mixed microflora or act on different metabolic elements within similar species or strains, which theoretically results in improved control compared with the use of one antimicrobial agent alone. However, actual proof of improved efficacy requires objective interpretation. Although testing of combined antibiotics for clinical use is well studied and relatively well standardized (Barry, 1976; Krogstad and Moellering, 1986; Eliopoulos and Moellering, 1991), application of such methodology to antimicrobials in food systems is poorly developed (Davidson and Parish, 1989; López-Malo et al., 2005b). The use of antibiotic combinations in medicine continues to be a subject of intensive investigation and a matter of great clinical relevance (Eliopoulos and Moellering, 1991).

Methods of testing combined antimicrobials usually involve agar diffusion, agar or broth dilution, or death time curves (NCCLS, 1999; 2002). Dilution methods yield quantitative data and are often conducted with various combined concentrations of two antimicrobials arranged in a "checkerboard" array. The checkerboard method is the technique used most frequently to assess antimicrobial combinations *in vitro*, presumably because a) its rationale is easy to understand, b) the mathematics necessary to calculate and interpret the results are simple, c) it can be readily performed in the laboratory using microdilution systems, and d) it has been the technique most frequently used in studies that have investigated synergistic interactions of antibiotics in clinical treatments (Eliopoulos and Moellering, 1991). The term checkerboard refers to the pattern (of tubes, plates or microtitre wells) formed by multiple dilutions of the two antimicrobials being tested in concentrations equal to, above, and below their MIC. Traditional clinical dilution testing uses two-fold dilutions of test compounds, but testing of food antimicrobials is often conducted with alternate dilution schemes (Rehm, 1959; Davidson and Parish, 1989; Eliopoulos and Moellering, 1991).

Combined studies are conducted to determine if specific types of interactions occur between the two combined antimicrobials. Traditionally, the terms "additive," "antagonistic," and "synergistic" were used to describe possible antimicrobial interactions. Additivity occurs when two combined antimicrobials give results that are equivalent to the sum of each antimicrobial acting independently, i.e. no enhancement or reduction in overall efficacy for the combined antimicrobials occurs compared to the individual results, and is also sometimes referred to as "indifference" (Krogstad and Moellering, 1986; Eliopoulos and Moellering, 1991). Antagonism refers to a reduced efficacy of the combined agents compared to the sum of the individual results. Synergism is an increase or enhancement of overall antimicrobial activity when two agents are combined compared to the sum of individual results (López-Malo et al., 2005b).

A conclusion that synergism occurs must be approached with caution since it implies that a reduction of overall antimicrobial concentration might be achieved in a food system without a reduction in efficacy. Gardner (1977) stated that true synergism is quite rare in relation to combined antibiotics. Other concerns about the misuse of the term "synergism" in relation to antimicrobials have been cited (Garrett, 1958; Davidson and Parish, 1989). Most commonly, additive interactions are misidentified as synergistic. A case in which an increase in antimicrobial activity is observed upon the addition of a second compound to a food system does not necessarily constitute synergy. A conclusion of synergism requires that the overall efficacy of the combination be significantly greater than the sum of the efficacies of the individual compounds.

Additive, synergistic, or antagonistic interactions can be interpreted with an MIC isobologram. Isobologram construction can be simplified using fractional inhibitory concentrations (FIC), which are MICs normalized to unity. The FIC is the concentration of a compound needed to inhibit growth (expressed as a fraction of its MIC) when combined with a known amount of a second antimicrobial compound. It is calculated as the ratio of the MIC of a compound when combined with a second compound divided by the MIC of the first compound alone. The FIC of two compounds in an inhibitory combination may be added to give a total FIC_{Index}. An FIC_{Index} near 1 indicates additivity, whereas an FIC_{Index} less than 1 indicates synergy and an FIC_{Index} greater than 1 indicates antagonism. The degree to which a result must be less than or greater than 1, to indicate synergism or antagonism is a matter of interpretation. Squires and Cleeland (1985) proposed that for antibiotic testing FIC_{Index} indicates additive results between 0.5 and 2.0. Synergism and antagonism are indicated by results $FIC_{Index} < 0.5$ and $FIC_{Index} > 2.0$, respectively. Research is needed to provide a database for proper interpretation of FIC and FIC_{Index} in relation to food antimicrobial systems. Data interpretation must be conducted conditionally and will depend upon a number of variables, such as specific test conditions, microbial strain, and target food system (López-Malo et al., 2005b). It should be noted that interpretations might also vary depending upon the specific concentrations of each antimicrobial used in combination. Parish and Carroll (1988) observed additivity between SO_2 and either sorbate or butylparaben when the inhibitory concentration contained less than 0.25 FIC of SO_2. However, for the same combination at higher SO_2, the calculated FIC indicated antagonistic results. Rehm (1959) observed similar anomalies when sodium sulphite was combined with formate or borate.

An examination of these results shows that it is not easy to anticipate the effects or to explain observed activity when considering binary mixtures of antimicrobials. Moreover, there is an increasing awareness that many combinations may result in antagonism. Four mechanisms of antimicrobial interaction that produce synergism are generally accepted: a) sequential inhibition of a common biochemical pathway, b) inhibition of protective enzymes, c) combinations of agents active against cell walls; and d) the use of cell wall active agents to enhance the uptake of other antimicrobials (Eliopoulos and Moellering 1991). Mechanisms of antimicrobial interaction that produce antagonism are less well understood and include: a) combinations of bactericidal/fungicidal and bacteriostatic/ fungistatic agents; b) use of compounds that act on the same target of the microorganism; and c) chemical (direct or indirect) interactions between the compounds (Larson, 1984). The specific modes of action of plant constituents on metabolic activities of microorganisms still need to be clearly defined, even when they are the only stress factor.

It is not easy to anticipate the effects of binary mixtures of antimicrobials, or to explain their activity, and it becomes even more difficult if more than two antimicrobials are mixed. Despite the lack of scientific knowledge about the mechanisms of interaction of natural and synthetic antimicrobials, synergistic combinations could be useful in reducing the amounts of antimicrobials needed, diminishing consumer concerns about the use of chemical preservatives. Our objective in this study was to evaluate the individual and combined (binary or ternary) effects of selected natural phenolic compounds such as eugenol, vanillin, thymol, and carvacrol, and a synthetic antimicrobial agent (potassium sorbate) as antifungal agent mixtures.

2. MATERIALS AND METHODS

2.1. Microorganisms and Preparation of Inocula

Aspergillus flavus (ATCC 16872) was cultivated on potato dextrose agar slants (PDA; Merck, Mexico City, Mexico) for 10 days at 25°C and the spores were harvested with 10 ml of 0.1% Tween 80 (Merck, Mexico City, Mexico) solution sterilized by membrane (0.45 μm) filtration. The spore suspension was adjusted with the same solution to give a final spore concentration of 10^6 spore/ml and used the same day.

2.2. Preparation of Agar Systems

PDA agar systems were prepared with the addition of commercial sucrose to produce a_w 0.99, sterilized for 15 min at 121°C, cooled and acidified with hydrochloric acid to attain pH 3.5. The amounts of sucrose and hydrochloric acid needed were previously determined. The sterile acidified agar was aseptically divided and the required amounts of thymol, carvacrol, vanillin, eugenol (0, 25, 50, 75, up to 1300 ppm), and/or potassium sorbate (0, 50, 100, 150 up to 1000 ppm) were added and mechanically incorporated aseptically. Test compounds were obtained from Sigma Chemical, Co., St. Louis, MO. Agar solutions were poured into sterile Petri dishes. The combinations tested are given in Table 1.

As examples of the checkerboard array employed to evaluate the effects of binary mixtures of antimicrobials, Tables 2 and 3 present the conditions and concentrations evaluated. In the case of ternary mixtures a general scheme proposed by Berenbaum et al. (1983) was used where the maximum concentrations (represented as 1) of each antimicrobial correspond to the MIC and the concentration of every agent in the mixture represents a fraction of the MIC (Table 4).

2.3. Inoculation and Incubation

Triplicate Petri dishes of each system were centrally inoculated by spotting 2 μl of the spore suspension ($\approx 2.0 \times 10^3$ spores/plate) to give

Table 1. Mixtures of phenolic and synthetic antimicrobial agents evaluated to inhibit *Aspergillus flavus* growth

Binary Mixtures	Ternary Mixtures
Vanillin -Eugenol	Thymol -Carvacrol -Potassium sorbate
Vanillin -Carvacrol	Thymol -Eugenol -Potassium sorbate
Vanillin -Thymol	Thymol -Vanillin -Potassium sorbate
Vanillin -Potassium sorbate	Carvacrol -Vanillin -Potassium sorbate
Eugenol -Carvacrol	Carvacrol -Eugenol -Potassium sorbate
Eugenol -Thymol	Eugenol -Vanillin -Potassium sorbate
Eugenol -Potassium sorbate	
Thymol -Potassium sorbate	

Table 2. Aspergillus flavus growth (G) or no growth (NG) response after one month of incubation in potato dextrose agar formulated with pH 3.5, a_w 0.99 and selected binary mixtures of vanillin and eugenol

	Eugenol (ppm)					
Vanillin (ppm)	100	200	300	400	500	600
100	G	G	G	G	NG	NG
200	G	G	G	NG	NG	NG
300	G	G	G	NG	NG	NG
400	G	G	G	NG	NG	NG
500	G	G	NG	NG	NG	NG
600	G	G	NG	NG	NG	NG
700	G	G	NG	NG	NG	NG
800	G	NG	NG	NG	NG	NG
900	NG	NG	NG	NG	NG	NG
1000	NG	NG	NG	NG	NG	NG
1100	NG	NG	NG	NG	NG	NG
1200	NG	NG	NG	NG	NG	NG

Table 3. Aspergillus flavus growth (G) or no growth (NG) response after one month of incubation in potato dextrose agar formulated with pH 3.5, a_w 0.99 and selected binary mixtures of carvacrol and thymol.

	Thymol (ppm)						
Carvacrol (ppm)	50	100	150	200	250	300	350
50	G	G	G	G	G	G	NG
100	G	G	G	G	NG	NG	NG
150	G	G	G	NG	NG	NG	NG
200	G	G	NG	NG	NG	NG	NG
250	G	NG	NG	NG	NG	NG	NG

Table 4. Experimental design utilized to evaluate ternary mixtures of antimicrobial agents[a]

Experiment	Agent A	Agent B	Agent C
1	1	0	0
2	0	1	0
3	0	0	1
4	0	1/2	1/2
5	1/2	0	1/2
6	1/2	1/2	0
7	1/3	1/3	1/3
8	1/6	1/6	1/6
9	1/6	1/6	2/3
10	2/3	1/6	1/6
11	1/6	2/3	1/6
12	1/12	1/12	1/3
13	1/3	1/12	1/12
14	1/12	1/3	1/12

[a]From Berenbaum et al., 1983

a circular inoculum of 1 mm diameter. Growth controls without antimicrobials were prepared and inoculated as above. Three plates of every system were maintained without inoculation for a_w and pH measurements. The inoculated plates and controls were incubated for 1 month at 25°C in hermetically closed plastic containers to avoid dehydration. Enough headspace was left in the containers to avoid anoxic conditions. Periodically, the inoculated plates were removed briefly to observe them and immediately re-incubated.

Water activity was measured with an AquaLab CX-2 instrument (Decagon Devices, Inc., Pullman, WA) calibrated and operated following the procedure described by López-Malo et al. (1994). pH was determined with a Beckman pH meter model 50 (Beckman Instruments, Inc., Fullerton, CA). Measurements were made in triplicate.

2.4. Calculation of the Mould Growth Responses

The inoculated systems were examined daily using a stereoscopic microscope (American Optical, model Forty), if no growth was observed the plates were re-incubated and if mould growth was detected plates were discarded. Inhibition was defined as no observable mould growth after one month of incubation.

Minimal inhibitory concentrations (MIC) for individual antimicrobials were defined as the minimal concentration of the compounds used required to inhibit mould growth. MIC data was transformed to fractional inhibitory concentration (FIC), as defined by Davidson and Parish (1989) for binary mixtures:

$$FIC_A = (MIC_{\text{compound A with compound B}}) / (MIC_{\text{compound A}}) \quad (1)$$

$$FIC_B = (MIC_{\text{compound B with compound A}}) / (MIC_{\text{compound B}}) \quad (2)$$

that can be extended to ternary mixtures (López-Malo et al., 2005b):

$$FIC_A = (MIC_{\text{compound A with compounds B and C}}) / (MIC_{\text{compound A}}) \quad (3)$$

$$FIC_B = (MIC_{\text{compound B with compounds A and C}}) / (MIC_{\text{compound B}}) \quad (4)$$

$$FIC_C = (MIC_{\text{compound C with compounds A and B}}) / (MIC_{\text{compound C}}) \quad (5)$$

FIC_{Index} was calculated with the FICs for individual antimicrobials as follows:

$$FIC_{Index} = FIC_A + FIC_B + FIC_C \quad (6)$$

3. RESULTS AND DISCUSSION

The pH and a_w of the PDA systems without inoculation determined at the beginning and at the end of incubation demonstrated that the desired values remained constant under the storage conditions. In control systems without antimicrobials, *A. flavus* grew at a_w 0.99 and pH 3.5.

Individual effects of the synthetic antimicrobial (potassium sorbate) and naturally occurring antimicrobials (thymol, carvacrol, vanillin, eugenol) on *A. flavus* growth response in PDA at a_w 0.99 and pH 3.5, represented as the minimal concentrations (MIC) that inhibit the mould growth for individual antimicrobials, are presented in Table 5. *A. flavus* exhibited higher sensitivity to thymol, eugenol, carvacrol and potassium sorbate than to vanillin. MICs varied from 300 ppm for carvacrol to 1300 ppm for vanillin.

3.1. Binary Mixtures

The combined effects of vanillin and eugenol and carvacrol and thymol resulted in the inhibitory conditions presented in Tables 2 and 3, respectively. In the same manner, we obtained growth/no growth responses for every binary mixture evaluated (data not shown). In many cases, combining antimicrobial agents resulted in no growth observations with lower inhibitory antimicrobial concentrations than when individually evaluated (Tables 2 and 3). These no growth results were transformed in fractional inhibitory concentrations to construct isobolograms and calculate each FIC_{Index}.

An isobologram may be thought of as an array of differing concentrations of two compounds, where one compound ranges from lowest to highest concentration on the x-axis and the other on the y-axis. All possible permutations of combined concentrations are reflected within the array. If those concentrations that inhibit the

Table 5. Minimal inhibitory concentrations[a] (MIC) of selected antimicrobials for *Aspergillus flavus* in potato dextrose agar formulated with a_w 0.99 and pH 3.5

Antimicrobial	MIC (ppm)
Vanillin	1300
Eugenol	600
Carvacrol	300
Thymol	400
Potassium sorbate	400

[a]Minimal concentration required for inhibiting mould growth for two months at 25°C

growth of the test organism fall on an approximately straight line that connects the individual MIC, or the FIC, on the x and y axes, the combined effect is additive. Deviation of linearity to the left or right of the additive line is interpreted as synergism or antagonism, respectively. As examples, FIC isobolograms are presented for combinations of vanillin and potassium sorbate (Figure 1), potassium sorbate and eugenol (Figure 2), and eugenol and carvacrol (Figure 3) that inhibited *A. flavus*. Depending on the antimicrobials used in the binary mixture, different isobolograms were obtained (Figures 1 to 3), indicating differences or similarities in the ability of the compounds to inhibit mould growth. Points representing no growth combinations do not always follow a well-defined pattern and in some cases the shape of the curve is concentration dependent. In other words, mould inhibition can be obtained with different combinations of two antimicrobials but the overall result (synergic, additive or antagonist) depends on the concentration of each antimicrobial in the mixture. This is also observed in Tables 6-10. Calculated FIC_{Index} for each binary mixture are also presented in Tables 6-10. In a similar way as isobolograms, a FIC_{Index} near 1 implicates additivity; FIC_{Index} < 1 implies synergy; and FIC_{Index} > 1 implies antagonism (López-Malo et al., 2005b).

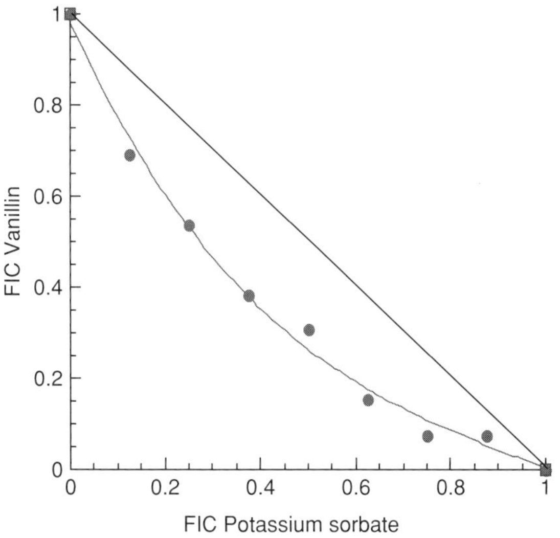

Figure 1. Fractional inhibitory concentration (FIC) isobologram for potassium sorbate and vanillin combinations to inhibit *Aspergillus flavus* in potato dextrose agar (a_w 0.99 and pH 3.5) after 30 days incubation at 25°C.

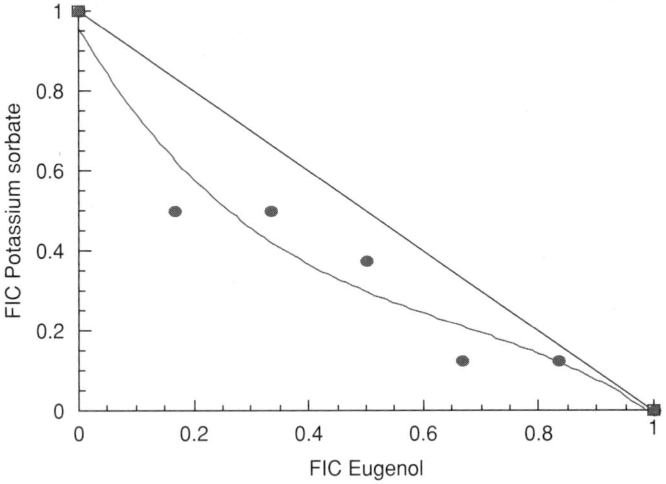

Figure 2. Fractional inhibitory concentration (FIC) isobologram for eugenol and potassium sorbate combinations to inhibit *Aspergillus flavus* in potato dextrose agar (a_w 0.99 and pH 3.5) after 30 days incubation at 25°C.

Several combinations of vanillin and potassium sorbate were synergistic (Figure 1, Table 6) as were some other binary mixtures (Tables 7-10). Others, such as combinations of thymol and carvacrol, demonstrated only additive effects (Table 10). In summary, the following antimicrobial

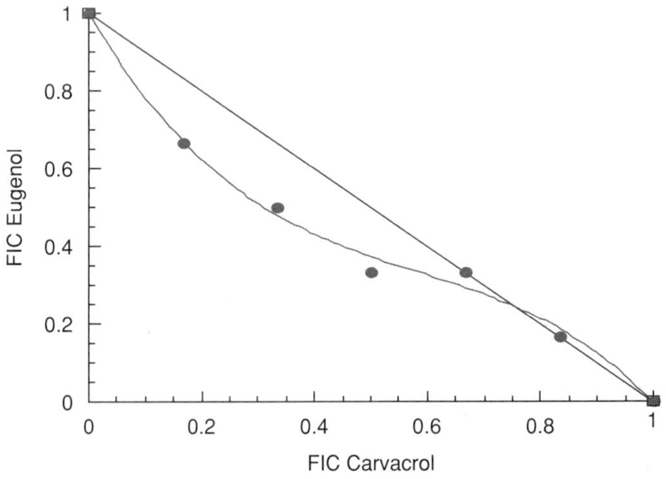

Figure 3. Fractional inhibitory concentration (FIC) isobologram for carvacrol and eugenol combinations to inhibit *Aspergillus flavus* in potato dextrose agar (a_w 0.99 and pH 3.5) after 30 days incubation at 25°C.

Table 6. Fractional inhibitory concentration index (FIC_{Index}) for *Aspergillus flavus* using binary mixtures of antimicrobials in potato dextrose agar formulated at a_w 0.99, pH 3.5.

Potassium sorbate (ppm)	Vanillin (ppm)	FIC_{Index}	Eugenol (ppm)	Vanillin (ppm)	FIC_{Index}
50	900	0.817	100	900	0.859
100	700	0.788	200	800	0.949
150	500	0.760	300	500	0.885
200	400	0.808	400	200	0.821
250	200	0.779	500	100	0.910
300	100	0.827			
350	100	0.952			

Table 7. Fractional inhibitory concentration index (FIC_{Index}) for *Aspergillus flavus* using binary mixtures of antimicrobials in potato dextrose agar formulated at a_w 0.99, pH 3.5.

Carvacrol (ppm)	Vanillin (ppm)	FIC_{Index}	Thymol (ppm)	Vanillin (ppm)	FIC_{Index}
50	800	0.782	50	900	0.817
100	700	0.872	100	800	0.865
150	500	0.885	150	600	0.837
200	500	1.051	200	500	0.885
250	200	0.987	250	200	0.779
			300	100	0.827

combinations were synergistic, in at least one combination, in inhibiting *A. flavus* for at least 30 days of incubation at a_w 0.99 and pH 3.5: vanillin-potassium sorbate (FIC_{Index} 0.76), vanillin-eugenol (FIC_{Index} 0.82), vanillin-thymol (FIC_{Index} 0.82), vanillin-carvacrol (FIC_{Index} 0.78); eugenol-carvacrol (FIC_{Index} 0.83), thymol-potassium sorbate (FIC_{Index} 0.75), eugenol-potassium sorbate (FIC_{Index} 0.67), and eugenol-thymol (FIC_{Index} 0.83). Antimicrobial combinations including potassium

Table 8. Fractional inhibitory concentration index (FIC_{Index}) for *Aspergillus flavus* using binary mixtures of antimicrobials in potato dextrose agar formulated at a_w 0.99, pH 3.5.

Eugenol (ppm)	Potassium sorbate (ppm)	FIC_{Index}	Carvacrol (ppm)	Potassium sorbate (ppm)	FIC_{Index}
100	200	0.667	50	250	0.792
200	200	0.833	100	200	0.833
300	150	0.875	150	200	1.000
400	100	0.917	200	150	1.042
500	50	0.958	250	100	1.083

Table 9. Fractional inhibitory concentration index (FIC_{Index}) for *Aspergillus flavus* using binary mixtures of antimicrobials in potato dextrose agar formulated at a_w 0.99, pH 3.5.

Thymol (ppm)	Potassium sorbate (ppm)	FIC_{Index}	Carvacrol (ppm)	Eugenol (ppm)	FIC_{Index}
50	250	0.750	50	400	0.833
100	200	0.750	100	300	0.833
150	200	0.875	150	200	0.833
200	200	1.000	200	200	1.000
250	150	1.000	250	100	1.000
300	100	1.000			
350	50	1.000			

sorbate and carvacrol or thymol were synergistic when small concentrations of phenolics were combined with > 150 ppm potassium sorbate.

Fungal inhibition can be achieved by combining spices (or their phenolic compounds) and traditional antimicrobials, reducing the concentrations needed to achieve the same effect than when using only one antimicrobial agent. In previous studies, Azzous and Bullerman (1982) reported that clove was an efficient antimycotic agent against *A. flavus, A. parasiticus* and *A. ochraceus* and four strains of *Penicillium*, delaying mould growth by more than 21 days. These authors also observed additive and synergic effects combining 0.1% clove with 0.1-0.3% potassium sorbate, delaying mould germination time. Sebti and Tantaoui-Elaraki (1994) reported that the combination of sorbic acid (0.75 g/kg) with an aqueous cinnamon extract (20 g/kg) inhibited growth of 151 mould and yeast strains isolated from a Moroccan bakery product. In contrast, when using

Table 10. Fractional inhibitory concentration index (FIC_{Index}) for *Aspergillus flavus* using binary mixtures of antimicrobials in potato dextrose agar formulated at a_w 0.99, pH 3.5.

Thymol (ppm)	Eugenol (ppm)	FIC_{Index}	Thymol (ppm)	Carvacrol (ppm)	FIC_{Index}
50	500	0.958	100	250	1.083
100	400	0.917	150	200	1.042
150	300	0.875	200	150	1.000
200	200	0.833	250	100	0.958
250	100	0.792	300	100	1.083
300	100	0.917	350	50	1.042
350	100	1.042			

only one antimicrobial agent to inhibit the studied microorganisms, 2000 ppm of sorbic acid was needed. Matamoros-León et al. (1999) evaluated individual and combined effects of potassium sorbate and vanillin concentrations on the growth of *Penicillium digitatum, P. glabrum* and *P. italicum* in PDA adjusted to a_w 0.98 and pH 3.5, and observed that 150 ppm potassium sorbate inhibited *P. digitatum* while 700 ppm were needed to inhibit *P. glabrum*. Using vanillin, inhibitory concentrations varied from 1100 ppm for *P. digitatum* and *P. italicum* to 1300 ppm for *P. glabrum*. When used in combination, minimal inhibitory concentration (MIC) isobolograms illustrated that curves deviated to the left of the additive line. Also, calculated FIC_{Index} values varied from 0.60 to 0.84. FIC_{Index} as well as isobolograms demonstrated synergistic effects on mould inhibition when vanillin and potassium sorbate were applied in combination (Matamoros-León et al., 1999).

3.2. Ternary Mixtures

Several of the tested combinations of two antimicrobials exhibited synergy in an experimental system. The question arises as to whether combinations of more than two agents might show even greater synergy. However it is not easy to answer this question (Berenbaum et al., 1983). An alternative approach to improving fungal inhibition could be to combine three antimicrobials agents, especially for resistant strains which cannot be inhibited with individual or binary mixtures of antimicrobials. As already mentioned, little is known about the interaction between antifungal agents against filamentous fungi. In order to determine the potential use of ternary combinations of antifungal agents to inhibit growth of *A. flavus* we decided to study the interactions among selected antimicrobials (Table 4).

Tables 11-16 present the growth/no growth results for the evaluated ternary mixtures, as well as FIC of each antimicrobial in the mixture and FIC_{Index}. The experimental design used, proposed by Berenbaum et al. (1983), has been used for antibiotics and is focused on determining synergistic mixtures and establishing if they are consistently synergistic, i.e. results are not dependent on concentration. Several experiments (mixtures) proposed in the design (Table 4) and evaluated (Tables 11-16) have by definition an FIC_{Index} equal to 1, and are included to corroborate individual and binary inhibitory effects, as well as ternary combinations in which the proportions of antimicrobials (fractions of MIC) add up to 1. Two thirds of the MIC of one agent and 1/6 of the MIC of the other two, or 1/3 of the MIC of each

Table 11. Fractional inhibitory concentration (FIC) and FIC_{Index} for *Aspergillus flavus* using ternary mixtures of potassium sorbate (KS), carvacrol and thymol in potato dextrose agar formulated at a_w 0.99, pH 3.5.

KS (ppm)	Carvacrol (ppm)	Thymol (ppm)	Growth Response	FIC KS	FIC Carvacrol	FIC Thymol	FIC_{Index}
400	0	0	NG	1.00	0.00	0.00	1.00
0	300	0	NG	0.00	1.00	0.00	1.00
0	0	400	NG	0.00	0.00	1.00	1.00
0	150	200	NG	0.00	0.50	0.50	1.00
200	0	200	NG	0.50	0.00	0.50	1.00
200	150	0	NG	0.50	0.50	0.00	1.00
132	99	132	NG	0.33	0.33	0.33	1.00
68	51	68	NG	0.17	0.17	0.17	0.50
68	51	268	NG	0.17	0.17	0.67	1.00
268	51	68	NG	0.67	0.17	0.17	1.00
68	201	68	NG	0.17	0.67	0.17	1.00
32	24	132	NG	0.08	0.08	0.33	0.50
132	24	32	G				
32	99	32	NG	0.08	0.33	0.08	0.50

agent are examples of these ternary mixtures. The rest of the experiments are used to test synergy by combining 1/6 MIC of each agent or 1/3 MIC of one antimicrobial with 1/12 MIC of the other two. Berenbaum et al. (1983) indicated that if no growth is obtained in every combination tested, the ternary mixture is synergic in a consistent way.

Table 12. Fractional inhibitory concentration (FIC) and FIC_{Index} for *Aspergillus flavus* using ternary mixtures of potassium sorbate (KS), eugenol and thymol in potato dextrose agar formulated at a_w 0.99, pH 3.5.

KS (ppm)	Eugenol (ppm)	Thymol (ppm)	Growth Response	FIC KS	FIC Eugenol	FIC Thymol	FIC_{Index}
400	0	0	NG	1.00	0.00	0.00	1.00
0	600	0	NG	0.00	1.00	0.00	1.00
0	0	400	NG	0.00	0.00	1.00	1.00
0	300	200	NG	0.00	0.50	0.50	1.00
200	0	200	NG	0.50	0.00	0.50	1.00
200	300	0	NG	0.50	0.50	0.00	1.00
132	198	132	NG	0.33	0.33	0.33	1.00
68	102	68	NG	0.17	0.17	0.17	0.50
68	102	268	NG	0.17	0.17	0.67	1.00
268	102	68	NG	0.67	0.17	0.17	1.00
68	402	68	NG	0.17	0.67	0.17	1.00
32	48	132	NG	0.08	0.08	0.33	0.50
132	48	32	G				
32	198	32	NG	0.08	0.33	0.08	0.50

Table 13. Fractional inhibitory concentration (FIC) and FIC_{Index} for *Aspergillus flavus* using ternary mixtures of potassium sorbate (KS), vanillin and thymol in potato dextrose agar formulated at a_w 0.99, pH 3.5.

KS (ppm)	Vanillin (ppm)	Thymol (ppm)	Growth Response	FIC KS	FIC Vanillin	FIC Thymol	FIC_{Index}
400	0	0	NG	1.00	0.00	0.00	1.00
0	1300	0	NG	0.00	1.00	0.00	1.00
0	0	400	NG	0.00	0.00	1.00	1.00
0	650	200	NG	0.00	0.50	0.50	1.00
200	0	200	NG	0.50	0.00	0.50	1.00
200	650	0	NG	0.50	0.50	0.00	1.00
132	429	132	NG	0.33	0.33	0.33	1.00
68	221	68	G				
68	221	268	G				
268	221	68	NG	0.67	0.17	0.17	1.00
68	871	68	NG	0.17	0.67	0.17	1.00
32	104	132	G				
132	104	32	G				
32	429	32	G				

Therefore, above the lowest concentration of every antimicrobial tested in the mixture, the combination will be synergistic.

In ternary mixtures including potassium sorbate-thymol-carvacrol (Table 11) and potassium sorbate-thymol-eugenol (Table 12), mould growth was observed only when 1/3 MIC of potassium sorbate was

Table 14. Fractional inhibitory concentration (FIC) and FIC_{Index} for *Aspergillus flavus* using ternary mixtures of potassium sorbate (KS), eugenol and carvacrol in potato dextrose agar formulated at a_w 0.99, pH 3.5.

KS (ppm)	Eugenol (ppm)	Carvacrol (ppm)	Growth Response	FIC KS	FIC Eugenol	FIC Carvacrol	FIC_{Index}
400	0	0	NG	1.00	0.00	0.00	1.00
0	600	0	NG	0.00	1.00	0.00	1.00
0	0	300	NG	0.00	0.00	1.00	1.00
0	300	150	NG	0.00	0.50	0.50	1.00
200	0	150	NG	0.50	0.00	0.50	1.00
200	300	0	NG	0.50	0.50	0.00	1.00
132	198	99	NG	0.33	0.33	0.33	1.00
68	102	51	NG	0.17	0.17	0.17	0.50
68	102	201	NG	0.17	0.17	0.67	1.00
268	102	51	NG	0.67	0.17	0.17	1.00
68	402	51	NG	0.17	0.67	0.17	1.00
32	48	99	G				
132	48	24	G				
32	198	24	NG	0.08	0.33	0.08	0.50

Table 15. Fractional inhibitory concentration (FIC) and FIC_{Index} for *Aspergillus flavus* using ternary mixtures of potassium sorbate (KS), vanillin and carvacrol in potato dextrose agar formulated at a_w 0.99, pH 3.5.

KS (ppm)	Vanillin (ppm)	Carvacrol (ppm)	Growth Response	FIC KS	FIC Vanillin	FIC Carvacrol	FIC_{Index}
400	0	0	NG	1.00	0.00	0.00	1.00
0	1300	0	NG	0.00	1.00	0.00	1.00
0	0	300	NG	0.00	0.00	1.00	1.00
0	650	150	NG	0.00	0.50	0.50	1.00
200	0	150	NG	0.50	0.00	0.50	1.00
200	650	0	NG	0.50	0.50	0.00	1.00
132	429	99	G				
68	221	51	G				
68	221	201	NG	0.17	0.17	0.67	1.00
268	221	51	G				
68	871	51	NG	0.17	0.67	0.17	1.00
32	104	99	G				
132	104	24	G				
32	429	24	G				

combined with 1/12 MIC of thymol and 1/12 MIC of carvacrol or eugenol. In both ternary mixtures when growth was observed the phenolic compounds represent the lowest MIC fraction tested (1/12). Combinations that result in synergism (FIC = 0.5) include at least one phenolic in a fraction higher than 1/12 MIC, Therefore, we can

Table 16. Fractional inhibitory concentration (FIC) and FIC_{Index} for *Aspergillus flavus* using ternary mixtures of potassium sorbate (KS), vanillin and eugenol in potato dextrose agar formulated at a_w 0.99, pH 3.5.

KS (ppm)	Vanillin (ppm)	Eugenol (ppm)	Growth Response	FIC KS	FIC Vanillin	FIC Eugenol	FIC_{Index}
400	0	0	NG	1.00	0.00	0.00	1.00
0	1300	0	NG	0.00	1.00	0.00	1.00
0	0	600	NG	0.00	0.00	1.00	1.00
0	650	300	NG	0.00	0.50	0.50	1.00
200	0	300	NG	0.50	0.00	0.50	1.00
200	650	0	NG	0.50	0.50	0.00	1.00
132	429	198	NG	0.33	0.33	0.33	1.00
68	221	102	G				
68	221	402	G				
268	221	102	G				
68	871	102	NG	0.17	0.67	0.17	1.00
32	104	198	G				
132	104	48	G				
32	429	48	G				

conclude that synergistic antifungal combinations of these agents must include: 1/12 MIC < phenolic concentrations < 1/6 MIC. Comparing binary and ternary mixture results, it can be observed that a considerable reduction of potassium sorbate concentration from 150-250 ppm to 32-68 ppm is possible. Maintaining a potassium sorbate concentration near 100 ppm, allows a considerable reduction in phenolic antimicrobial concentrations in the inhibitory ternary mixtures. Figure 4 presents a representation of FIC of potassium sorbate-thymol-eugenol inhibitory combinations that inhibit *A. flavus*, this 3-D isobologram as well as FIC_{Index} illustrate the synergistic combinations.

Ternary mixtures where vanillin is included (Tables 13, 15 and 16) have in common that growth was observed in those combinations where synergism was expected. However, some ternary combinations presented an additive result ($FIC_{Index} = 1$). Observing binary results of those mixtures that include vanillin, synergism was anticipated in ternary mixtures but the results cannot be predicted, as can be seen in Tables 13, 15 and 16. Monzón et al. (2001) reported that individual or binary antimicrobial agents that exhibit synergistic results do not necessary generate similar outcomes in ternary antimicrobial mixtures.

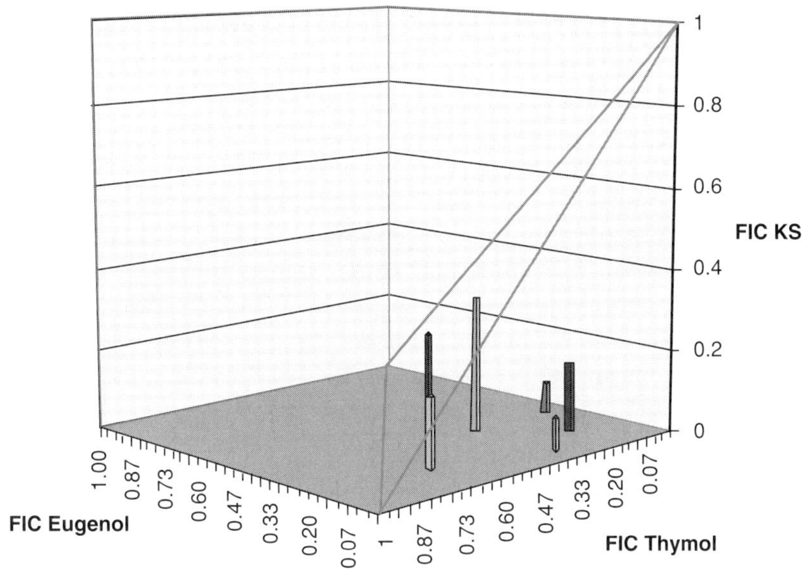

Figure 4. Fractional inhibitory concentration (FIC) isobologram for potassium sorbate (KS), eugenol and thymol combinations to inhibit *Aspergillus flavus* in potato dextrose agar (a_w 0.99 and pH 3.5) after 30 days incubation at 25°C.

4. CONCLUSIONS

Combinations of antimicrobials can be selected when identified microorganisms are resistant to inhibition and/or inactivation by legal levels of single, conventional antimicrobials. The combination may exert the desired antimicrobial activity. Also, and more frequently, antimicrobial combinations can be selected to provide broad-spectrum preservation (Alzamora et al., 2003). Relatively little work has been reported to date on the combined action of mixtures of conventional and natural antimicrobials against microorganisms, even in model systems. Moreover, little attention has been given to the study of the mechanisms underlying their toxicity and modes of resistance, particularly for microorganisms of concern in fruit products, among them moulds and yeasts. This lack of understanding of the relative contribution of factors to obtain safe, high quality foods is surprising, because the necessity for an adequate database on which to develop safe multifactorial preservation systems was pointed out some time ago (Roberts, 1989).

Further work to combine natural antimicrobials with conventional ones in foods is still required. Several combinations could be exploited for mild food preservation techniques in the near future. However, their mode of action in model systems and in food matrices is still not well understood and represents a barrier to their application. Results of the study reported here, and various others in the literature, raise certain questions about the use of antimicrobial mixtures (Alzamora and López-Malo, 2002; Alzamora et al., 2003). Conversely, a better understanding of microbial ecology and the physiological response of microorganisms to individual preservation factors as well as to combinations of natural and conventional antimicrobials in different food environments will offer new opportunities and provide greater precision for a rational selection of antimicrobial combinations. The answer to these points can be established by appropriate scientific and experimental inquiry and it is part of the challenge for the future of antifungal combinations.

5. ACKNOWLEDGMENTS

We acknowledge financial support from CONACyT -Mexico (Projects 32020-B, 33405-B and 44088), Universidad de las Américas-Puebla, and CYTED Program (Project XI.15).

6. REFERENCES

Al-Khayat, M. A., and Blank, G., 1985, Phenolic spice components sporostatic to *Bacillus subtilis*, *J. Food Sci.* **50**:971-974, 980.

Alzamora, S. M., López-Malo, A., Guerrero, S., and Palou, E., 2003, Plant antimicrobials combined with conventional preservatives for fruit products, in: *Natural Antimicrobials for the Minimal Processing of Foods*, S. Roller, ed., Woodhead Publishing, Ltd., London, pp. 235-249.

Alzamora, S. M., and López-Malo, A., 2002, Microbial behavior modeling as a tool in the design and control of minimally processed foods, in: *Engineering and Food for the 21st Century*, J. Welti-Chanes, G. V. Barbosa-Canovas, and J. M. Aguilera, ed., CRC Press, Boca Raton, FL, pp. 631-650.

Azzous, M. A., and Bullerman, L. B., 1982, Comparative antimycotic effects of selected herbs, spices, plant components and commercial antifungal agents, *J. Food Prot.* **45**:1298-1301.

Barry, A. L., 1976, *The Antimicrobic Susceptibility Test: Principles and Practices*, Lea & Febiger, Philadelphia, PA.

Berenbaum, M. C, Yu, V. L., and Felegie, T. P., 1983, Synergy with double and triple antibiotic combinations compared, *J. Antimicrob. Chemother.* **12**:555-563.

Beuchat, L. R., 1994, Antimicrobial properties of spices and their essential oils, in: *Natural Antimicrobial Systems and Food Preservation*, V. M. Dillon, R. G. Board, ed., CAB International, Wallingford, England, pp. 167-179.

Beuchat, L. R., and Golden, D. A., 1989, Antimicrobials occurring naturally in foods, *Food Technol.* **43**(1):134-142.

Beuchat, L. R., 1976, Sensitivity of *Vibrio parahaemolyticus* to spices and organic acids, *J. Food Sci.* **41**:899-902.

Board, R. G., 1995, Natural antimicrobials from animals, in: *New Methods of Food Preservation*, G. W. Gould, ed., Blackie Academic & Professional, Glasgow, UK, pp. 40-57.

Branen, A. L., Davidson, P. M., and Katz, B., 1980, Antimicrobial properties of phenolic antioxidants and lipids, *Food Technol.* **34**(5):42-53.

Buchanan, R. L., and Shepherd, A. J., 1981, Inhibition of *Aspergillus parasiticus* by thymol, *J. Food Sci.* **46**:976-977.

Bullerman, L. B., Lieu, F. Y., and Seier, S. A., 1977, Inhibition of growth and aflatoxin production by cinnamon and clove oils. Cinnamic aldehyde and eugenol, *J. Food Sci.* **42**:1107-1109, 1116.

Bullerman, L. B., 1974, Inhibition of aflatoxin production by cinnamon, *J. Food Sci.* **39**:1163-1167.

Busta, F. F., and Foegeding, P. M., 1983, Chemical food preservatives, in: *Disinfection, Sterilization and Preservation,* 3rd edition, S. S. Block, ed., Lea & Febiger, Philadelphia, PA, p. 656.

Cerrutti, P., and Alzamora, S. M., 1996, Inhibitory effects of vanillin on some food spoilage yeasts in laboratory media and fruit purées, *Int. J. Food Microbiol.* **29**:379-386.

Cerrutti, P., Alzamora, S. M., and Vidales, S. L., 1997, Vanillin as an antimicrobial for producing shelf-stable strawberry purée, *J. Food Sci.* **62**:608-610.

Chikindas, M. L., and Montville, T. J., 2002, Perspectives for application of bacteriocins as food preservatives, in: *Control of Foodborne Microorganisms*, V. K. Juneja, and J. N. Sofos, eds, Marcel Dekker, New York, pp. 303-322.

Conner, D. E., 1993, Naturally occurring compounds, in: *Antimicrobials in Foods*, 2nd edition, P. M. Davidson, A. L. Branen, eds, Marcel Dekker, New York, pp. 441-467.

Conner, D. E., and Beuchat, L. R., 1984a, Effects of essential oils from plants on growth of food spoilage yeasts, *J. Food Sci.***49**:429-434.

Conner, D. E., and Beuchat, L. R., 1984b, Sensitivity of heat-stressed yeasts to essential oils of plants, *Appl. Environ. Microbiol.* **47**:229-233.

Davidson, P. M., 2001, Chemical preservatives and naturally antimicrobial compounds, in: *Food Microbiology. Fundamentals and Frontiers*, 2nd edition, M. P. Doyle, L. R. Beuchat, and T. J. Montville, eds, ASM Press, Washington, D.C., pp. 593-628.

Davidson, P. M., and Branen, A. L., 1993, *Antimicrobials in Foods*, Marcel Dekker, New York.

Davidson, P. M., and Parish, M. E., 1989, Methods for testing the efficacy of food antimicrobials, *Food Technol.* **43**(1):148-155.

Eliopoulos, G. M., and Moellering, R. C., 1991, Antimicrobial combinations, in: *Antibiotics in Laboratory Medicine*, 3rd edition, V. Lorian, ed., Williams & Wilkins, Baltimore, MD, p. 432.

Farag, R. S., Daw, Z. Y., Hewedi, F. M., and El-Baroty, G. S. A., 1989, Antimicrobial activity of some Egyptian spice essential oils, *J. Food Prot.* **52**:665-667.

Farrell, K. T., 1990, *Spices, Condiments and Seasonings*, 2nd edition, Van Nostrand Reinhold, New York.

Fung, D. Y. C., Taylor, S., and Kahan, J., 1977, Effects of butylated hydroxyanisole (BHA) and butylated hydroxitoluene (BHT) on growth and aflatoxin production of *Aspergillus flavus*, *J. Food Safety* **1**:39-51.

Gardner, J. R., 1977, Principles of antimicrobial activity, in: *Disinfection, Sterilization and Preservation*, 2nd edition, S. S. Block, ed., Lea & Febiger, Philadelphia, PA.

Garrett, E. R., 1958, Classification and evaluation of combined antibiotic activity, *Antibiot. Chemother.* **8**:8.

Gould, G. W., 2002. Control with naturally occurring antimicrobial systems including bacteriolytic enzymes, in: *Control of Foodborne Microorganisms*, V. K. Juneja, and J. N. Sofos, eds, Marcel Dekker, New York, pp. 281-302.

Gould, G. W., 1995, Overview, in: *New Methods of Food Preservation*, G. W. Gould, ed., Blackie Academic & Professional, Glasgow, UK, pp. xv-xix.

Hill, C., 1995, Bacteriocins: natural antimicrobials from microorganisms, in: *New Methods of Food Preservation*, G. W. Gould, ed., Blackie Academic & Professional, Glasgow, UK, pp. 22-39.

Hitokoto, H., Morozumi, S., Wauke, T., and Sakai, S., 1978, Inhibitory effects of spices on growth and toxin production of toxigenic fungi, *Appl. Environ. Microbiol.* **39**:818-822.

Juven, B. J., Kanner, J., Schved, F., and Weisslowicz, H., 1994, Factors that interact with the antibacterial action of thyme essential oil and its active constituents, *J. Appl. Bacteriol.* **76**:626-631.

Kabara, J. J., and Eklund, F., 1991, Organic acids and esters, in: *Food Preservatives*, N. J. Russel, and G. W. Gould, eds, Blackie Academic & Professional, Glasgow, UK, pp. 44-71.

Katayama, T., and Nagai, I., 1960, Chemical significance of the volatile components of spices in the food preservative view point. VI. Structure and antibacterial activity of terpenes, *Bull. Japan Soc. Sci. Fisheries* **26**:29-32.

Krogstad, D. J., and Moellering, R. C., 1986, Combinations of antibiotics, mechanisms of interaction against bacteria, in: *Antibiotics in Laboratory Medicine*, 2nd edition, V. Lorian, ed., Williams & Wilkins, Baltimore, MD, p. 537.

Larson, E., 1984, *Clinical and Infection Control*, Blackwell Scientific, Boston, MA.

López-Malo, A., Alzamora, S. M., and Argaiz, A., 1998, Vanillin and pH synergistic effects on mold growth, *J. Food Sci.* **63**:143-146.

López-Malo, A., Alzamora, S. M., and Argaiz, A., 1997, Effect of vanillin concentration, pH and incubation temperature on *Aspergillus flavus*, *A. niger*, *A. ochraceus* and *A. parasiticus* growth, *Food Microbiol.* **14**:117-124.

López-Malo, A., Alzamora, S. M., and Argaiz, A., 1995, Effect of natural vanillin on germination time and radial growth of moulds in fruit-based agar systems, *Food Microbiol.* **12**:213-219.

López-Malo, A., Alzamora, S. M., and Guerrero, S., 2000, Natural antimicrobials from plants, in: *Minimally Processed Fruits and Vegetables. Fundamentals Aspects and Applications*, S. M. Alzamora, M. S. Tapia, and A. López-Malo, ed., Aspen Publishers, Gaithersburg, MD, p. 237.

López-Malo, A., Alzamora, S. M., and Palou, E., 2005a, Naturally occurring compounds – Plant sources, in: *Antimicrobials in Food*, 3rd edition, P. M. Davidson, J. N. Sofos, and A. L. Branen, eds, CRC Press, New York, pp. 429-451.

López-Malo, A., Palou, E., and Argaiz, A., 1994, Measurement of water activity of saturated salt solutions at various temperatures, in: *Proceedings of the Poster Session: International Symposium on the Properties of Water, Practicum II*, A. Argaiz, A. López-Malo, E. Palou, and P. Corte, eds, Universidad de las Américas-Puebla, Mexico, pp. 113-116.

López-Malo, A., Palou, E., Parish, M. E., and Davidson, P. M., 2005b, Methods for activity assay and evaluation of results, in: *Antimicrobials in Food*, 3rd edition, P. M. Davidson, J. N. Sofos, and A. L. Branen, eds, CRC Press, New York, pp 659-680.

Matamoros-León, B., Argaiz, A., and López-Malo, A., 1999, Individual and combined effects of vanillin and potassium sorbate on *Penicillium digitatum*, *P. glabrum* and *P. italicum* growth, *J. Food Prot.* **62**:540-542.

Monzón, M., Oteiza, C., Leiva, J., and Amorena, B., 2001, Synergy of different antibiotic combinations in biofilms of *Staphylococcus epidermis*, *J. Antimicrob. Chemother.* **48**:793-801.

NCCLS, 1999, *Methods for Determining Bactericidal Activity of Antimicrobial Agents; Approved Guideline*, National Committee for Clinical Laboratory Standards, Wayne, PA.

NCCLS, 2002, *Methods for Dilution Antimicrobial Susceptibility Tests for Bacteria That Grow Aerobically; Approved Standard*, 6th edition, National Committee for Clinical Laboratory Standards, Wayne, PA.

Nychas, G. J. E., 1995, Natural antimicrobials from plants, in: *New Methods of Food Preservation*, G. W. Gould, ed., Blackie Academic & Professional, Glasgow, UK, pp. 58-89.

Parish, M. E., and Carroll, D. E., 1988, Effects of combined antimicrobial agents on fermentation initiation by *Saccharomyces cerevisae* in a model broth system, *J. Food Sci.* **53**:240.

Paster, N., Juven, B. J., Shaaya, E., Menasherov, M., 1990, Inhibitory effect of oregano and thyme essential oils on moulds and foodborne bacteria, *Lett. Appl. Microbiol.* **11**:33-37.

Prindle, R. F., and Wright, E. S., 1977, Phenolic compounds, in: *Disinfection, Sterilization and Preservation*, 2nd edition, S. S. Block, ed., Lea & Febiger, Philadelphia, PA.

Rehm, H., 1959, Untersuchung zur Sirkung von Konservierungsmittelkombinationen, *Z. Lebesm. Untersuch. Forsch.* **110**:356.

Roberts, T. A., 1989, Combinations of antimicrobials and processing methods, *Food Technol.* **42**:156-163.

Sebti F., and Tantaoui-Elaraki, A., 1994, In vitro inhibition of fungi isolated from "Pastilla" papers by organic acids and cinnamon, *Lebensm.-Wiss. u-Technol.* **27**:370-74.

Shelef, L.A., 1983, Antimicrobial effects of spices, *J. Food Safety* **6**:29-44.

Smid, E. J., and Gorris, L. G. M., 1999, Natural antimicrobials for food preservation, in: *Handbook of Food Preservation*, M. S. Rahman, ed., Marcel Dekker, New York, pp. 285-308.

Sofos, J. N., Beuchat, L. R., Davidson, P. M., and Johnson, E. A., 1998, *Naturally Occurring Antimicrobials in Food*, Council for Agricultural Science and Technology, Task Force Report No 132.

Squires, E., and Cleeland, R., 1985, *Methods of Testing Combinations of Antimicrobial Agents*, Hoffman-LaRoche, Nutley, NJ.

Wilkins, K. M., and Board, R. G., 1989, Natural antimicrobial systems, in: *Mechanisms of Action of Food Preservation Procedures*, G. W. Gould, ed., Elsevier, New York, pp. 285-362.

Zaika, L. L., 1988, Spices and herbs: their antimicrobial activity and its determination, *J. Food Safety* **6**:29-44.

PROBABILISTIC MODELLING OF *ASPERGILLUS* GROWTH

*Enrique Palou and Aurelio López-Malo**

1. INTRODUCTION

Filamentous fungi are of concern to the food industry as potential spoilage micro-organisms (Pitt, 1989; Samson, 1989) and mycotoxin producers (Smith and Moss, 1985). It is important to understand the growth kinetics of these fungi in the food context, in order to control product quality from formulation to storage. This is especially applicable to long shelf life products, but also in those food products where formulation ingredients could be source of fungal contamination which may cause spoilage during processing or storage. In low water activity foods, mould spoilage is controlled by controlling a_w, either by drying or the addition of solutes (NaCl, sucrose, glucose or fructose). However mould growth can also occur at a_w values, especially when the preservation factors inhibit bacteria and allow fungal growth (Gould, 1989; Pitt and Miscamble, 1995). Mould growth in these products depends on the pH, a_w and antimicrobial agents, which are directly a function of the product formulation (Rosso and Robinson, 2001), the solutes used (ICMSF, 1980; Pitt and Hocking, 1977), and the storage temperature. Several models describing the effect of a_w or solute concentration on the growth of moulds have been published (Gibson et al., 1994; Cuppers et al., 1997; Valik et al., 1999; Rosso and Robinson, 2001).

*Ingeniería Química y Alimentos, Universidad de las Américas, Puebla. Cholula 72820, Mexico. Correspondence to epalou@mail.udlap.mx

Predictive modelling as defined by the US Advisory Committee on Microbiological Criteria for Foods is the use of mathematical expressions to describe the likely behaviour of biological agents. Mathematical modelling of microbial growth or decline (also called "predictive microbiology") is receiving a great deal of attention because of its enormous potential within the food industry. Predictive modelling provides a fast and relatively inexpensive way to get reliable first estimates on microbial growth and survival (McMeekin et al. 1993). Predictive microbiology is gaining importance as a powerful tool in food microbiology. Predictive modelling can be used for describing behaviour of microorganisms under different conditions, as well as assisting process design and optimization for production and distribution chains, based on microbial safety and shelf-life (Alavi et al., 1999; Alzamora and López-Malo, 2002). The main driving forces for the impressive progress in microbial modelling have been: the advent of reasonably priced microcomputers that has facilitated multifactorial data analysis, and the great improvement in techniques to establish mathematical models in the area of predictive microbiology.

Predictive microbiology involves the use of mathematical expressions to describe microbial behaviour. These include functions that relate microbial density to time, and growth rate to environmental conditions such as temperature, pH, a_w, and presence of antimicrobial agents. Predictive models in food microbiology can be divided, according to their aim, into two main categories: kinetic models and probability models. Kinetic models that predict growth of foodborne microorganisms are effective under a wide range of conditions; however, they are less useful close to the boundary between growth and no growth. Probabilistic models are useful where the objective is to determine whether or not microbial growth can occur under specific conditions. Much of the effort spent on generating predictive microbiology databases has focused on kinetic data, in which growth rates of microorganisms are determined in the normal temperature range and in combination with a_w, pH and nutrient levels that do not prevent growth of the modelled organism (Salter et al., 2000). This strategy is adequate when the desired information is the extent of growth of food spoilage organisms, or of pathogens for which some tolerance of growth is acceptable. However, in many situations it is important to ensure that microorganisms do not contaminate foods (Ross and McMeekin, 1994; Tienungoon et al., 2000).

The goal of analysis using any statistical model-building technique is to find the best suited and most parsimonious, yet biologically reasonable, model to describe the relationship between a dependent variable and a set of independent variables. However, while kinetic models make possible the calculation of the food shelf-life or the prediction of the time span in which significant microbial growth might occur, the probabilistic models focus their attention towards deciding whether a microorganism might or might not grow. Consequently, probability modelling is particularly useful when pathogenic or mycotoxin-producing species are involved. In this case, the growth rate of a microorganism is of lesser importance than the fact that it is present, and potentially able to multiply up to infectious dose or toxic levels. Ratkowsky and Ross (1995) proposed the application in food microbiology of the logistic regression model, which enables modelling of the boundary between growth and no growth for selected microbial species when one or more growth controlling factors are used. This approach was subsequently used by Presser et al. (1998), Bolton and Frank (1999), López-Malo et al. (2000), López-Malo and Palou (2000a, b), McMeekin et al. (2000), Lanciotti et al. (2001), and Palou and López-Malo (2003). In recent years, there has been a continuing interest in the development of predictive microbiology models describing microbial responses in food. The benefits of their application in the food industry could be substantial and various, such as prediction of shelf life, or as an aid to the elaboration of minimally processed foods (Alzamora and López-Malo, 2002). Several researchers have indicated that a need exists for predictive models with advantageous mathematical characteristics such as parsimony, robustness and stability (McMeekin et al., 1992; McMeekin et al., 1993; Ratkowsky, 1993; Whiting and Call, 1993; Baranyi and Roberts, 1994; Massana and Baranyi, 2000a). These properties would decrease the error of predictions and would increase the confidence in using predictive models. Multivariate polynomials are commonly used in predictive microbiology to summarise experimental results on the effect of environmental conditions on fungal growth (López-Malo and Palou, 2000b). They allow the use of linear regression for curve fitting procedures, which results in ease of computation and well established statistical analyses.

The growth of microorganisms in food can be fully described only by a combination of the two kinds of models, kinetic and probabilistic. An integrated description of the microbial response could be given by first establishing the likelihood of growth through a probability model (growth/no growth boundary model), and then predicting the growth parameters; specific rate and lag time, if growth is expected (López-Malo and Palou, 2000a, b; Massana and Baranyi, 2000b;

Palou and López-Malo, 2003). Boundary models will then help to define the range of applicability of kinetic models and may also be important for establishing food safety regulations as highlighted by Schaffner and Labuza (1997). Boundary models can predict the most suitable combinations of factors to prevent microbial growth, thus giving a significant degree of quality and safety from spoilage or food borne disease. This was also the aim of the hurdle approach proposed by Leistner (1985). Despite their potential importance, until now there have been only a few attempts to model the growth/no growth boundary for vegetative microorganisms.

In this study, selected experimental designs, i.e. central composite or factorials, and the combined effects of incubation temperature, a_w, pH and concentration of antimicrobial agent (vanillin or sodium benzoate) were incorporated into laboratory media to evaluate the growth/no growth response of three important mycotoxigenic *Aspergillus* species, namely *Aspergillus flavus*, *A. ochraceus* and *A. parasiticus*.

2. MATERIALS AND METHODS

2.1. Microorganisms and Preparation of Inocula

Aspergillus flavus ATCC 16872, *A. ochraceus* ATCC 22947 and *A. parasiticus* ATCC 26691 were cultivated on potato dextrose agar (PDA; Merck, Mexico) slants for 10 days at 25°C and the spores harvested with 10 ml of 0.1% Tween 80 (Merck, Mexico) solution sterilized by membrane (0.45 μm) filtration. Spore suspensions were adjusted with the same solution to give a final spore concentration of 10^6 spores/ml and were used the same day. Depending on the experimental design, a cocktail of these three species (*A. flavus*, *A. parasiticus* and *A. ochraceus*) or *A. flavus* alone were used.

2.2. Experimental Designs

A three level central composite design (Montgomery, 1984) was employed in a first study to assess the effects of pH, incubation temperature and vanillin concentration on mould growth response at 0.98 a_w. Independent variable levels are presented in Table 1. The results of this first set of experiments were used to select levels, including a_w, of each variable in a range where fungi might or might not grow. A facto-

Table 1. Central composite design utilized to evaluate growth response of a cocktail of conidia of *Aspergillus flavus*, *A. parasiticus* and *A. ochraceus* in laboratory media formulated with a_w 0.98, selected pH, vanillin concentration and incubated at different temperatures.

pH	Incubation Temperature (°C)	Vanillin Concentration (ppm)
3.0	10.0	0
4.0	10.0	0
3.0	25.0	0
4.0	25.0	0
2.7	17.5	0
4.3	17.5	0
3.5	4.9	0
3.5	30.1	0
3.5	17.5	0
3.0	10.0	500
4.0	10.0	500
3.0	25.0	500
4.0	25.0	500
3.0	10.0	1000
4.0	10.0	1000
3.0	25.0	1000
4.0	25.0	1000
2.7	17.5	750
4.3	17.5	750
3.5	4.9	750
3.5	30.1	750
3.5	17.5	330
3.5	17.5	1170
3.5	17.5	750

rial design was employed to assess the effects of a_w, pH, antimicrobial concentration, antimicrobial type (sodium benzoate or vanillin) and incubation temperature, on *Aspergillus flavus* growth response, the levels of every variable are presented in Table 2. Triplicate systems were prepared with the resulting variable combinations (Tables 1 and 2).

2.3. Laboratory Media

Following the experimental designs, PDA systems were prepared with commercial sucrose to reach a_w 0.98, 0.96 or 0.94, sterilized for 15 min at 121°C, cooled and acidified with hydrochloric acid to the desired pH. The amounts of sucrose and hydrochloric acid needed in every case had been previously determined. The sterilized and acidified agar

Table 2. Factorial design utilized to evaluate the growth response of *Aspergillus flavus* in laboratory media formulated with selected water activity, pH, sodium benzoate or vanillin concentration and incubated at different temperatures.

0.98 a_w			0.96 a_w			0.94 a_w		
pH	Incubation Temp (°C)	Conc[a] (ppm)	pH	Incubation Temp (°C)	Conc (ppm)	pH	Incubation Temp (°C)	Conc (ppm)
3	15	0	3	15	0	3	15	0
4	15	0	4	15	0	4	15	0
5	15	0	5	15	0	5	15	0
3	25	0	3	25	0	3	25	0
4	25	0	4	25	0	4	25	0
5	25	0	5	25	0	5	25	0
3	15	100	3	15	100	3	15	100
4	15	100	4	15	100	4	15	100
5	15	100	5	15	100	5	15	100
3	25	100	3	25	100	3	25	100
4	25	100	4	25	100	4	25	100
5	25	100	5	25	100	5	25	100
3	15	200	3	15	200	3	15	200
4	15	200	4	15	200	4	15	200
5	15	200	5	15	200	5	15	200
3	25	200	3	25	200	3	25	200
4	25	200	4	25	200	4	25	200
5	25	200	5	25	200	5	25	200
.	.	.	.	.	.	.	.	.
.	.	.	.	.	.	.	.	.
.	.	.	.	.	.	.	.	.
3	15	1000	3	15	1000	3	15	1000
4	15	1000	4	15	1000	4	15	1000
5	15	1000	5	15	1000	5	15	1000
3	25	1000	3	25	1000	3	25	1000
4	25	1000	4	25	1000	4	25	1000
5	25	1000	5	25	1000	5	25	1000

[a] Concentration of sodium benzoate or vanillin

solutions were aseptically divided and depending on the experimental design, the necessary amount of vanillin or sodium benzoate (Sigma Chemical, Co., St. Louis, MO, was added and mechanically incorporated under sterile conditions, then poured into sterile Petri dishes.

2.4. Inoculation and Incubation

Triplicate Petri dishes of every system were centrally inoculated by pouring 2 μl of the spore suspension (≈ 2.0×10^3 spores/plate) to give a circular inoculum (1 mm diameter). For every tested pH and a_w, growth controls without antimicrobial were prepared and inoculated as above, including one control without pH adjustment (pH = 5.5) or a_w change (a_w = 0.998). Three plates of each system were maintained without inoculation for a_w and pH measurement. The inoculated plates and controls were incubated for 1 month at selected temperatures (Tables 1 and 2) in hermetically closed plastic containers to avoid dehydration. A sufficient headspace was left in the containers to avoid anoxic conditions. Periodically, inoculated plates were removed briefly to observe them and determine if growth had occurred and immediately re-incubated. Water activity was measured with a Decagon CX-1 instrument (Decagon Devices, Inc., Pullman, WA) calibrated and operated following the procedure described by López-Malo et al. (1993). pH was determined with a Beckman pH meter model 50 (Beckman Instruments, Inc., Fullerton, CA). Measurements were made by triplicate. The pH and a_w of the PDA systems without inoculation determined at the beginning and at the end of incubation demonstrated that the desired values remained constant under incubation conditions.

2.5. Mould Growth Response

The inoculated systems were examined daily using a stereoscopic microscope (American Optical, model Forty). A diameter of approximately 2 mm was defined as a positive sign of growth (López-Malo et al., 1998) and registered as "1." If no growth was observed during the incubation period (one month), the response was registered as "0".

2.6. Model Construction

A logistic regression model relates the probability of occurrence of an event, Y, conditional on a vector, x, of explanatory variables (Hosmer and Lemeshow, 1989). The quantity p (x) = E (Y | x) represents the conditional mean of Y given x when the logistic distribution is used. The specific model of the logistic regression is as follows:

$$p\,(x) = [\exp\,(\Sigma\, b_i\, x_i)] \,/\, [1+ \exp\,(\Sigma\, b_i\, x_i)] \quad (1)$$

The logit transformation of p (x) is defined as:

$$\text{logit}\,(p) = g\,(x) = \ln\,\{[p\,(x)\,] \,/\, [1\text{-}p\,(x)]\} = \Sigma\, b_i\, x_i \quad (2)$$

For our particular case, a_w, pH, incubation temperature (T), antimicrobial type (A), antimicrobial concentration (C) and their interactions are the independent variables and the outcome or dependent variable is mould response. In order to fit the logistic model the following equations were selected:

Central composite design – Aspergillus flavus, A. ochraccus *and* A. parasiticus *growth response*

$$g(x)=\beta_0+\beta_1 pH+\beta_2 pH^2+\beta_3 T+\beta_4 T^2+\beta_5 C+\beta_6 C^2+\beta_7 pHT+ \beta_8 pHC+\beta_9 TC+\beta_{10} pHTC \quad (3)$$

Factorial design -Aspergillus flavus *growth response*

$$g(x)=\beta_0+\beta_1 T+\beta_2 a_w+\beta_3 pH+\beta_4 C+\beta_5 A+\beta_6 Ta_w+\beta_7 TpH+ \beta_8 TC+\beta_9 TA+\beta_{10} a_w pH+\beta_{11} a_w C+\beta_{12} a_w A+\beta_{13} pHC+ \beta_{14} pHA+\beta_{15} CA+\beta_{16} Ta_w pH+\beta_{17} TpHC+ \beta_{18} TCA+\beta_{19} Ta_w C+\beta_{20} TpHA+\beta_{21} a_w pHC+ \beta_{22} a_w CA+\beta_{23} pHCA+\beta_{24} Ta_w pHC+\beta_{25} Ta_w pHA+ \beta_{26} a_w pHCA+\beta_{27} Ta_w pHCA \quad (4)$$

where the coefficients (β_i) are the parameters to be estimated by fitting the models to our experimental data.

If an independent variable is discrete, then it is inappropriate to include it in the model as if it was interval scaled. In this situation, the method of choice is to use a collection of design or dummy variables (Hosmer and Lemeshow, 1989). For antimicrobial type (A) which is a discrete variable the codification we used was as follows: "1" for vanillin and "0" for sodium benzoate. Logistic regression was performed with the logistic subroutine in SPSS 10.0 (SPSS Inc., Chicago, IL). A forward stepwise selection procedure was performed to fit the logistic regression equation. The significance of the coefficients was evaluated and were eliminated from the model if the probability of being zero was greater than 0.1.

After fitting the logistic regression equation, predictions of the growth/no growth interface were made at probability levels of 0.50, 0.10 and 0.05, by substituting the value of logit (*p*) in the model and finding the value of one independent variable maintaining fixed the other independent variables. Also probability of growth was calculated using the logistic equation for the evaluated conditions.

3. RESULTS AND DISCUSSION

For every tested pH and a_w, controls prepared without antimicrobials produced growth when the incubation temperature was 15°C or higher, but at temperatures of 10°C or lower no growth was observed after one month of incubation (Table 3). Growth/no growth results for the conditions of central composite design (Table 1) and factorial design (Table 2) are summarized in Tables 3 and 4, respectively. Results demonstrate that mould inhibition (no observable growth) can be obtained with several pH, incubation temperature and vanillin concentration combinations (Table 3) or with combinations of a_w, pH, incubation temperature and antimicrobial (sodium benzoate or vanillin) concentration (Table 4). In most cases once a replicate from a combination produced growth eventually all the others also grew. Therefore, the probabilities observed were, with some exceptions, either close to 1 or 0. This observation agrees with Ratkowsky et al. (1991), they reported higher variance of the microbial response under more stressful conditions. Some synergistic combinations can be

Table 3. Response (growth=1, no growth=0) of an *Aspergillus flavus A. parasiticus* and *A. ochraceus* cocktail inoculated in potato dextose agar formulated at a_w 0.98, selected pH values and different concentrations of vanillin incubated at different temperatures.

pH	Incubation Temperature (°C)	Vanillin Concentration (ppm)	Mould Cocktail Growth Response
2.7	17.5	0	1
2.7	17.5	750	0
3.0	10.0	0	0
3.0	10.0	500	0
3.0	10.0	1000	0
3.0	25.0	0	1
3.0	25.0	500	1
3.0	25.0	1000	0
3.5	4.9	0	0
3.5	4.9	750	0
3.5	17.5	0	1
3.5	17.5	330	1
3.5	17.5	750	1
3.5	17.5	1170	0
3.5	30.1	0	1
3.5	30.1	750	1
4.0	10.0	0	0
4.0	10.0	500	0
4.0	10.0	1000	0
4.0	25.0	0	1
4.0	25.0	500	1
4.0	25.0	1000	1
4.3	17.5	0	1
4.3	17.5	750	1

Table 4. Response (growth=1, no growth=0) of *Aspergillus flavus* inoculated in potato dextrose agar formulated with selected a_w, pH values and different concentrations of sodium benzoate or vanillin incubated at 25 or 15°C.

			0.94 a_w		0.96 a_w		0.98 a_w	
Temp (°C)	pH	Conc.[a] (ppm)	Sodium Benzoate	Vanillin	Sodium Benzoate	Vanillin	Sodium Benzoate	Vanillin
25	3	0	1	1	1	1	1	1
		200	1	1	1	1	1	1
		400	0	1	0	1	1	1
		600	0	1	0	1	0	1
		800	0	1	0	1	0	1
		1000	0	0	0	0	0	0
	4	0	1	1	1	1	1	1
		200	1	1	1	1	1	1
		400	1	1	0	1	1	1
		600	0	1	0	1	1	1
		800	0	1	0	1	0	1
		1000	0	1	0	1	0	1
	5	0	1	1	1	1	1	1
		200	1	1	1	1	1	1
		400	1	1	1	1	1	1
		600	1	1	1	1	1	1
		800	1	1	1	1	1	1
		1000	1	0	1	1	1	1
15	3	0	1	1	1	1	1	1
		200	0	1	0	1	1	1
		400	0	0	0	0	0	0
		600	0	0	0	0	0	0
		800	0	0	0	0	0	0
		1000	0	0	0	0	0	0
	4	0	1	1	1	1	1	1
		200	0	1	1	1	1	1
		400	0	0	0	0	0	1
		600	0	0	0	0	0	0
		800	0	0	0	0	0	0
		1000	0	0	0	0	0	0
	5	0	1	1	1	1	1	1
		200	0	1	1	1	1	1
		400	0	0	0	0	0	1
		600	0	0	0	0	0	1
		800	0	0	0	0	0	0
		1000	0	0	0	0	0	0

[a]Concentration of antimicrobial compound (sodium benzoate or vanillin)

detected in Tables 3 and 4 where fungal growth was inhibited at relatively high pH values or incubation temperatures when combined with reduced a_w and selected antimicrobial concentrations. This is the principle of the multifactorial preservation (or hurdle technology) approach: the search for those "minimal" combinations of factors that inhibit microbial growth.

Fitting Eqs. (3) and (4) to the growth/no growth data by logistic regression and eliminating non-significant (p>0.10) terms resulted in the reduced models presented in Table 5. Variables included in the models were statistically significant (p>0.005). The goodness of fit of every model was tested by log likelihood ratio and Chi-square tests, both being significant, which indicates that models are useful to predict the outcome variable (growth or no growth). The models' goodness of fit was also evaluated comparing predicted values and experimental observations. A predicted probability of growth (cut value) ≥ 0.50 was considered as a growth prediction. Using this criterion, the overall correct observation was 99.2% for the central composite design; with only 3 misclassified predictions from a total of 171 observations. In only one of these three disagreements was growth predicted when no growth was observed, and in the other two cases the model predicted no growth when growth was observed. For the factorial experimental design, the overall correct observation was

Table 5. Reduced logistic model coefficients utilized to predict mould growth/no-growth for the evaluated experimental designs.

Factorial design			Central composite design		
Coefficient	Term	Estimate	Coefficient	Term	Estimate
β_0	Constant	–9876.684	β_0	Constant	–201.673
β_1	T	744.823	β_5	C	1070.559
β_2	a_w	9985.567	β_7	pH*T	4.624
β_3	pH	2182.488	β_9	T*C	–109.451
β_5	A	–7002.668			
β_6	T*a_w	–753.751			
β_7	T*pH	–156.855			
β_8	T*C	–0.336			
β_9	T*A	–19.818			
β_{10}	a_w*pH	–2201.049			
β_{12}	a_w*A	7619.495			
β_{13}	pH*C	–1.523			
β_{15}	C*A	–1.274			
β_{16}	T*a_w*pH	158.516			
β_{17}	T*pH*C	0.126			
β_{18}	T*C*A	0.051			
β_{19}	T*a_w*C	0.340			
β_{20}	T*pH*A	63.797			
β_{21}	a_w*pH*C	1.541			
β_{24}	T*a_w*pH*C	–0.128			
β_{25}	T*a_w*pH*A	–64.832			

98.0%, with only 13 misclassified predictions from a total of 648 observations. In ten of these 13 discrepancies, growth was predicted when no growth was observed, and in the other three cases the model predicted no growth when growth occurred.

Probabilistic microbial models based on logistic regression have been reported for *Shigella flexneri* (Ratkowsky and Ross, 1995), *Escherichia coli* (Presser et al., 1998), *Saccharomyces cerevisiae* (López-Malo et al., 2000), *Listeria monocytogenes* (Bolton and Frank, 1999) and *Zygosaccharomyces bailii* (Cole et al., 1987; López-Malo and Palou, 2000a, b). These reports illustrate the flexibility of logistic regression in constructing the model, taking into account square root type kinetic models (Ratkowsky and Ross, 1995; Presser et al., 1998) or polynomial type models (Cole et al., 1987; Bolton and Frank, 1999; López-Malo et al., 2000; López-Malo and Palou, 2000a, b). However, in every case growth/no growth observations were recorded at fixed storage times, which limited probabilistic models to predict microbial response during that specific period of time. Polynomial type models do not contribute to the understanding of the mechanism involved in microbial growth inhibition. However, they are useful to determine independent variable effects and their interactions. As reported for *Z. bailii,* the probability of growth depends on individual effects of pH, °Brix, sorbic acid, benzoic acid and sulfite concentrations as well as upon complex interactions among preservation factors (Cole et al., 1987). For *Saccharomyces cerevisiae*, polynomial probabilistic models predicted the boundary between growth/no growth interface of as a function of a_w, pH and potassium sorbate concentration (López-Malo et al., 2000).

The predicted probabilities of growth for a cocktail of *Aspergillus flavus*, *A. ochraceus* and *A. parasiticus* are given in Figures 1 and 2, and for *Aspergillus flavus* after one month of incubation in Figures 3 and 4. pH reduction gradually increased the number of combinations of incubation temperature and antimicrobial concentration with probabilities >0.05 for inhibition of mould growth. An important shift of probability of growth curves is also observed with increasing vanillin concentration (Figures 1 and 2). From Figures 3 and 4, the magnitude of the shift in the predicted boundary depends on the combination of temperature, antimicrobial type and water activity considered.

The transition from "likely to grow" conditions ($p > 0.90$, or > 90% likelihood of growth) to "unlikely to grow" conditions ($p < 0.10$, or < 10% likelihood of growth), as predicted from the fitted models, was abrupt as can be seen graphically for combinations of pH and temperatures (Figures 1 and 2) as well as for pH and antimicrobial

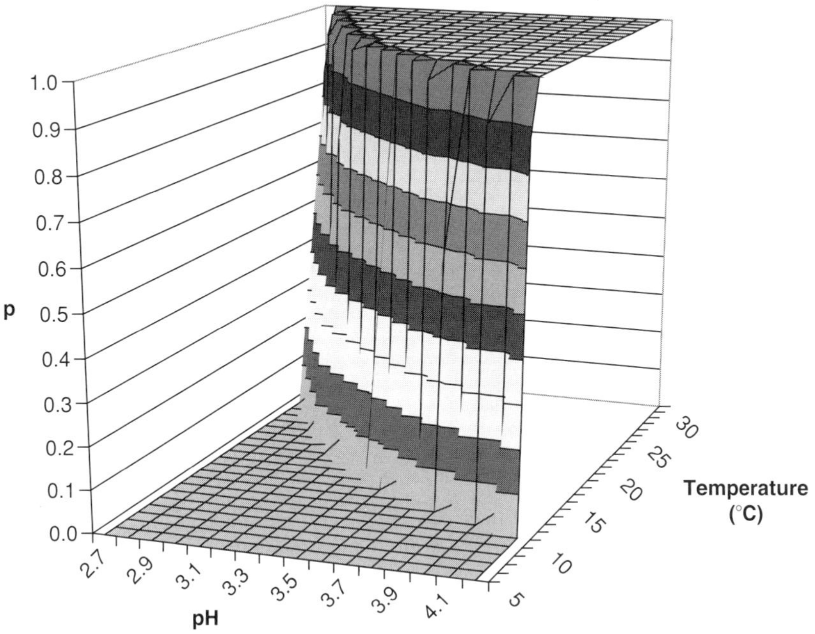

Figure 1. Aspergillus flavus, A. ochraceus and *A. parasiticus* cocktail probability of growth (p) in potato dextrose agar formulated at a_w 0.98, selected pH values and 750 ppm vanillin after one month of incubation at different temperatures.

concentration (Figures 3 and 4). The abruptness of the transition between growth or no growth conditions influenced by pH can be as little as 0.1 to 0.2 pH units, which is close to the limit of reproducibility for pH measurements. For temperature the transition is much less abrupt, occurring over increments of temperature that exceed that of measurement or experimental error. For 648 factor combinations for *Aspergillus flavus*, only 15 of these combinations gave a response different from "all grew" or "none grew". Thus, the experimental data showed an abrupt transition between growth and no growth. This abruptness does indicate a microbiological reality in which small changes in environmental factors within an experiment may have a strong influence on the position of the interface (Tienungoon et al., 2000; Masana and Baranyi, 2000a, b).

An important feature of the generated probabilistic models is that the level of probability can be set, depending on the level of stringency required, to calculate critical values of selected variables. To illustrate, three sets of model predictions using factorial design results were compared, $p = 0.50$, $p = 0.10$ and $p = 0.05$. More stringent values

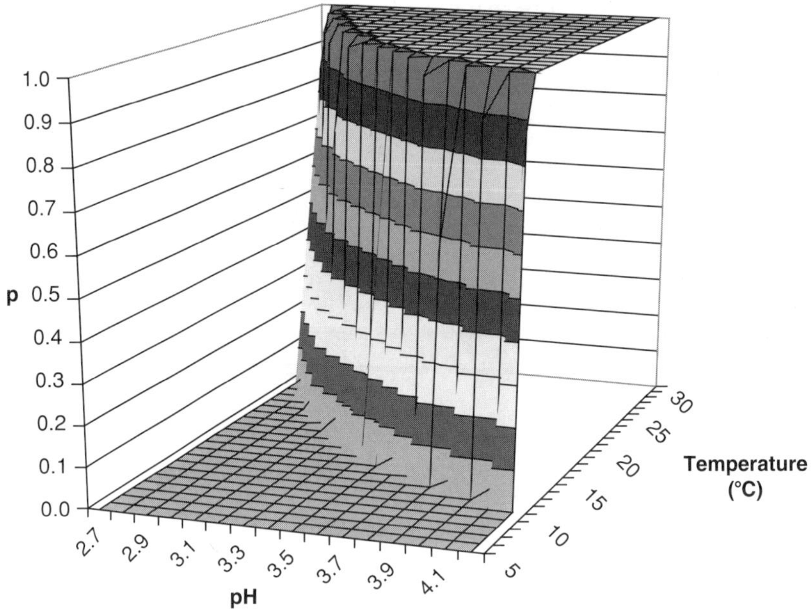

Figure 2. Aspergillus flavus, A. ochraceus and *A. parasiticus* cocktail probability of growth (p) in potato dextrose agar formulated with a_w 0.98, selected pH values and 1000 ppm vanillin concentration after one month of incubation at different temperatures.

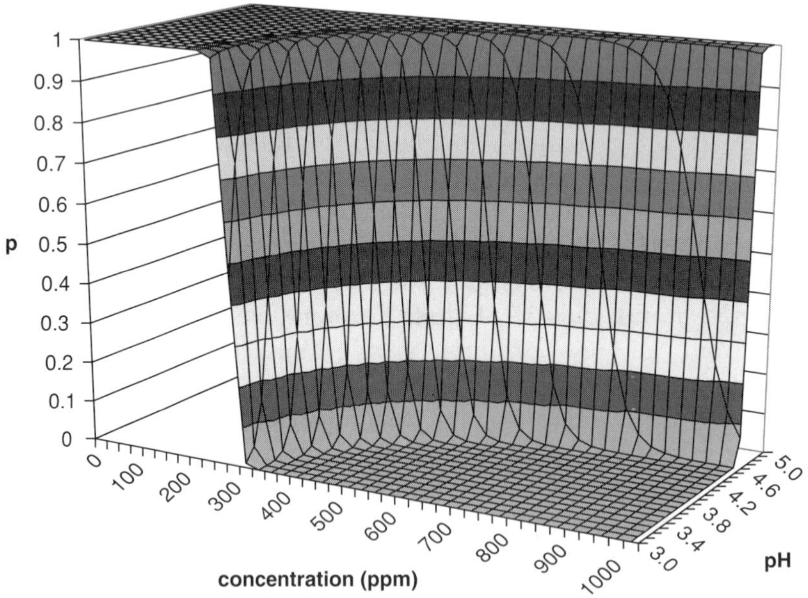

Figure 3. Aspergillus flavus probability of growth (p) in potato dextrose agar formulated with selected pH, sodium benzoate concentration (ppm) and a_w 0.94 after one month of incubation at 25°C.

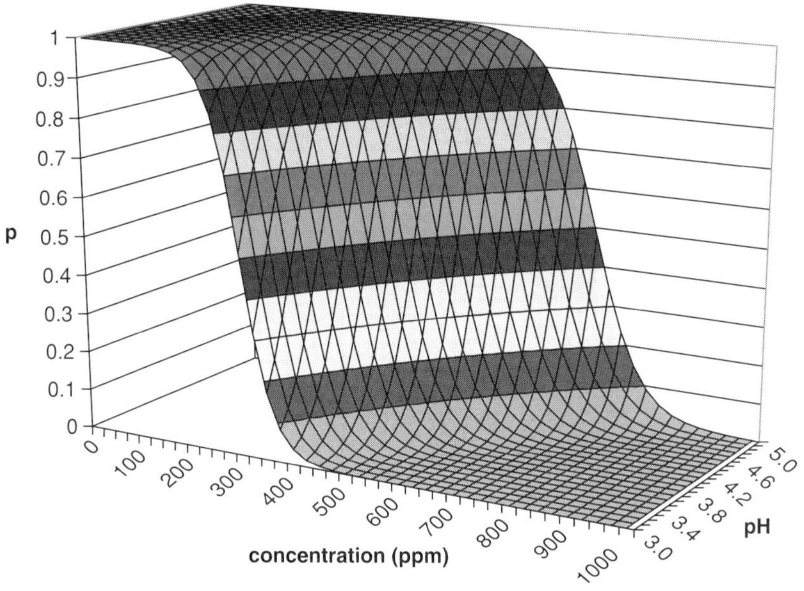

Figure 4. Aspergillus flavus probability of growth (p) in potato dextrose agar formulated with selected pH, vanillin concentration (ppm) and a_w 0.98 after one month of incubation at 15°C.

(p = 0.01 or 0.001) may be necessary in some instances. Critical antimicrobial concentration predictions (Tables 6 and 7) made at p = 0.50 (50:50 chance that *A. flavus* will grow) represent a relatively conservative series of estimates, with the predicted response on the boundary being no better than a coin toss. Model predictions were made more stringent by making p = 0.10 or 0.05 (10 or 5% chance of a false prediction) which causes a shift in the predicted critical antimicrobial concentration or, in other words, in the growth/no growth boundary to a lower temperature, pH and water activity. As a_w decreases from 0.98 (Table 6) to 0.94 (Table 7), the critical concentration of sodium benzoate or vanillin decreases. Also as incubation temperature increases higher antimicrobial concentrations are needed to achieve no growth results after one month of incubation.

4. CONCLUSIONS

Traditional preservation and storage procedures to produce safe and stable food product are generally based on microbial control. If

Table 6. Critical antimicrobial concentrations (ppm) for selected probabilities (p) to inhibit *Aspergillus flavus* growth in laboratory media formulated at a_w 0.98 and selected pH values during one month of incubation at various temperatures.

		Incubation Temperature (°C)				
p	pH	15.0	17.5	20.0	22.5	25.0
Sodium Benzoate						
0.05						
	3.0	353	357	364	384	566
	3.5	360	365	372	393	656
	4.0	368	373	381	403	860
	4.5	377	381	390	415	> 1000
	5.0	385	389	399	427	> 1000
0.10						
	3.0	351	355	361	379	542
	3.5	359	363	370	388	623
	4.0	367	371	378	398	806
	4.5	375	379	387	409	> 1000
	5.0	383	387	396	422	> 1000
0.50						
	3.0	347	349	354	365	471
	3.5	355	357	362	374	525
	4.0	363	365	370	383	646
	4.5	371	374	379	393	> 1000
	5.0	379	382	388	405	> 1000
Vanillin						
0.05						
	3.0	392	532	682	842	> 1000
	3.5	473	595	741	921	> 1000
	4.0	569	679	834	> 1000	> 1000
	4.5	685	799	1004	> 1000	> 1000
	5.0	829	984	> 1000	> 1000	> 1000
0.10						
	3.0	369	509	658	817	988
	3.5	448	567	711	888	> 1000
	4.0	542	647	796	> 1000	> 1000
	4.5	655	760	950	> 1000	> 1000
	5.0	795	934	> 1000	> 1000	> 1000
0.50						
	3.0	302	439	586	743	911
	3.5	375	488	624	790	998
	4.0	462	553	683	880	> 1000
	4.5	567	646	790	> 1000	> 1000
	5.0	696	790	> 1000	> 1000	> 1000

Table 7. Critical antimicrobial concentrations (ppm) for selected probabilities (p) to inhibit *Aspergillus flavus* growth in laboratory media formulated at a_w 0.94 and selected pH values during one month of incubation at various temperatures.

		Incubation Temperature (°C)				
p	pH	15.0	17.5	20.0	22.5	25.0
Sodium Benzoate						
0.05						
	3.0	50	68	98	152	291
	3.5	79	94	120	173	341
	4.0	108	122	146	200	436
	4.5	139	151	175	236	701
	5.0	170	183	209	287	> 1000
0.10						
	3.0	49	67	96	150	287
	3.5	78	93	119	171	335
	4.0	107	120	144	197	428
	4.5	137	150	173	233	685
	5.0	169	182	207	282	> 1000
0.50						
	3.0	46	63	91	143	274
	3.5	75	89	113	163	318
	4.0	104	116	138	188	403
	4.5	134	146	167	222	638
	5.0	165	177	201	269	> 1000
Vanillin						
0.05						
	3.0	232	410	599	799	> 1000
	3.5	233	383	558	767	> 1000
	4.0	235	349	502	714	> 1000
	4.5	236	307	417	614	> 1000
	5.0	238	251	275	345	> 1000
0.10						
	3.0	228	406	595	795	1007
	3.5	229	379	554	761	> 1000
	4.0	230	344	496	707	> 1000
	4.5	232	301	410	604	> 1000
	5.0	233	244	265	326	> 1000
0.50						
	3.0	217	394	582	782	994
	3.5	217	365	539	746	995
	4.0	217	330	479	687	997
	4.5	218	284	388	574	1001
	5.0	219	225	237	271	> 1000

the microbial hazard or spoilage cannot be totally eliminated from the food, microbial growth and toxin production must be inhibited. Microbial growth can be inhibited by combining intrinsic food characteristics with extrinsic storage and packaging conditions. Accurate quantitative data about the effects of combined factors on growth or survival of selected microorganisms are needed. Predictive models can provide decision support tools for the food industry. In many cases models are empirical, interpreting only the response of the microorganism without understanding the mechanism of the response. However, if the models are used properly, predictive probabilistic models are helpful tools for evaluating microbial responses which can in turn identify potential problems for a product, process or storage conditions. Logistic regression is a useful tool for modelling the boundary between growth and no growth. The probabilistic microbial modelling approach can provide a practical means of evaluating the combined effects of food formulation, processing and storage conditions.

5. ACKNOWLEDGMENTS

We acknowledge financial support from CONACyT -Mexico (Projects 32405-B and 44088), Universidad de las Américas-Puebla, and CYTED Program (Project XI.15).

6. REFERENCES

Alavi, S. H., Puri, V. M., Knabel, S. J., Mohtar, R. H., and Whiting, R. C., 1999, Development and validation of a dynamic growth model for *Listeria monocytogenes* in fluid whole milk, *J. Food Prot.* **62**:170-176.

Alzamora, S.M., and López-Malo, A., 2002, Microbial behavior modeling as a tool in the design and control of minimally processed foods, in: *Engineering and Food for the 21st Century*, J. Welti-Chanes, G. V. Barbosa-Canovas, and J. M. Aguilera, eds, CRC Press, Boca Raton, FL, pp. 631-650.

Baranyi, J., and Roberts, T. A., 1994, A dynamic approach to predicting bacterial growth in food, *Int. J. Food Microbiol.* **23**:277-294.

Bolton, L. F., and Frank, J. F., 1999, Defining the growth/no growth interface for *Listeria monocytogenes* in Mexican-style cheese based on salt, pH and moisture content, *J. Food Prot.* **62**:601-609.

Cole, M. B., Franklin, J. G., and Keenan, M. H. J., 1987, Probability of growth of the spoilage yeast *Zygosaccharomyces bailii* in a model fruit drink system, *Food Microbiol.* **4**:115-119.

Cuppers, H. G. M., Oomes, S., and Brul, S., 1997, A model for the combined effects of temperature and salt concentration on growth rate of food spoilage moulds, *Appl. Environ. Microbiol.* **63**:3764-3769.

Gibson, A. M., Baranyi, J., Pitt, J. I., Eyles, M. J., and Roberts, T. A., 1994, Predicting fungal growth: effect of water activity on *Aspergillus flavus* and related species, *Int. J. Food Microbiol.* **23**:419-431.

Gould, G. W., 1989, *Mechanisms of Action of Food Preservation Procedures*, Elsevier, London.

Hosmer, D. W., and Lemeshow, S., 1989, *Applied Logistic Regression*, John Wiley and Sons, New York, p. 307.

ICMSF (International Commission on the Microbiological Specifications for Foods), 1980, *Microbial Ecology of Foods, Vol. 1. Factors Affecting Life and Death of Microorganisms. ICMSF*, Academic Press, New York.

Lanciotti, R., Sinigaglia, M., Gardini, F., Vannini, L., and Guerzoni, M. E., 2001, Growth/no growth interfaces of *Bacillus cereus*, *Staphylococcus aureus* and *Salmonella enteritidis* in model systems based on water activity, pH, temperature and ethanol concentration, *Food Microbiol.* **18**:659-668.

Leistner, L., 1985, Hurdle technology applied to meat products of shelf stable and intermediate moisture food types, in: *Properties of Water in Foods in Relation to Quality and Stability*, D. Simatos, and J. L. Multon, ed., Martinus Nihof Publishers, Dordrecht, The Netherlands, pp. 309-329.

López-Malo, A., and Palou E., 2000a, Growth/no growth interface of *Zygosaccharomyces bailii* as a function of temperature, water activity, pH, potassium sorbate and sodium benzoate concentration, Presented at Predictive Modeling in Foods, Leuven, Belgium, September 12-15.

López-Malo, A., and Palou E., 2000b, Modeling the growth/no growth interface of *Zygosaccharomyces bailii* in mango puree, *J. Food Sci.* **65**:516-520.

López-Malo, A., Alzamora, S. M., Argaiz, A., 1998, Vanillin and pH synergistic effects on mold growth, *J. Food Sci.* **63**:143-146.

López-Malo, A., Guerrero, S., and Alzamora, S. M., 2000, Probabilistic modeling of *Saccharomyces cerevisiae* inhibition under the effects of water activity, pH and potassium sorbate, *J. Food Prot.* **63**:91-95.

López-Malo, A., Palou, E., and Argaiz, A., 1993, Medición de la actividad de agua con un equipo electrónico basado en el punto de rocío, *Información Tecnológica.* **4**(6): 33-37.

Masana, M. O., and Baranyi, J., 2000a, Adding new factors to predictive models: the effect on the risk of extrapolation, *Food Microbiol.* **17**:367-374.

Masana, M. O., and Baranyi, J., 2000b, Growth/no growth interface of *Brochothrix thermosphacta* as a function of pH and water activity, *Food Microbiol.* **17**:485-493.

McMeekin, T. A., Olley, J., Ross, T., and Ratkowsky, D. A., 1993, *Predictive Microbiology: Theory and Application*, Research Studies Press, Tauton, UK.

McMeekin, T. A., Presser, K., Ratkowsky, D. A., Ross, T., Salter, M., and Tienungoon, S., 2000, Quantifying the hurdle concept by modelling the growth/no growth interface. A review, *Int. J. Food Microbiol.* **55**:93-98.

McMeekin, T. A., Ross, T., and Olley, J., 1992, Application of predictive microbiology to assure the quality and safety of fish and fish products, *Int. J. Food Microbiol.* **15**:13-32.

Montgomery, D. C., 1984, *Design and Analysis of Experiments*, John Wiley and Sons, New York.

Palou, E., and López-Malo, A., 2004, Growth/no-growth interface modeling and emerging technologies, in: *Novel Food Processing Technologies*, G. V. Barbosa-Canovas, M. S. Tapia, and P. Cano, eds, Marcel Dekker, New York, pp. 629-651.

Pitt, J. I., 1989, Food mycology – an emerging discipline, *Soc. Appl. Bacteriol. Symp. Suppl.* 1989:1S-9S.

Pitt, J. I., and Hocking, A. D., 1977, Influence of solute and hydrogen ion concentration on the water relations of some xerophilic fungi, *J. Gen. Microbiol.* **101**:35-40.

Pitt, J. I., and Miscamble, B. F., 1995, Water relations of *Aspergillus flavus* and closely related species, *J. Food Prot.* **58**:86-90.

Presser, K. A., Ross, T., and Ratkowsky, D. A., 1998, Modelling of the growth limits (growth/no-growth) of *Escherichia coli* as a function of temperature, pH, lactic acid concentration, and water activity, *Appl. Environ. Microbiol.* **64**:1773-1779.

Ratkowsky, D. A., and Ross, T., 1995, Modelling the bacterial growth/no-growth interface, *Lett. Appl. Microbiol.* **20**:29-33.

Ratkowsky, D. A., 1993, Principles of modelling, *J. Indust. Microbiol.* **12**:195-199.

Ratkowsky, D. A., Ross, T., McMeekin, T. A., and Olley, J., 1991, Comparison of Arrhenius-type and Belehradek-type models for prediction of bacterial growth in foods, *J. Appl. Bacteriol.* **71**:452-459.

Ross, T., and McMeekin, T. A., 1994, Predictive microbiology, *Int. J. Food Microbiol.* **23**:241-264.

Rosso, L., and Robinson, T. P., 2001, Cardinal model to describe the effect of water activity on the growth of moulds, *Int. J. Food Microbiol.* **63**:265-273.

Salter, M. A., Ratkowsky, D. A., Ross, T., and McMeekin, T. A., 2000, Modelling the combined temperature and salt (NaCl) limits for growth of a pathogenic *Escherichia coli* strain using nonlinear logistic regression, *Int. J. Food Microbiol.* **61**:159-167

Samson, R. A., 1989, Filamentous fungi in food and feed, *Soc. Appl. Bacteriol. Symp. Suppl.* 1989:27S-35S.

Schaffner, D. W., and Labuza, T. P., 1997, Predictive microbiology: where are we, and where are we going?, *Food Technol.* **51**:95-99.

Smith, J. E., and Moss, M .O., 1985, *Mycotoxins, Formation, Analysis and Significance*, John Wiley and Sons, Chichester, UK.

Tienungoon, S., Ratkowsky, D. A., McMeekin, T. A., and Ross, T., 2000, Growth limits of *Listeria monocytogenes* as a function of temperature, pH, NaCl, and lactic acid, *Appl. Environ. Microbiol.* **11**:4979-4987.

Valik, L., Baranyi, J., and Gorner, F., 1999, Predicting fungal growth: the effect of water activity on *Penicillium roqueforti*, *Int. J. Food Microbiol.* **47**:141-146.

Whiting, R. C., and Call, J. E., 1993, Time of growth model for proteolytic *Clostridium botulinum*, *Food Microbiol.* **10**:295-301.

ANTIFUNGAL ACTIVITY OF SOURDOUGH BREAD CULTURES

*Lloyd B. Bullerman, Marketa Giesova, Yousef Hassan, Dwayne Deibert and Dojin Ryu**

1. INTRODUCTION

Many strategies have been studied for control of mould growth and reduction in mycotoxin production in foods. The most effective strategy for controlling the presence of mycotoxins in foods is prevention of growth of the mycotoxin-producing fungi in foods and field crops in the first place. Mycotoxin contamination may occur prior to harvest of crops and is often the dominant reason for the occurrence of mycotoxins in foods and feeds. However, fungal growth on stored foods and commodities is also a serious and continuing problem. In recent years increased public concern over chemical food additives and fungicides in foods has prompted searches for safe naturally occurring biological agents with antifungal potential. One source of such compounds are the lactic acid bacteria.

While only a relatively limited number of studies have reported the inhibitory effects of lactic acid bacteria on fungal growth and mycotoxin production, it is generally believed that it is safe for humans to consume lactic acid bacteria and has been known for many years that lactic acid bacteria may positively influence the gastrointestinal tract

* Lloyd B. Bullerman, Yousef Hassan, Dwayne Deibert and Dojin Ryu, Department of Food Science and Technology, University of Nebraska, Lincoln, NE 68583-0919, USA. Marketa Giesova, Department of Dairy and Fat Technology, Institute of Chemical Technology, Prague, Czech Republic. Correspondence to lbbuller@unl-notes.unl.edu

of humans and other mammals (Sandine, 1996). Lactic acid bacteria have been used to ferment foods for centuries, which suggests the non-toxic nature of metabolites produced by these bacteria (Garver and Muriana, 1993; Klaenhammer, 1998). Indigenous lactic acid bacteria are commonly found in retail foods, which suggests that the public consumes viable lactic acid bacteria in many ready-to-eat products (Garver and Muriana, 1993). Thus the metabolic activity of lactic acid bacteria that may contribute in a number of ways to the control of bacterial pathogens and might also have applications for preventing fungal growth (Gourama and Bullerman, 1995; Holzapfel et al., 1995; Klaenhammer, 1998).

The potential of lactic acid bacteria for use as biological control agents of moulds in barley and thus improve the quality and safety of malt during the malting process has been studied in Finland (Haikara et al., 1993; Haikara and Laitila, 1995). The latter authors found a group of lactic acid bacteria with antagonistic activities against *Fusarium. Lactobacillus planarum* and *Pediococcus pentosaceum* inhibited *Fusarium avenaceum* obtained from barley kernels. The preliminary characterization of these starter cultures revealed new types of antimicrobial substances with low molecular mass and features not previously reported for lactic acid bacteria microbiocides (Haikara and Niku-Paavola, 1993; Niku-Paavola et al., 1999).

Studies of *Lactobacillus* strains that possess antifungal properties carried out in the Czech Republic showed that *Lactobacillus rhamnosus* VT1 exhibited strong antifungal properties (Stiles et al., 1999; Plockova et al., 2000, 2001). Further research has shown that *L. rhamnosus* is capable of inhibiting the growth of many spoilage and toxigenic fungi including species in the genera *Aspergillus, Penicillium* and *Fusarium* (Stiles et al., 2002).

The use of sourdough bread cultures has been reported to increase the shelf life of baked goods by delaying mould growth due to the presence of lactic acid bacteria (Gobbetti, 1998). This activity has been attributed to the presence of organic acids, particularly lactic and acetic acids (Spicher, 1983; Rocken, 1996). Further studies have shown that although acetic acid contributes to the antifungal activity of lactic acid bacteria, other bacterial metabolites also have antifungal activity and may contribute to the inhibition of mould growth (Corsetti et al., 1998; Gourama, 1997; Niku-Paavola et al., 1999). Strains of *Lactobacillus plantarum* in particular seem to possess strong antifungal activity (Gobetti et al., 1994a,b; Karunaratne et al., 1990). Lavermicocca et al. (2000) found that *L. plantarum* from sourdough was fungicidal to *F. graminearum* in wheat flour hydrolysate and reported the purification and characterization of novel antifungal compounds from this strain. The compounds that they reported to have strong antifungal activity were phenyllactic acid and 4-hydroxyphenyllactic acid. Two strains of *L. plantarum* isolated from sour-dough cultures produced these compounds and were inhibitory to a number of fungi isolated from baked products (Lavermicocca et al., 2003).

The objective of this work was to test *L. plantarum* and *L. paracasei* isolated from sourdough bread cultures for antifungal activity against several mycotoxigenic molds and to test the ability of an intact sourdough bread culture to inhibit mold growth.

2. MATERIALS AND METHODS

The fungi used were as follows: *Aspergillus flavus* NRRL 1290, *Aspergillus ochraceus* NRRL 3174, *Penicillium verrucosum* NRRL 846, *Penicillium roqueforti* NRRL 848, and *Penicillium commune* NRRL 1899. Bacterial cultures isolated and identified from the sourdough cultures included, *Lactobacillus plantarum* 01 and 011, *Lactobacillus paracasei* 02, 03 and 05 and *Lactobacillus paracasei* SF1, SF2 and SF21. Isolates 01, 02, 03, 05 and 011 were obtained from an old original household sourdough from West Texas, USA which has been kept active for about 100 years. Isolates SF1, SF2 and SF21 were obtained from a commercial San Francisco type sourdough culture.

Sourdough starter cultures and other bacterial isolates were screened for antifungal activity using a dual agar plate assay in which 1.0% of the activated sourdough culture or isolate was added to 15 ml of wheat flour hydrolysate agar (WFH) formulated according to Gobbetti (et al., 1994b), commercial deMan-Rogosa-Sharpe (MRS) agar (Oxoid, Cat. CM0361) and modified MRS (mMRS) agar in Petri dishes. Modified MRS agar was produced by making it without sodium acetate. The sourdough and bacterial isolate agar cultures were overlaid with soft (0.75% agar) yeast extract sucrose agar (YES) or potato dextrose agar (PDA). The centres of the YES or PDA agars were then inoculated at a single point with a mould spore suspension containing 10^3 spores. The plates were incubated at 30°C for 21 days. Colony diameters of the growing mould cultures were measured in mm and recorded daily.

After initial studies the work concentrated on using *L. plantarum* 01 and *L. paracasei* SF1 as they appeared to be the most active antifungal cultures and were representative of the main isolates. These cultures were then grown and compared on MRS and mMRS agars with PDA overlay.

3. RESULTS AND DISCUSSION

The influence of bacterial cultures on fungal growth (colony diameter) was plotted as a functional of time (Figures 1-5). Fungal cultures growing on an underlay of uninoculated bacterial media were used as controls.

Growth of *Aspergillus flavus* was delayed in the presence of *L. paracasei* SF1 and *L. plantarum* 01 (Figure 1). *L. plantarum* was more inhibitory than *L. paracasei* and caused a greater delay in growth. Both bacterial strains appeared to be more inhibitory when grown on MRS agar than mMRS agar, although there was no apparent difference in the growth of the control fungal cultures on the two media. *Aspergillus ochraceus* was inhibited to a greater degree than was *A. flavus* (Figure 2). Both *L. plantarum* and *L. paracasei* on MRS agar caused complete inhibition of growth of *A. ochraceus*. On mMRS

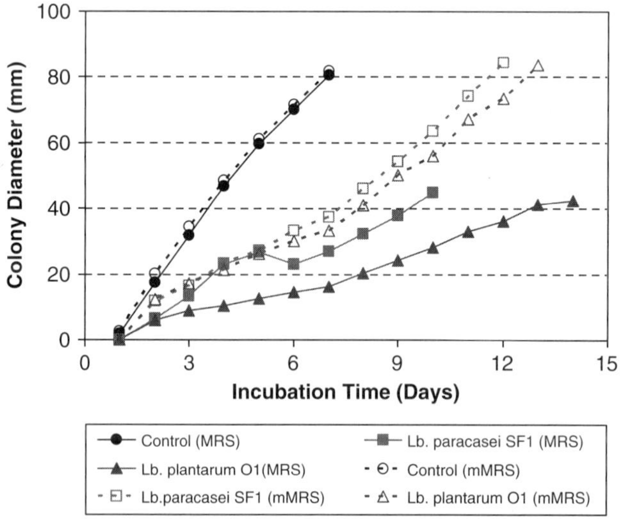

Figure 1. Inhibitory effect of *Lactobacillus plantarum* O1 and *Lactobacillus paracasei* SF1 grown in MRS and mMRS agars on growth of *Aspergillus flavus* NRRL 1290 on a yeast extract sucrose agar overlay.

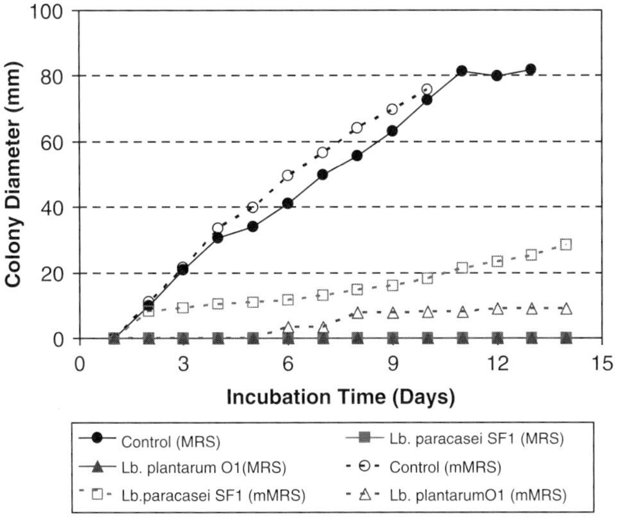

Figure 2. Inhibitory effect of *Lactobacillus plantarum* O1 and *Lactobacillus paracasei* SF1 grown in MRS and mMRS agars on growth of *Aspergillus ochraceus* NRRL 3174 on a yeast extract sucrose agar overlay.

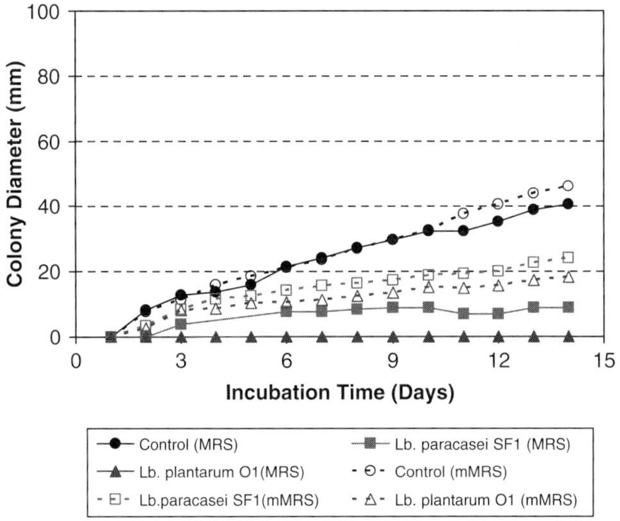

Figure 3. Inhibitory effect of *Lactobacillus plantarum* O1 and *Lactobacillus paracasei* SF1 grown in MRS and mMRS agars on growth of *Penicillium verrucosum* NRRL 846 on a potato dextrose agar overlay.

agar limited growth of *A. ochraceus* occurred in the presence of *L. plantarum*, with growth being delayed until the fifth day of incubation. In the presence of *L. paracasei* on mMRS growth of *A. ochraceus* began by the second day of incubation, but growth was limited, though more growth occurred than in the presence of *L. plantarum*.

Growth of *Penicillium verrucosum* was completely inhibited by *L. plantarum* on MRS agar and *L. paracasei* was also more inhibitory on MRS agar than on the mMRS agar (Figure 3). Growth occurred sooner on mMRS, but colonies were smaller than for the control. *Penicillium roqueforti* grew more rapidly in the presence of the two bacteria than the other mould species (Figure 4). *L. plantarium* showed a greater inhibitory effect on both MRS and mMRS. *P. roquefortii* was slightly inhibited by *L. paracasei* compared with the controls. Growth of *P. commune* was completely inhibited by *L. plantarum* on MRS medium and strongly inhibited by *L. plantarum* on mMRS agar (Figure 5). *L. paracasei* was less inhibitory than *L. plantarum* on both media but with no real difference in inhibitory effect between either medium.

With all fungal species, differences in growth were seen due to the medium in which the *Lactobacillus* species were grown. The lactobacilli were more inhibitory when grown in MRS agar than in mMRS. The original MRS medium and the Commercial MRS medium contain 5 g of sodium acetate per litre (0.5%) of medium (de Man et al., 1960). Stiles et al. (2002) reported that *L. rhamnosus* had greater inhibitory action against 40 different fungal strains when grown in MRS medium than when grown in modified MRS in which the sodium acetate was omitted. They concluded that the acetate was also exerting an inhibitory effect either independently or in addition to inhibitory substances produced by *L. rhamnosus. P. roqueforti* was least inhibited by either bacterium and did not seem to be inhibited to any greater degree on the MRS over the mMRS by either bacterium, although *L. plantarum* tended to be more inhibitory than *L. paracasei. Penicillium roqueforti* is known to have resistance to preservatives such as acetic acid (Gravesen et al., 1994).

It appeared possible that MRS may be a better growth medium for lactobacilli from dairy environments, but sourdough cultures may grow better in a medium that has more cereal based ingredients. Therefore, it was decided to add a medium developed by Gobbetti et al. (1994b) called Wheat Flour Hydrolysate Agar (WFH). In this study a complete sourdough mixed culture, not individual bacterial isolates, was added to the base agar layers of MRS, mMRS and WFH in Petri dishes. Base agar layers were then overlaid with YES agar

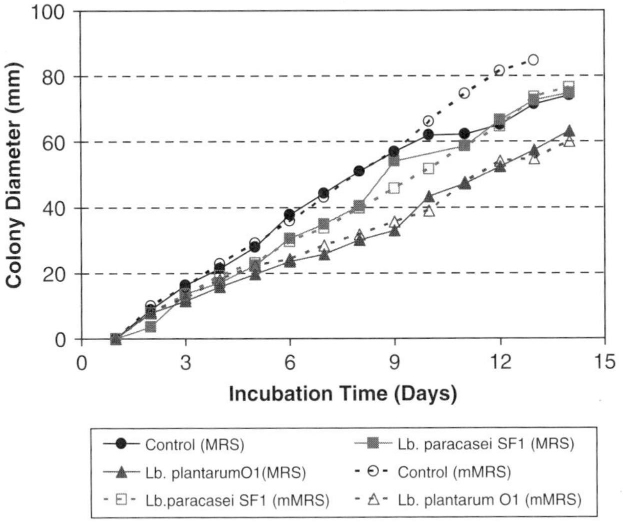

Figure 4. Inhibitory effect of *Lactobacillus plantarum* O1 and *Lactobacillus paracasei* SF1 grown in MRS and mMRS agars on growth of *Penicillium roqueforti* NRRL 848 on a potato dextrose agar (PDA) overlay.

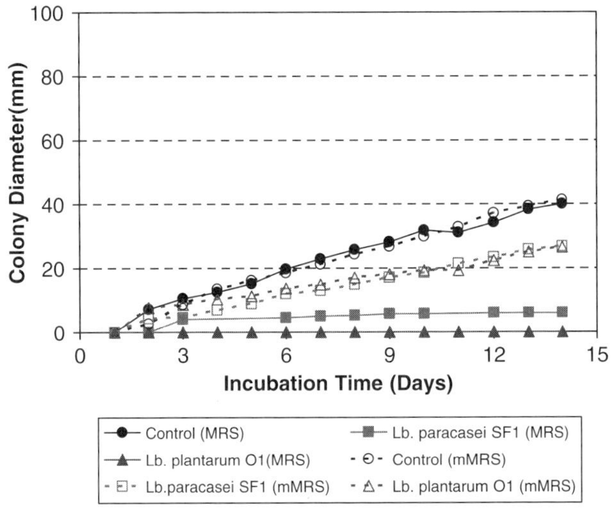

Figure 5. Inhibitory effect of *Lactobacillus plantarum* O1 and *Lactobacillus paracasei* SF1 grown in MRS and mMRS agars on growth of *Penicillium commune* NRRL 1899 on a potato dextrose agar overlay.

which was inoculated with a single point of *A. flavus* NRRL 1290 spores in the centre of the plate as previously described. The growth of *A. flavus* was delayed and reduced on all three media by the intact or complete sourdough culture (Figure 6). Essentially no difference in inhibition was observed in treatments where the sourdough culture was grown in the MRS, mMRS and WFH agars. Thus in this study with the complete sourdough culture there did not appear to be an effect from the medium in which the sourdough culture was grown as was observed with the individual bacterial cultures.

Overall this study has shown that *L. plantarum* and *L. paracasei* isolated from sourdough bread cultures and an intact sourdough bread culture were inhibitory to several species of mycotoxigenic fungi. The inhibition was manifested both as delay of growth and suppression of the growth rate. The inhibitory effects were influenced by the culture medium or substrate in which the individual sourdough bacteria, but not the complete sourdough culture were grown. The presence of 0.5% sodium acetate in MRS medium resulted in a stronger inhibitory effect by *L. plantarum* and *L. paracasei* than mMRS medium from which the sodium acetate was removed. The inhibitory effect was about 20% less when sodium acetate was excluded from the medium. Additional studies are in progress to further evaluate the antifungal

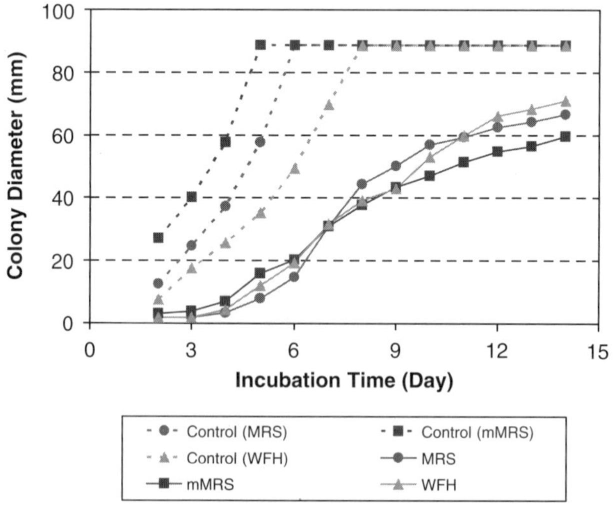

Figure 6. Inhibitory effect of an intact sourdough bread culture grown in MRS, mMRS and WFH agars on growth of *Aspergillus flavus* NRRL 1290 on a yeast extract sucrose agar overlay.

activities of the complete sourdough cultures, and various *L. plantarum* and *L. paracasei* isolates.

4. ACKNOWLEDGEMENTS

This manuscript is published as Paper No. 14941, Journal Series. This research was carried out under Project 16-097, Agricultural Research Division, University of Nebraska-Lincoln and was supported in part by a research grant from the Anderson Research Fund of the NC-213 Multistate Research Project.

5. REFERENCES

Corsetti, A., Gobbetti M., and Damiani, P., 1998, Antimould activity of sourdough lactic acid bacteria: identification of a mixture of organic acids produced by *Lactobacillus sanfrancisco* CB1, *Appl. Microbiol. Biotechnol.* **50**:253-256.

deMan, J. C., Rogosa, M., and Sharpe, M. E., 1960, A medium for the cultivation of lactobacilli, *J. Appl. Bacteriol.* **23**:130-135.

Garver, K. I., and Muriana, P. M., 1993, Detection, identification and characterization of bacteriocin-producing lactic acid bacteria from retail food products, *Int. J. Food Microbiol.* **19**:241-258.

Graveson, S., Frisvad J. C., and Samson, R. A., 1994, *Microfungi*, Munksgaard. Copenhagen, Denmark.

Gobbetti, M., 1998, The sourdough microflora: interaction of lactic acid bacteria and yeasts, *Trends Food Sci. Technol.* **9**:267-274.

Gobbetti, M., Corsetti, A., and Rossi, I. 1994a, The sourdough microflora: interactions between lactic acid bacteria and yeasts: metabolism of amino acids, *World J. Microbiol. Biotechnol.* **10**:275-279.

Gobbetti, M., Corsetti, A. and Rossi, I., 1994b, The sourdough microflora: interactions between lactic acid bacteria and yeasts: metabolism of carbohydrates, *Appl. Microbiol. Biotechnol.* **41**:456-460.

Gourama, H., 1997, Inhibition of growth and mycotoxin production of *Penicillium* by *Lactobacillus* species, *Lebensm. Wiss. Technol.* **30**:279-283.

Gourama, H., and Bullerman, L. B., 1995, Antimycotic and antiaflatoxigenic effect of lactic acid bacteria. A Review, *J. Food Prot.* **57**:1275-1280.

Haikara, A., and Laitila, A., 1995, Influence of lactic acid starter cultures on the quality of malt and beer, in: *Proceedings of the 26th Congress of the European Brewers Convention, Brussels*, IRL Press, Oxford, UK, pp. 249-256.

Haikara, A., and Niku-Paavola, M., 1993, Fungicidic substances produced by lactic acid bacteria, *FEMS Microbiol. Rev.* **12**:120.

Haikara, A., Uljas, H., and Suurnakki, A., 1993, Lactic starter cultures in malting: a novel solution to gushing problems, in: *Proceedings of the 25th Congress of the European Brewers Convention, Oslo*, IRL Press, Oxford, UK, pp. 163-172.

Holzapfel, W. H., Geisen, R., and Schillinger, U., 1995, Biological preservation of foods with reference to protective cultures, bacteriocins and food grade enzymes, *Int. J. Food Microbiol.* **24**:343-362.

Karunaratne, A., Wezenberg, E., and Bullerman, L. B., 1990, Inhibition of mold growth and aflatoxin production by *Lactobaciillus* spp., *J. Food Prot.* **53**:230-236.

Klaenhammer, T. R., 1998, Functional activities of *Lactobacillus* probiotics: genetic mandate, *Int. Dairy J.* **8**:497-505.

Lavermicocca, P., Valerio, F., Evidente, A., Lazzaroni, S., Corsetti, A., and Gobbetti, M., 2000, Purification and characterization of novel antifungal compounds from the sourdough *Lactobacillus plantarum* Strain 21B, *Appl. Environ. Microbiol.* **66**:4084-4090.

Lavermicocca, P., Valerio, F., and Visconti, A., 2003, Antifungal activity of phenyllactic acid against molds isolated from bakery products, *Appl. Environ. Microbiol.* **69**:634-640.

Niku-Paavola, M.–L., Laitila, A., Mattila-Sandholm, T., and Haikara, A., 1999, New types of antimicrobial compounds produced by *Lactobacillus plantarum*, *J. Appl. Microbiol.* **86**:29-35.

Plockova, M., Stiles, J., and Chumchalova, J., 2000, Evaluation of antifungal activity of lactic acid bacteria by the milk agar plate method, *Czech Dairy J.* **62**:19-19.

Plockova, M., Stiles, J., Chumchalova, J., and Halfarova, R., 2001, Control of mould growth by *Lactobacillus rhamnosus* VT1 and *Lactobacilus reuteri* CCM 3625 on milk agar plates, *Czech J. Food Sci.* **19**:46-50.

Rocken, W., 1996, Applied aspects of sourdough fermentation, *Adv. Food Sci.* **18**: 212-216.

Sandine, W. E., 1996, Commercial production of dairy starter cultures, in: *Dairy Starter Cultures,* T. M. Cogan and J.-P. Accolas, eds., VCH Publishers, New York, pp. 191-206.

Spicher, G., 1983, Baked goods, in: *Biotechnology, Vol. 5: Food and Feed Production with Microorganisms*, G. Reed, ed., Verlag Chemie, Weinheim, Germany.

Stiles, J., Plockova, M., Toth, V., and Chumchalova, J., 1999, Inhibition of *Fusarium* sp. DMF 0101 by *Lactobacillus* strains grown in MRS and Elliker broths, *Adv. Food Sci.* **21**:117-121.

Stiles, J., Penkar, S., Plockova, M., Chumchalova, J., and Bullerman, L. B., 2002, Antifungal activity of sodium acetate, a component of MRS medium, *J. Food Prot.* **65**:1188-1191.

PREVENTION OF OCHRATOXIN A IN CEREALS IN EUROPE

Monica Olsen[1], Nils Jonsson[2], Naresh Magan[3], John Banks[4], Corrado Fanelli[5], Aldo Rizzo[6], Auli Haikara[7], Alan Dobson[8], Jens Frisvad[9], Stephen Holmes[10], Juhani Olkku[11], Sven-Johan Persson[12] and Thomas Börjesson[13]

1. INTRODUCTION

This paper describes objectives and activities of a major European Community project (OTA PREV) aimed at understanding sources of contamination of ochratoxin A in European cereals and related foodstuffs, and the development of strategies to minimise ochratoxin A in the food supply. The project ran from February 2000 to July 2003.

[1] National Food Administration, PO Box 622, SE-751 26 Uppsala, Sweden
[2] Swedish Institute of Agricultural and Environmental Engineering, PO Box 7033, SE-750 07 Uppsala, Sweden
[3] Cranfield Biotechnology Centre, Cranfield University, Barton Road, Silsoe, Bedfordshire MK45 4DT, UK
[4] Central Science Laboratory, Sand Hutton, York YO41 1LZ, UK
[5] Laboratorio di Micologia, Univerisità "La Sapienza", Largo Cristina di Svezia 24, I-00165 Roma, Italy
[6] National Veterinary and Food Res. Inst., PO Box 45, FIN-00581, Helsinki, Finland
[7] VTT Biotechnology, PO Box 1500, FIN-02044 Espoo, Finland
[8] Microbiology Department, University College Cork, Cork, Ireland
[9] BioCentrum-DTU, Building 221, DK-2800 Kgs. Lyngby, Denmark
[10] ADGEN Ltd, Nellies Gate, Auchincruive, Ayr KA6 5HW, UK
[11] Oy Panimolaboratorio-Bryggerilaboratorium AB, P.O. Box 16, FIN-02150 Espoo, Finland
[12] Akron maskiner, SE-531 04 Järpås, Sweden
[13] Svenska Lantmännen, Östra hamnen, SE-531 87 Lidköping, Sweden. Correspondence to: mool@slv.se

1.1. The Objectives

The over-all objective for the OTA PREV project is the protection of the consumer's health by describing measures for decreasing the amount of ochratoxin A in cereals produced in Europe. This has been achieved by identifying the key elements in an effective HACCP programme for ochratoxin A for cereals, and by providing tools for preventive and corrective actions. A summary of the tools provided by this project is presented in Table 1. The project included 11 work packages covering the whole food chain from primary production to the final processed food product (Table 2). The objectives and expected achievements were divided into four tasks, all important steps in a HACCP managing programme for ochratoxin A in cereals: 1. Identification of the critical control points (CCP); 2. Establishment of critical limits for the critical control points; 3. Developing rapid monitoring methods, and 4. Establishment of corrective actions in the event of deviation of a critical limit. The outcome will serve as a pool of knowledge for HACCP-based management programmes, which will increase food safety and support the European cereal industry.

1.2. Why Ochratoxin A?

EC legislation and Codex Alimentarius are currently addressing the problem of ochratoxin A in food commodities and raw materials. Ochratoxin A can be found in cereals, wine, grape juice, dried vine fruits, coffee, spices, cocoa, and animal derived products such as pork products. The current EC legislation includes unprocessed cereals, cereal products and dried vine fruits (Commission regulation 472/2002) and limits for other commodities are being discussed, including baby food. JECFA (the FAO/WHO Joint Expert Committee on Food Additives) evaluated ochratoxin A at its 56th meeting in 2001 (JECFA, 2001). Ochratoxin A is nephrotoxic in all tested animal species and may cause renal carcinogenicity, but the mechanism of action is still being debated. Both genotoxic and non-genotoxic mechanisms have been proposed.

JECFA retained the previously established provisional tolerable weekly intake (PTWI) at 100 ng/kg bodyweight (b.w.), corresponding to approximately 14 ng/kg b.w. per day. Estimates of tolerable daily intake for ochratoxin A, based on non-threshold mathematical modelling approaches or a safety factor/threshold approach, have ranged from 1.2 to 14 ng/kg b.w. per day. The Scientific Committee for Food of the European Commission (SCF, 1998) considered that "it would

Table 1. Summary of tools to prevent ochratoxin A in the cereal production chain as provided by the EC project known as OTA PREV[a]

Site	Control type	Tools provided	Comments (possible % reduction of OTA)
Harvest	GAP	Recommendation: Keep machinery and areas in contact with the harvested grain, clean. Remove old grain and dust. (WP1)	(% prevention not possible to estimate, but significant)
Buffer storage before drying and during drying (in near-ambient dryers)	CCP	Mathematical model which can predict safe storage time (critical limits). (WP4)	(up to 100 % prevention possible)
		Rapid monitoring methods for OTA and producing fungi. (WP8)	Monitoring tools: LFDs and ELISAs.
		Data on environmental conditions conducive to growth and OTA production. (WP3)	(% prevention not possible to estimate, but useful tools in DSS)
Storage	GSP/ CCP	Recommendations on silo design and maintenance. (WP5)	(% prevention not possible to estimate, but significant)
		Critical limits for remoistening. (WP5)	(up to 100 % prevention possible)
		Food grade antioxidants and natural control measures to prevent OTA formation in wet grain. (WP 6)	(>80 % prevention but not yet economically feasible)
Intake at cereal processing industry	CCP	Rapid monitoring methods for OTA* and OTA producing fungi in grain. (WP8)	LFD (with reader for ochratoxin A) and ELISA
		Critical limit: less than 1000 cfu/g *P. verrucosum* in wheat. (WP4)	Indicating risk of OTA levels above 5 μg/kg
		Monitoring method for *P. verrucosum*. (WP1, WP8, and WP9)	Monitoring tools: DYSG, LFD, ELISA, and PCR
Milling industry	GMP	Reductive measures during milling. (WP10)	(cleaning 2-3%, scouring 3-44%, milling up to 60%)

Continued

Table 1. Summary of tools to prevent ochratoxin A in the cereal production chain as provided by the EC project known as OTA PREV[a]–cont'd

Site	Control type	Tools provided	Comments (possible % reduction of OTA)
Cereal processing industry	GMP	Reductive measures during extrusion and baking. (WP10)	(baking up to 5-10%, extrusion up to 40%)
Intake at malting industry	CCP	Critical limits: <3% internal infection or <400 cfu/g with *P. verrucosum* in barley. (WP11)	(up to 100 % prevention possible)
Malting industry	GMP	Recommendation: effect of temperature on OTA formation during malting. (WP11)	(decrease temperature from 16-18 to 12-14°C reduces OTA formation 4 times)
Intake at brewing industry	CCP	Rapid monitoring methods for OTA* in malt. (WP8)	LFD (with reader) and ELISA
Brewing industry	GMP	Fate of OTA during brewing. (WP 11)	(up to 80 % reduction)
Official control	CCPs	Rapid monitoring methods for OTA*. (WP8)	LFD (with reader) and ELISA

[a] Abbreviations used: GAP, Good Agricultural Practice; GSP, Good Storage Practice; GMP, Good Manufacturing Practice; CCP, Critical Control Point; DSS, decision support systems; cfu, colony forming units; OTA, ochratoxin A; LFD, lateral flow device; WP, project work package.

* the critical limits at these points are the same as the legislative limits (currently 5 and 3 μg/kg for the unprocessed cereals and products, respectively)

Table 2. Work packages included in the OTA PREV project

WP no.	Work Package Title
1	Which are the ochratoxin A producing fungi and what are the sources of inoculum
2	Differences depending on various cereal species, farming methods and climate.
3	The effect of temporal environmental factors on fungal growth, patterns of colonisation and ochratoxin production
4	Modelling the growth of *Penicillium verrucosum* in cereal grain during aerobic conditions
5	Grain silos – hygienic design
6	Control of ochratoxin production in cereals using food grade antioxidants and natural control measures to prevent entry into the food chain
7	Development of rapid biosensor for ochratoxin detection using molecular imprinted polymers
8	Development of rapid ELISA system for ochratoxin producing fungi
9	Molecular characterisation of the ochratoxin biosynthetic genes in the mycotoxigenic fungi *Aspergillus ochraceus* and *Penicillium verrucosum*
10	Establishing corrective actions during milling and processing
11	Establishing corrective actions during malting and brewing

be prudent to reduce exposure to ochratoxin A as much as possible, ensuring that exposures are towards the lower end of the range of tolerable daily intakes of 1.2-14 ng/kg b.w. per day which have been estimated by other bodies, e.g. below 5 ng/kg b.w. per day". In the most recent assessment of ochratoxin A intake by European consumers (SCOOP, 2002) and in earlier investigations, cereals have been found to be the most important dietary source of ochratoxin A, contributing from 50 to 80% of the intake. Consequently, prevention of ochratoxin A formation by specific moulds in cereals would have a significant impact on the consumer intake of ochratoxin A in Europe.

Maximum permissible levels of 5 μg/kg in raw cereals and 3 μg/kg in cereal products have been set for ochratoxin A within the EC. Surveys of stored cereals in Europe over many years (e.g. Olsen et al. 1993; Scudamore et al., 1999; Wolff, 2000; Puntaric et al., 2001) show that samples examined sometimes exceed this level. Even if the percentage of samples contaminated at the statutory limits within the Community were as low as 3% this would represent a large tonnage of grain (approximately 6 million tonnes) and a potential serious economic loss. In monetary terms this would equate to a loss of 800-1000 million euros assuming that no alternative use for the grain was available. In addition, the high cost of monitoring programmes (roughly 0.3 and 100 million euros for official and internal control, respectively) for preventing contaminated grain entering the food chain must also be considered.

The food and brewing industries are increasingly demanding high quality cereals for food and drink products and require grain at least conforming to the statutory limits set for ochratoxin A. Thus there is a major incentive for European cereal industry to minimise ochratoxin A (and other mycotoxins) in grain to enable it to remain competitive worldwide and to reduce consumer risk as far as possible. Clearly, there is an urgent need to understand the factors that encourage mycotoxin formation both pre and post harvest as this will assist in developing strategies to minimise mycotoxin formation.

Application of HACCP-like assessment of the cereal food chain supports the established knowledge that drying grain quickly at harvest is the most crucial factor in avoiding mould and mycotoxin formation during subsequent storage. Thus, instigation of a suitable monitoring check system to ensure that grain is dried rapidly to a safe moisture content, together with regular inspection and monitoring of grain moisture, should eliminate the risk of ochratoxin A formation during storage. In addition, compliance with Good Agricultural Practice (GAP) will assist in the production of good quality grain. However, the continued occurrence of ochratoxin A in grain shown by surveys suggests that either good practice is not or cannot always be fully followed or that all the factors involved in the formation of ochratoxin A are not completely understood. In other words, if grain cannot always be dried quickly enough it becomes important that all of the factors that affect the potential for formation of ochratoxin A are fully understood. Only by obtaining this information can sound and effective advice be made available in the field on how to minimise this risk.

One of the tasks within this project (WP2; see Table 2), aims to establish how cereal production, harvesting and storage procedures vary across the EC by compiling a small data base from the information provided in a questionnaire sent to cereal experts. The cooperation of 'grass root' experts means that the information is their view and not necessarily an official or government view, especially where answers required to a particular question may be sensitive or subjective, e.g. 'is ochratoxin A considered a problem in your Country?' The data and this appraisal are intended to complement the results from the scientific studies of within this project.

1.3. Fungi Producing Ochratoxin A

Penicillium verrucosum has been found to be the ochratoxin A producer in several national investigations of cereal grain in Europe.

Frisvad and Viuf (1986) investigated 70 samples of barley from Denmark containing up to 7400 μg/kg ochratoxin A and all ochratoxin A producers from these samples were identified as *Penicillium viridicatum* II as defined by Ciegler et al. (1973). Taxonomists subsequently concluded that all known ochratoxin A producers in *Penicillium* belonged to *P. verrucosum* (Frisvad 1985; Pitt 1987; Frisvad 1989). Larsen et al. (2001) found that *P. viridicatum* III (Ciegler et al., 1973) was a distinct species, which was named *Penicillium nordicum,* and that it should be separated from *P. verrucosum. P. nordicum* was found to be associated with meat and cheese products while *P. verrucosum* isolates are usually derived from plants (Ciegler et al., 1973; Dragoni and Cantoni, 1979; Larsen et al., 2001).

Several *Aspergillus* species and *Aspergillus* teleomorphs have been reported to produce ochratoxin A including *A. ochraceus, A. sulphureus, A. niger, A. carbonarius, Neopetromyces muricatus* and *Petromyces alliaceus* (Frisvad and Samson, 2000). However, ochratoxin A producing isolates of *Aspergillus* species have never been found on European cereals. One of the objectives with this study was to investigate the occurrence of all potential ochratoxin A producers in cereals produced in Europe (WP1).

Dichloran Rose Bengal Yeast Extract Sucrose agar (DRYES, Frisvad et al., 1992) is recommended for the enumeration of *P. verrucosum* in food (NMKL Method no. 152, 1995), however, Dichloran Yeast extract Sucrose 18% Glycerol agar (DYSG, Frisvad et al., 1992) may be a more selective and better diagnostic medium. The efficiency of DYSG needs to be compared with DRYES and with DG18 (Hocking and Pitt, 1980), which is a general purpose medium suitable for detecting *P. verrucosum.* The objective of the work, presented in this report, was to examine whether the numbers of *P. verrucosum,* recovered on the different media are statistically different, by investigating a large number of cereal samples containing ochratoxin A. In addition, the three different media were validated in a collaborative study (WP1).

To be able to find contamination sources and critical control points, new molecular fingerprinting methods, such as AFLP (Vos et al., 1995; Jannsen et al., 1996), have been used with success in several studies of fungal populations and clones of different fungi (Majer et al., 1996; Arenal et al., 1999; Bakkeren et al., 2000; Tooley et al., 2000; Kothera, 2003; Kure et al., 2003; Lund et al., 2003; Schmidt et al., 2003). The AFLP method has been considered superior to RAPD fingerprinting (Lund and Skouboe, 1998; Bakkeren et al., 2000). The aim of one study within this project was to examine whether the AFLP fingerprinting method could be used to find critical control points in the process from soil to table in order to prevent or minimize ochratoxin A formation (WP1).

1.4. Critical Limits for Fungal Growth and Production of Ochratoxin A

Fungal growth and ochratoxin A production during grain storage are influenced by a wide variety of complex interactions between abiotic and biotic factors. Water availability and temperature are the two most important abiotic factors. They interact to determine the range of microorganisms that can colonise a given substrate, their growth and contribution to spontaneous heating.

During respiration of damp grain, oxygen is utilized and carbon dioxide is produced. Concentrations of oxygen and carbon dioxide in the inter-granular atmosphere are important in determining the pattern of fungal colonization of grain during storage. As most moulds including *P. verrucosum* and *A. ochraceus* are considered obligate aerobes, increased levels of CO_2 should inhibit their growth. However, relatively few studies have been carried out regarding the relationship between various carbon dioxide levels, water activity (a_w) and temperatures on fungal growth, particularly *P. verrucosum* and *A. ochraceus,* the main ochratoxin A producers during wheat storage.

Changing either temperature or a_w affects growth and may affect the ability of species to compete (Magan and Lacey, 1985a,b) which could effect ochratoxin A production (Ramakrishna et al., 1996). A range of interactions can occur between species, which can be scored to compare competitiveness of different species in various culture conditions (Magan and Lacey, 1984). On maize grain, interactions and competition have been shown to have a marked influence on ochratoxin A production by *A. ochraceus* (Ramakrishna et al., 1993, 1996; Marin et al., 1998; Lee and Magan, 2000). There have been no previous studies to examine the competitiveness of *P. verrucosum* against other fungi on wheat grain and the influence these interactions may have on ochratoxin A production. It has been suggested that the co-existence of microorganisms may be mediated via nutritional resources partitioning. Wilson and Lindow (1994a,b) determined niche overlap indices (NOI) for epiphytic bacteria and the level of ecological similarity for isolating effective bacterial control agents. Niche overlap values greater than 0.9 have been suggested to indicate co-existence between species in an ecological niche, whereas scores less than 0.9 indicate occupation of separate niches.

In this project the effect of a_w and temperature was examined on 1) growth and ochratoxin A production by *Penicillium verrucosum* over time on irradiated wheat grain 2) growth and ochratoxin A production at various gas compositions and 3) *in vitro* interactions on growth and toxin production between an ochratoxin A producing isolate of *P. verrucosum* and other wheat spoilage fungi. The effect of environmental factors on *in vitro* carbon source utilisation patterns and NOIs for the ochratoxin A producing strain of *P. verrucosum* in relation to all species was determined (WP3).

The minimum moisture content (m.c.) in wheat that allows growth of *P. verrucosum* is about 16-17%, or approximately 0.80 a_w (Northolt et al., 1979). To produce ochratoxin A, the fungus probably needs about 1% higher m.c. A study of the grain quality on farms indicates that the occurrence of ochratoxin A in grain is attributed to insufficient drying or to excessively long pre-drying storage (Jonsson and Pettersson, 1992). Studies on the safe storage period for cereal grains are few, and are based on visible moulding (Kreyger, 1972), dry-matter loss (Steel et al., 1969; White et al., 1982) or loss of seed germination (Kreyger, 1972; White et al., 1982). For maize, maximum allowable storage time has been estimated based on the time before dry matter loss exceeds 0.5% (Steel et al., 1969). This loss is estimated to correspond to the loss of one US grade, which is based on visible inspection. Visible mould may be an unreliable criterion, because considerable losses can occur before moulding is visible, depending on whether or not the conditions favour fungal growth and sporulation (Seitz et al., 1982).

Measurement of respiration is a widely used method for estimating fungal growth, biomass and dry-matter losses. In soil the rate of CO_2 production has been used to estimate total living microbial biomass (Anderson and Domsch, 1975). CO_2 production was highly correlated with ergosterol (r = 0.98) content when *Eurotium repens* colonised maize (Martin et al., 1989).

The aim of one study in this project was to develop a mathematical model, which describes the effect of a_w and temperature on safe storage time before obvious growth of *P. verrucosum* and formation of ochratoxin A in cereal grain. Data from a respirometer on CO_2 production during the storage was compared with data on the growth of *P. verrucosum* and production of ochratoxin A (WP4).

1.5. Prevention of Ochratoxin A Formation

After cereal grain is dried and cooled in a high temperature drier and placed into storage at 13-14% moisture content, its temperature is frequently well above the average ambient temperature in temperate climates (Brooker et al., 1992). Because dry grain is an effective thermal insulator, the periphery of the bulk changes temperature faster than the less exposed grain in the centre. This is especially valid for grain masses over 50 tons, which fail to cool or warm uniformly during seasonal thermal changes (Foster and Tuite, 1992). The resulting temperature gradient causes moisture to move from warmer to colder parts of the grain bulk. Moisture migration is particularly a problem in grain with larger size, such as maize, stored in uninsulated outdoor steel silos in areas with large seasonal changes in air temperature.

Factors responsible for heat and mass transfer during storage without aeration are conduction, diffusion and natural convection. However, depending on the geometry of the storage structure, material properties of the grain bulk and loading practices, the effect of natural convection may be negligible (Smith and Sokhansanj, 1990; Maier and Montross, 1998). Small grain, such as wheat, offers more resistance to air movement within the grain mass and is usually stored at a lower moisture content than maize (Foster and Tuite, 1992). It has been shown that mass transfer is not significant in wheat stored without aeration during the summer in North Dakota, with a maximum increase of the moisture content of 0.45 % just below the surface. (Hellevang and Hirning, 1988). Similar moisture increases were measured in simulated tests in the laboratory. However, if the plenum openings are not fully sealed, wind-induced air currents may increase the heat and mass transfer (Montross et al., 2002).

Thermal properties of the construction material determine the extent and frequency of fluctuations of silo temperature, and research has shown that approximately 90 % of the environmental temperature changes were transmitted to the silo gas space with amplification of 2.5-3.0 times in an uninsulated steel silo (Meiering, 1986). This amplification is markedly reduced when the steel silo is placed indoors protected from direct solar radiation. A concrete silo does not follow the daily fluctuations of solar radiation and environmental air temperature due to its approximately 20 times higher resistance to heat flow, lower thermal conductivity and larger thermal inertia.

For smaller amounts of grain stored in uninsulated steel silos outdoors, the headspace may be the most critical factor due to large diurnal temperature variations, causing water to condense on to the cold grain (Bailey, 1992) and also on the inside of the roof, dripping back into the grain. The spoilage of grain close to the headspace may go undetected because of mixing that takes place when the silo is unloaded (Lundin, 1998). Uninsulated steel silos have become common for outdoor storage of dried cereal grain in Europe because they require lower investment costs compared with those placed indoors, or reinforced concrete silos.

Numerous investigators have developed mathematical models to describe the heat and mass transfer in stored bulk grain although mainly for maize. Most models have only considered heat transfer during storage but there are more comprehensive models which also include mass transfer (Metzger and Muir, 1983; Tanaka and Yoshida, 1984; Smith and Sokhansanj, 1990; Maier, 1992; Casada and Young, 1994; Chang et al., 1994; Khankari et al., 1995). Most numerical models assume overly simplistic boundary conditions and do not accurately model the headspace and plenum conditions during storage (Maier and Montross, 1998). In a validation of a comprehensive model using realistic boundary conditions, it was concluded that the available literature values for a number of critical variables used to model headspace and plenum conditions proved to be inaccurate and needed more research (Montross et al., 2002).

The influence of different systems for air exchange between the headspace and the ambient air, on the temperature and relative humidity in the grain close to the headspace were studied in this project. A mathematical model of dew point temperature and collected data were used to describe the risk of moisture condensation in the headspace of the silos during the storage. Preventive strategies to avoid rehydration of the grain and the risk for mould growth and production of ochratoxin A were also suggested (WP5).

There are two main strategies for reducing the occurrence of mycotoxins in cereals: prevention of contamination or detoxification of mycotoxins already present. One of the approaches in this project is based on the preventive strategy during post harvest storage. The use of grain protectant chemicals has been widely applied during the last few years. This has achieved good results in the short-term, but the widespread use of chemical compounds has resulted in resistance in the target organisms.

Storage fungi are commonly controlled using preservatives such as organic acids, however the use of preservatives has met with increasing consumer resistance in recent years (Foegeding and Busta, 1991; Basilico and Basilico, 1999). In the past decade research has focused on the use of natural preservatives (Foegeding and Busta, 1991), which are perceived as raising few concerns among consumers, regulatory agencies or within the food industry (Dillon and Board, 1994; Nychas, 1995; Lopez-Malo et al., 1997; Basilico and Basilico, 1999; Etcheverry et al., 2002).

Essential oils are volatile products of plant secondary metabolism which in many cases are biologically active, with antimicrobial, antioxidant and bioregulatory properties (French, 1985; Caccioni et al., 1998). Essential oils of oregano, thyme, basil, garlic, onion and cinnamon have been reported having the greatest antimicrobial effectiveness (Paster et al., 1995; Basilico and Basilico, 1999; Cosentino et al., 1999; Yin and Tsao, 1999). More specifically, cloves and cinnamon have been found to be strong antifungal agents against *Penicillium* and *Aspergillus* species. Many of the essential oils that have been tested have been shown to have an antagonistic effect against aflatoxigenic *Aspergillus* spp. However, few studies have investigated the effects of essential oils on growth and ochratoxin A production by ochratoxigenic strains of *A. ochraceus* (Basilico and Basilico, 1999). Furthermore, many studies have not taken account of different environmental factors such as temperature and a_w on the effectiveness of essential oils in control of fungal growth and ochratoxin A production.

Resveratrol is a polyphenolic compound, which exhibits antimicrobial and antioxidant properties. It is a phytoalexin in grapes and is produced in response to various kinds of stress including attack by fungal pathogens. Resveratrol also has anticancer activity (Pervaiz, 2001) and antimicrobial activity against some dermatophytes (Man-Ying, 2002). The effect of resveratrol in combination with environmental variables on growth and ochratoxin A production by *P. verrucosum* and *A. ochraceus* has not been investigated.

Lactic acid bacteria have been used as biocontrol agents in malting trials (Haikara et al., 1993). *Lactobacillus plantarum* (VTT E-78076) was reported to have a fungistatic effect against *Fusarium* species *in vitro* and in laboratory-scale malting (Laitila et al., 1997, 2002).

One of the objectives in the present investigation was to determine the effect of a range of food-grade essential oils, antioxidants and resveratrol under different interacting a_w and temperature regimens on (a) growth rate, (b) ochratoxin A production by *P. verrucosum* and *A. ochraceus* isolates from wheat grain and (c) the concentration of essential oil needed for control of *P. verrucosum*. In addition, for barley and malting, the aims were to first investigate the effectiveness of lactic acid bacteria against the growth of *P. verrucosum* and subsequent production of ochratoxin A *in vitro* and in a mini-scale storage experiment. A pilot-scale experiment was also carried out to study mould growth and ochratoxin A formation during storage of barley at different moisture levels (WP6).

1.6. Monitoring Methods

Establishment of regulatory limits for ochratoxin A in Europe has led to the requirement for rapid monitoring methods for the presence of ochratoxin A. Conventional analysis of ochratoxin A by HPLC is expensive and does not offer a turn round time that is compatible with the needs of the industry. However, rapid immunodiagnostic methods offer a real alternative.

The concept of using immunological systems to detect mycotoxins is not new and was first reported by Aalund et al. (1975). Commercial kits to detect ochratoxin A were first introduced in the late 1980s and clearly demonstrate that assays can be made user friendly and rapid (around 20 minutes). At the start of the OTA PREV project, currently available immunoassays had generally been optimised to meet the USA requirements of around 20 ppb and at best were only sensitive down to about 8 ppb, which far exceeds the EU limits <3 or <5 ppb. In addition, the only assays that were available were in ELISA format, which are generally more suited to laboratory use.

An alternative to ELISA based detection systems is molecular imprinting. A template molecule is "imprinted" into a functional polymeric matrix, which maintains its steric and chemical recognition for the template once it has been removed, by chemical elution. The resulting material can be used to selectively rebind the template. These materials are extremely robust and can be used in a number of analytical methods including purification, separation and pre-concentration steps and in sensor design. In a part of this project, the aim was to imprint ochratoxin A, test the viability of the system and integrate the imprinted polymer into a suitable sensor platform (WP7).

There is a need for specific monoclonal antibodies (Mab) to ochratoxin A producing fungi. Generally suitable antibodies are developed in to ELISA systems but in certain applications, other formats such as lateral flow devices (LFD) may be more suitable. These are one step rapid systems that only take a few minutes to complete, can be performed by unskilled personnel and have been developed and are commercially available for a number of plant pathogens (Danks and Barker, 2000). One part of this project describes efforts to develop a specific Mab to the main ochratoxin A producing fungi (i.e. *P. verrucosum* and *A. ochraceus*) and the incorporation of a pre-existing Mab, specific to *Aspergillus* and *Penicillum* species, into an ELISA and LFD (WP8).

Ochratoxin A comprises a chlorinated isocoumarin derivative linked to *L*-phenyl-alanine. The biosynthetic pathway for ochratoxin

A has not yet been completely established; however, the isocoumarin group is a pentaketide skeleton formed from acetate and malonate via a polyketide synthesis pathway with the *L*-phenylalanine being derived from the shikimic acid pathway (Moss, 1996, 1998). No information currently is available on the enzymes or the genes responsible for any of these biosynthetic steps. In this project two studies have been performed with the objectives to clone and characterise genes involved in the biosynthesis of ochratoxin A and to develop nucleic acid sequence based methods to detect the presence of *A. ochraceus* and *P. verrucosum*. As ochratoxin A has a polyketide backbone, a polyketide synthase gene was targeted (WP9).

1.7. Reductive Measures During Processing

Drying and subsequently keeping grain under safe storage conditions should reduce or eliminate the risk of occurrence of ochratoxin A. However, this has often been found difficult to achieve in practice and surveys continue to find contaminated grain and cereal food products (e.g. Scudamore et al., 1999; Wolff, 2000). The extent to which ochratoxin A is degraded or lost during processing and whether it can be, at least partially, removed during the preparation of cereal products such as bread and breakfast cereals is uncertain. Relatively few studies have been carried out on the reduction of ochratoxin A during processing. Cooking of polished wheat using a procedure common in Egypt only removed 6% of ochratoxin A (El-Banna and Scott, 1984). A similar result was shown by Osborne et al. (1996) when whole wheat (both hard and soft) containing ochratoxin A at about 60 μg/kg was milled. There was a reduction in the ochratoxin A content in the white flour compared with the original grain, although subsequently only a small further reduction occurred when this was baked into bread. The fate of ochratoxin A during bread making (Subirade, 1996), malting and brewing (Baxter, 1996), the effects of processing on the occurrence of ochratoxin A in cereals (Alldrick, 1996), and animal feed (Scudamore, 1996) and effects of processing and detoxification treatments on ochratoxin A (Scott, 1996a) have all been reported.

Ochratoxin A is relatively stable once formed but, some breakdown occurs under high temperature, acid or alkaline conditions or in the presence of enzymes. Ochratoxin A tends to be concentrated in the outer bran layers of cereals, so that both reduction and increase in concentration can occur depending on the milled fraction examined (Osborne et al., 1996). However, in contrast, studies in Poland showed that cleaning and milling wheat and barley did not remove ochratoxin A in naturally contaminated samples and levels in flour and bran were the same as in the whole grains (Chelkowski et al., 1981).

Extrusion processing is used for producing a range of cereal products such as breakfast foods, snacks and animal feedstuffs. Commodities such as wholemeal flour are forced through a die under pressure, and the process causes mechanical shearing stresses, elevated temperatures and rapid expansions (Riaz, 2000; Guy, 2001). During this process, chemical bonds in polymers may break and free radicals may form. Therefore, a contaminant such as ochratoxin A may be subjected both to high temperatures (up to 200°C) and to chemical reactions by free radical mechanisms. Information about the destruction of ochratoxin A during extrusion may assist in developing strategies for reducing consumer exposure to this mycotoxin.

Boudra et al. (1995) studied the decomposition of ochratoxin A under different moisture and temperature conditions. In the presence of 50% water, ochratoxin A loss increased at 100°C and 150°C in comparison to drier grain, but decreased at 200°C. Little work has been published on the effect of extrusion on ochratoxin A in wheat although unpublished studies in the authors' laboratory suggested that loss below 150°C is relatively small. Extrusion processing has been shown to have variable effects on other mycotoxins. For example, approximately 15% of aflatoxins B_1 and B_2 survived extrusion at 150°C when spiked maize dough was treated (Martinez and Monsalve, 1989) although in a study by Cazzaniga et al., (2001) in which the effects of flour moisture content, temperature and addition of sodium metabisulphite were examined, the reduction of aflatoxin B_1 was only between 10 and 25%.

A part of this project was carried out to confirm and clarify these earlier results, to establish the change in concentration of ochratoxin A during milling, abrasive scouring of the grain and baking and to assess the pattern of the distribution of ochratoxin A in all milled fractions. In addition, the stability of ochratoxin A was examined during extrusion of wheat under a range of extrusion cooker conditions (WP10).

Malt is produced from barley by germinating the seed under regulated conditions of moisture and temperature. The barley moisture content is raised from 14-16% (storage level) to roughly 45% by steeping for 1-2 days at 12-20°C. The seed is then germinated at 15-20°C for 4-6 days. The germinated barley, or green malt, is dried to about 4-5% moisture in a kilning step where the temperature is increased to 85°C within 21 h.

The malting process can provide conditions favourable for the growth of toxigenic fungi. Moist conditions during germination and the initial stages of kilning are conducive to the growth of many fungi. The extent of the development of mycotoxin producing fungi depends largely on the initial contamination of the barley and on the vitality of the organisms present. Moreover problems of mould growth during malting have often been associated with elevated temperatures (Flannigan, 1996). If *P. verrucosum* is present on barley it could proliferate during malting and produce ochratoxin A especially at temperatures near 20°C. However, no published data is available on the formation of ochratoxin A during the malting process. The European malting industry is now facing the new challenges caused by the established legislation for maximum levels of ochratoxin A for raw cereal grains and products derived from cereals.

There is little data available on the fate of ochratoxin A during the brewing process. In experiments where ochratoxin A was added at various stages during malting and brewing or malt containing high levels of ochratoxin A was used, reduced amounts (13-32%) were always found in beer (Chu et al., 1975; Nip et al., 1975; Baxter et al., 2001). Losses to spent grains, uptake by the yeast and degradation, especially during mashing, were the main reasons for reduced amounts recovered in beer (Baxter et al.; 2001). Due to the thermo stable nature of ochratoxin A, it is not destroyed to any significant extent during kilning or wort boiling (Scott, 1996b; Baxter et al.; 2001). Surveys of ochratoxin A in commercial beer are regularly carried out in several countries. Very low concentrations of ochratoxin A have ever been detected, the maximum levels ranging from 0.026 to 0.33 µg/l (Vanne and Haikara, 2001). The final part of this project aimed at studying the formation of ochratoxin A during malting and the subsequent fate during the brewing process (WP11).

2. SUMMARY OF RESULTS FROM THE OTA PREV PROJECT

2.1. Determination of the Critical Control Points

Investigation of grain samples has revealed that *Penicillium verrucosum* is the main, if not the only, producer of ochratoxin A in European cereals (Lund and Frisvad, 2003). It was concluded that *P. verrucosum* infection was best detected on DYSG media after seven days at 20°C. Numbers of *P. verrucosum* found on DYSG and ochratoxin A content in cereals were correlated. Kernel infection with *P. verrucosum* of more than 7% indicated likely ochratoxin A contamination. In the action to identify critical control points for infection, it was found that the AFLP fingerprinting technique developed did not generate additional important information over that gained by the detection of *P. verrucosum* at species level by traditional taxonomic methods (Frisvad et al., 2005).

The sources of infection of the grain were the contaminated environments of combines, dryers, and silos. Prompt and effective drying of cereals at harvest is the major CCP for preventing the formation of ochratoxin A. In regions of Europe where the cereal harvest is at greatest risk, measures to avoid mould and toxin problems are often most effective, while areas normally at less risk may not be the best prepared to avoid storage problems when unusual conditions occur. It may not be economic to have expensive drying machinery idle some years while in others the supply of damp grain may exceed the drying capacity available. Delays in drying may then put the grain at risk. Another problem arises when the infrastructure is such that sufficient funds and expertise are unavailable to advise on and ensure best storage practice (Scudamore, 2003).

2.2. Specification of Critical Limits and Establishing Preventive Actions

The studies of the effect of temporal environmental factors on fungal growth, patterns of colonisation and ochratoxin A production revealed interesting characteristics, which may explain why *P. verrucosum* is the main ochratoxin A producer in cereal grain in Europe. Generally, *P. verrucosum* was dominant at lower a_w and 15°C, whereas *Aspergillus ochraceus* was dominant at higher a_w at 25°C. Furthermore, results indicated that *P. verrucosum* was less sensitive to higher concentrations of CO_2 than *A. ochraceus*, which may also be a competitive advantage during storage (Cairns et al., 2003). A mathematical model for safe storage time before onset of significant growth of *P. verrucosum* and ochratoxin A production has been developed, describing the effect of water activity and temperature on the rate of growth of *P. verrucosum* in cereal grain. The model is valid for aerobic conditions, for instance when drying grain in near-ambient dryers or cooling grain by aeration prior to high-temperature drying. The model is described in the final report for the OTA PREV project, which is available at www.slv.se/otaprev.

The probability, of ochratoxin A levels above the EC maximum limit of 5 μg/kg at different concentration of *P. verrucosum* in the grain, clearly increased when the levels of *P. verrucosum* were above 1000 colony forming units/gram (Lindblad et al., 2004). A mathematical model was developed which describes the risk for condensation in the headspace of a silo during storage of cereal grain. The model has been used to identify the conditions which cause moistening of the grain, and to develop control strategies to reduce this and the risk for mould growth and ochratoxin A production (see www.slv.se/otaprev). Essential oils, resveratrol and lactic acid bacteria (LAB) can control growth and ochratoxin A production by both *P. verrucosum* and *A. ochraceus* on grain (Ricelli et al., 2002; Cairns and Magan, 2003; Fanelli et al., 2003). However, in small scale storage experiments and experimental maltings, the inhibitory effect of the selected LAB strain could not be shown clearly. Of twenty four essential oils tested the most effective were found to be thyme, cinnamon leaf and clove bud.

2.3. Establishing Monitoring Systems

New diagnostic tools have become available that will provide the means for rapid determination of ochratoxin A in cereals. This will enable the effective implementation of the European legislation and facilitate future internal control and scientific studies. Immunoassays in ELISA format, sensitive enough to meet the EU legislation for ochratoxin A, have been developed where hundreds of samples can be analysed in a few hours. A lateral flow device (LFD) taking less than five minutes to perform, which can be used on-site, has also been developed (Danks et al., 2003, and unpublished results). These assays are in the prototype stage but it is anticipated that a full validation and comparison with CEN or the criteria mentioned in directive 2002/26/EC can be carried out in the near future. However, it has been shown that these techniques are sensitive enough for the EU legislation for ochratoxin A.

A number of genes involved in ochratoxin A biosynthesis have been cloned, among them a polyketide synthase gene. PCR primer pairs have been developed which appear to be highly specific for *A. ochraceus* and *P. verrucosum* (O'Callaghan et al., 2003). The primers may find use in the development of rapid identification protocols for ochratoxigenic fungi.

Several advances have been made towards a molecularly imprinted polymer specific for ochratoxin A and its integration into a solid phase extraction (SPE) and sensor systems. Several polymers have been designed using a computational method and tested using SPE. The materials demonstrate a high affinity and specificity for the target molecule in aqueous model samples, however integration in real samples with complex biological matrices (grain samples) has proved difficult as interfering compounds affect binding and measurements of ochratoxin A. Attempts to isolate and remove these interfering materials were unsuccessful and consequently the detection limits were not at the level required to meet the legislative requirements (Turner et al., 2003).

2.4. Establishing Corrective or Reductive Measures

This project (OTA PREV) has contributed tools and recommendations for the cereal processing industry. These will facilitate decisions to be made to enable the dual maximum levels for ochratoxin A described in the Commission Regulation (EC) No 472/2002 of 12 March 2002 setting maximum levels for ochratoxin A in foodstuffs to be followed.

Examining the fate of ochratoxin A during milling revealed white flour having the most significant reduction of ochratoxin A of about 50%. An initial cleaning stage and scouring (1-2%) prior to milling, removed small amounts of ochratoxin A. Baking resulted in only a small fall in concentration. However, an overall reduction of about 80% is achievable for white bread with scouring included and up to 35% for wholemeal bread (Scudamore et al., 2003; 2004).

The increase of ochratoxin A concentration during malting was 2-4-fold in 75 % of the samples studied and process temperature had a pronounced effect. At the higher temperatures of 16-18°C ochratoxin A formation was 20-fold compared to 5-fold at the temperatures of 12-14°C. During the brewing process approximately 20% of the original ochratoxin A from the malt remained in the beer (Lehtonen and Haikara, 2002; Haikara et al., 2003).

3. ACKNOWLEDGEMENT

This work was supported by the European Commission, Quality of Life and Management of Living Resources Programme (contract no. QLK1-CT-1999-00433). The authors wish to thank all Project Participants and the Scientific Officer Achim Boenke for their work and constructive criticism during the progress of project. The help provided and valuable discussion with Claudine Vandemeulebrouke

(EUROMALT), Guislaine Veron Delor (IRTAC), Hans de Keijzer (GAM/COCERAL), Esko Pajunen (European Brewery Convention) and Roger Williams (Home Grown Cereals Authority, UK) have been highly appreciated. The contributions and support of Keith Scudamore (KAS Mycotoxins), who has been subcontracted to this project, have been very valuable.

4. REFERENCES

Aalund, O., Brunfeld, K., Hald, B., Krogh, P., and Poulsen, K., 1975, A radio immunoassay for ochratoxin A: a preliminary investigation, *Acta Path. Microbiol. Scand. Sect. C* **83**:390-392.

Alldrick, A. J., 1996, The effect of processing on the occurrence of ochratoxin A in cereals, *Food Addit. Contam.* **13** (suppl.):27-28.

Anderson, J. P. E., and Domsch K. H., 1975, Measurement of bacterial and fungal contributions to respiration of selected agricultural and forest soils, *Can. J. Microbiol.* **21**:314-322.

Arenal, F., Platas, G., Martín, J., Salazar, O., and Peláez, F., 1999, Evaluation of different PCR-based DNA fingerprinting techniques for assessing the genetic variability of isolates of the fungus *Epicoccum nigrum*, *J. Appl. Microbiol.* **87**:898-906.

Bailey J. E., 1992, Whole grain storage, in: *Storage of Cereal Grains and their Products*, 4th edition, D. B. Sauer, ed., American Association of Cereal Chemists, Inc., St Paul, MN, pp. 157-182.

Bakkeren, G., Kronstad, J. W., and Lévesque, A. C., 2000, Comparison of AFLP fingerprints and ITS sequences as phylogenetic markers in Ustilaginomycetes, *Mycologia* **92**:510-521.

Basilico, M. Z. and Basilico, J. C., 1999, Inhibitory effects of some spice essential oils on *Aspergillus ochraceus* NRRL 3174 growth and ochratoxin A production, *Lett. Appl. Microbiol.* **29**:238-241.

Baxter E. D., 1996, The fate of ochratoxin A during malting and brewing, in: *Occurrence and Significance of Ochratoxin A in Food*, Workshop held 10-12 January, 1996, in Aix-en-Provence, France, *Food Addit. Contam.* **13** (Suppl.):23-24.

Baxter, E. D., Slaiding, I. R., and Kelly, B., 2001, Behavior of ochratoxin A in brewing, *J. Am. Soc. Brew. Chem.* **59**:98-100.

Boudra, H., Le Bars, P., and Le Bars, J., 1995, Thermostability of ochratoxin A in wheat under two moisture conditions, *Appl. Environ. Microbiol.* **61**:1156-1158.

Brooker D. B. F. W., Bakker-Arkema, and Hall C. W., 1992, *Drying and Storage of Grains and Oilseeds*, Van Nostrand Reinhold, New York.

Caccioni, D. R. L., Guizzardi, M., Biondi, D. M., Renda, A., and Ruberto, G., 1998, Relationships between volatile components of citrus fruits, essential oils and antimicrobial action on *Penicillium digitatum* and *P. italicum*, *Int. J. Food Microbiol.* **43**:73-79.

Cairns, V., Hope, R., and Magan, N., 2003, Environmental factors and competing mycoflora affect growth and ochratoxin production by *Penicillium verrucosum* on wheat grain, *Aspects Appl. Biol.* **68**:81-90.

Cairns, V. and Magan, N., 2003, Impact of essential oils on growth and ochratoxin A production by *Penicillium verrucosum* and *Aspergillus ochraceus* on a wheat-based substrate, in: *Advances in Stored Product Protection*, P. Credland, D. M. Armitage, C. H. Bell, and P. M. Cogan, eds, CABI International, Wallingford, U.K., pp. 479-485.

Casada M. E., and Young J. H., 1994, Model for heat and moisture transfer in arbitrarily shaped two-dimensional porous media, *Trans. ASAE* **37**(6):1927-1938.

Cazzaniga, D., Basilico, J. C., Gonzalez, R. J., Torres, R. L., and Degreef, D. M., 2001, Mycotoxins inactivation by extrusion cooking of corn flour, *Lett. Appl. Microbiol.* **33**:144-147.

Chang C. S., Converse H. H. and Steele J. L., 1994, Modelling of moisture content of grain during storage with aeration, *Trans ASAE* **37**(6):1891-1898.

Chelkowski, J., Golinski, P., and Szebiotko, K., 1981, Mycotoxins in cereal grain. II. The fate of ochratoxin A after processing of wheat and barley grain, *Nahrung,* **25**:423-426.

Chu, F. S., Chang, C. C., Ashoor, S. H. and Prentice, N., 1975, Stability of aflatoxin B and ochratoxin A in brewing, *Appl. Microbiol.* **29**:313-316.

Ciegler, A., Fennell, D. I., Sansing, G. A., Detroy, R. W., and Bennett, G. A., 1973, Mycotoxin producing strains of *Penicillium viridicatum*: classification into subgroups, *Appl. Microbiol.* **26**:271-278.

Cossentino, S., Tuberoso, C. I. G., Pisano, B., Satta, M., Masca, U., Arzed, E., and Palmas, F., 1999, In vitro antimicrobial activity and chemical composition of Sardinian thyme essential oils, *Lett. Appl. Microbiol.* **29**:130-135.

Danks, C.. and Barker, I., 2000, On-site detection of plant pathogens using lateral-flow devices, *OEPP/EPPO Bulletin*, **30**: 421-426.

Danks, C., Ostoja-Starzewska, Flint, J., and Banks, J. N., 2003, The development of a lateral flow device for the discrimination of OTA producing and non-producing fungi, *Aspects Appl. Biol.* **68:**21-28.

Dillon, V. M., and Board, R. G., 1994, Future prospects for antimicrobial food preservation systems, in: *Natural Antimicrobial Systems and Food Preservation*. V. M. Dillon and R. G. Board, eds, Technomic Publishing Co. Inc., Lancaster, pp. 397-410.

Dragoni, I. and Cantoni, C., 1979, Le muffe negli insaccati crudi stagionati, *Ind. Aliment*. **19:**281-284.

El-Banna, A. A., and Scott, P. M., 1984, Fate of mycotoxins during processing. III. Ochratoxin A during cooking of faba beans (*Vicia faba*) and polished wheat, *J. Food Prot*. **47**:189-192.

Etcheverry, M., Torres, A., Ramirez, M. L., Chulze, S. and Magan, N., 2002, *In vitro* control of growth and fumonisin production by *F. verticillioides* and *F. proliferatum* using anti-oxidants under different water availability and temperature regimes, *J. Appl. Microbiol.* **92**:624-632.

European Commission (EC), 2002, Commission regulation No 472/2002 *Official Journal L 075, 16/03/2002, 18-20.*

Fanelli C., Taddei F., Trionfetti Nisini P., Jestoi M., Ricelli A., Visconti A., and Fabbri, A. A., 2003, Use of resveratrol and BHA to control fungal growth and mycotoxin production in wheat and maize seeds, *Aspects Appl. Biol.* **68**:63-71.

Flannigan, B., 1996, Mycotoxins in malting and brewing, in: *Mycotoxins in Cereals. An emerging Problem?* J. P. F. D'Mello, ed., Handbook for Fourth SAC Conference, October, 1996, pp. 45-55.

Foegeding, P. M., and Busta, F. F., 1991, Chemical food preservatives, in: *Disinfections, Sterilisation, and Preservation*, S. S. Block, ed., Lea and Febiger, Malvern, PA, pp. 802-832.

Foster G. H., and Tuite J., 1992, Aeration and stored grain management, in: *Storage of Cereal Grains and their Products*, 4th edition, D. B. Sauer, ed., Am. Ass. of Cereal Chem., Inc., St Paul, Minnesota, pp. 220-222.

French, R.C., 1985, The bio-regulatory action of flavour compounds on fungal spores and other propagules, *Ann. Rev. Phytopathol.* **23**:173-199.

Frisvad, J. C., 1985, Profiles of primary and secondary metabolites of value in classification of *Penicillium viridicatum* and related species, in: *Advances in* Penicillium *and* Aspergillus *Systematics,* R. A. Samson, and J. I. Pitt, eds, Plenum Press, New York, pp. 311-325.

Frisvad, J. C., 1989, The connection between the Penicillia and Aspergilli and mycotoxins with special emphasis on misidentified isolates, *Arch. Environ. Contam. Toxicol.* **18**:452-467.

Frisvad, J. C., Filtenborg, O., Lund, F., and Thrane, U., 1992, New selective media for the detection of toxigenic fungi in cereal products, meat and cheese, in: *Modern Methods in Food Mycology*, R. A. Samson, A. D. Hocking, J. I. Pitt, and A. D. King, eds, Elsevier Science Publishers, Amsterdam, Netherlands, pp. 275-284.

Frisvad, J. C., Lund, F., and Elmholt, S., 2005, Ochratoxin A producing *Penicillium verrucosum* isolates from cereals reveal large AFLP fingerprinting variability, *J. Appl. Microbiol.* **98**:684-692.

Frisvad, J. C., and Samson, R. A., 2000, *Neopetromyces* gen. nov. and an overview of teleomorphs of *Aspergillus* subgenus *Circumdati, Stud. Mycol.* **45**:201-207.

Frisvad, J. C. and Viuf, B. T., 1986, Comparison of direct and dilution plating for detecting *Penicillium viridicatum* in barley containing ochratoxin, in: *Methods for the Mycological Examination of Food,* A. D. King, J. I. Pitt, L. R. Beuchat, and J. E. L. Corry, eds, Plenum Press, New York, pp. 45-47.

Guy, R. C. E., 2001, *Extrusion Cooking: Technologies and Applications,* Woodhead Publishing, Cambridge, 206 pp.

Haikara, A., Lehtonen, S., Sarlin,T., Vanne, L., Jestoi, M., and Rizzo, A., 2003, The growth of *Penicillium verrucosum* and formation of ochratoxin A during malting, *Bulletin of EBC Symposium*, Mycotoxins and other Contaminants in the Malting and Brewing Industries, Brussels 26-28 January 2003 (*Bulletin is not an official publication*).

Haikara, A., Uljas, H. and Suurnäkki, A., 1993, Lactic starter cultures in malting a novel solution to gushing problems, *Proc. 24th Congr. Eur. Brew. Conv.*, Oslo, pp. 163-172.

Hellevang, K. J., and Hirning, H. J., 1988, Moisture movement in stored grain during summer, *Am. Soc. Agric. Eng. Tech. Pap.* 88-6052.

Hocking, A. D. and Pitt, J. I., 1980, Dichloran-glycerol medium for enumeration of xerophilic fungi from low-moisture foods, *Appl. Environ. Microbiol.* **39**:488-492.

Jannsen, P., Coopman, R., Huys, G., Swings, J., Bleeker, M., Vos, P., Zabeau, M., and Kersters, K., 1996, Evaluation of the DNA fingerprinting method AFLP as a new tool in bacterial taxonomy, *Microbiology (UK)* **142**:1881-1893.

JECFA (Joint FAO/WHO Expert Committee on Food Additives), 2001, Ochratoxin A, in: *Safety Evaluation of Certain Mycotoxins in Food.* Prepared by the Fifty-sixth meeting of the JECFA. FAO Food and Nutrition Paper 74, Food and Agriculture Organization of the United Nations, Rome, Italy.

Jonsson, N., and Pettersson, H., 1992, Comparison of different preservation methods for grain, in: *Cereals in the Future Diet*, Proc. 24 Nordic Cereal Congress, Stockholm 1990, H. Johansson, ed., Nordic Cereal Association, Lund, pp. 357-364.

Khankari, K. K., Morey, R. V., and Patankar, S. V., 1995, Application of a numerical model for prediction of moisture migration in stored grain, *Trans. ASAE* **38**(6):1789 – 1804.

Kothera, R. T., Keinath, A. P., Dean, R. A., and Franham, M. W., 2003, AFLP analysis of a worldwide collection of *Didymella bryoniae*, *Mycol. Res.* **107**:297-304.

Kreyger J., 1972, *Drying and Storing Grains, Seeds and Pulses in Temperate Climates,* Institute for Storage and Processing of Agricultural Produce, (IBVL), publication 205, Wageningen, Netherlands.

Kure, C. F., Skaar, I., Holst-Jensen, A., and Abeln, E. C. A., 2003, The use of AFLP to relate cheese-contaminating *Penicillium* strains to specific points in the production plant, *Int. J. Food Microbiol.* **83**: 195-204.

Laitila, A., Alakomi, H.-L., Raaska, L., Mattila-Sandholm, T., and Haikara, A., 2002, Antifungal activities of two *Lactobacillus plantarum* strains against *Fusarium* moulds *in vitro* and in malting of barley, *J. Appl. Microbiol.* **93**:566-576.

Laitila, A., Tapani, K-M., and Haikara, A., 1997, Prevention of formation of *Fusarium* mycotoxins with lactic acid starter cultures during malting, Proc. 26th Congr. Eur. Brew. Conv., Maastricht, pp. 137-144.

Larsen, T. O., Svendsen, A., and Smedsgaard, J., 2001, Biochemical characterization of ochratoxin A producing strains of the genus *Penicillium*, *Appl. Environ. Microbiol.* **67**:3630-3635.

Lee, H. B., and Magan, N., 2000, Environmental factors influence *in vitro* interspecific interactions between *A. ochraceus* and other maize spoilage fungi, growth and ochratoxin production, *Mycopathologia* **146**:43-47.

Lehtonen, S., and Haikara, A., 2002, The growth of *Penicillium verrucosum* and formation of ochratoxin A during malting, *Mallas ja Olut* **3**:65-75 (in Finnish, summary in English).

Lindblad, M., Johnsson, P., Jonsson, N., Lindqvist, R., and Olsen, M., 2004, Predicting non-compliant levels of ochratoxin A in cereal grain from *Penicillium verrucosum* counts, *J. Appl. Microbiol.* **97**:609-616.

Lopez-Malo, A., Alzamora, S. M., and Argaiz, A., 1997, Effect of vanillin concentration, pH and incubation temperature on *Aspergillus flavus, Aspergillus niger, Aspergillus ochraceus* and *Aspergillus parasiticus* growth, *Food Microbiol.* **14**: 117-124.

Lund, F., and Frisvad, J. C., 2003, *Penicillium verrucosum* in wheat and barley indicates presence of ochratoxin A, *J. Appl. Microbiol.* **95**:1117-1123

Lund, F., Nielsen, A. B., and Skouboe, P., 2003, Distribution of *Penicillium commune* isolates in cheese dairies mapped using secondary metabolite profiles, morphotypes, RAPD and AFLP fingerprinting, *Food Microbiol.* **20**:725-734.

Lund, F., and Skouboe, P., 1998. Identification of *Penicillium caseifulvum* and *P. commune* isolates related to specific cheese and rye bread factories using RAPD fingerprinting, *J. Food Mycol.* **1**:131-139.

Lundin, G., 1998, Influence of store design on grain quality. Sampling from eight farm stores, Report No. 250 from JTI – Swedish Institute of Agricultural and Environmental Engineering, Uppsala, Sweden.

Magan, N., and Lacey, J. 1984, Effect of gas composition and water activity on growth of field and storage fungi and their interactions, *Trans. Br. Mycol. Soc.* **82**:305-314.

Magan, N., and Lacey, J., 1985a, Interaction between field and storage fungi on wheat grain, *Trans. Br. Mycol. Soc.* **83**:29-37.

Magan, N., and Lacey, J., 1985b, The effect of water activity and temperature on mycotoxin production by *Alternaria alternata* in culture and on wheat grain, in: *Trichothecenes and Other Mycotoxins,* J. Lacey, ed., John Wiley and Sons, London, pp. 243-250.

Maier, D. E., 1992, The chilled aeration and storage of cereal grains, Ph. D. Thesis, Michigan State University, East Lansing, MI.

Maier D. E., and Montross, M. D., 1998, Modelling aeration and storage management strategies, paper presented at 7th International Working Conference on Stored-Product Protection, Beijing China 1998.

Majer, D., Mithen, R., Lewis, B. G., Vos, P., and Oliver, R. P., 1996, The use of AFLP fingerprinting for the detection of genetic variation in fungi, *Mycol. Res.* **100**: 1107-1111.

Man-Ying Chan, M., 2002, Antimicrobial effect of resveratrol on dermatophytes and bacterial pathogens of the skin, *Biochem. Pharmacol.* **63**:99-104.

Marin, S., Sanchis, V., Arnau, F., Ramos, A. J., and Magan, N., 1998, Colonisation and competitiveness of *Aspergillus* and *Penicillium* species on maize grain in the presence of *Fusarium moniliforme* and *Fusarium proliferatum*, *Int. J. Food Microbiol.* **45**:107-117.

Martin, S., Tuite, J., and Diekman, M. A., 1989, Inhibition radioimmunoassay for *Aspergillus repens* compared with other indices of fungal growth in stored corn, *Cereal Chem.* **66** (3): 39-144.

Martinez, A. J., and Monsalve, C., 1989, Aflatoxin occurrence in 1985-86 corn from Venezuela and its destruction by the extrusion process, in: *Biodeterioration Research 2*, C. E. O'Rear and G. C. Llewellyn, eds, Plenum Press, New York, pp. 251-259.

Meiering A. G., 1986, Oxygen control in sealed silos, *Trans. ASAE* **29**: 218-222.

Metzger, J. F., and Muir, W. E., 1983, Computer model of two-dimensional conduction and forced convection in stored grain, *Can. Agric. Eng.* **25**:119-125.

Montross, M. D., Maier, D. E., and Haghighi, K., 2002, Validation of a finite-element stored grain ecosystem model, *Trans. ASAE* **45**:1465-1474.

Moss, M. O., 1996, Mode of formation of ochratoxin A, *Food Addit. Contam.* **13** (suppl.):5-9.

Moss, M .O., 1998, Recent studies of mycotoxins, *J. Appl. Microbiol.* **84**:62S-76S.

Nip, W. K., Chang, F. C., Chu, F. S. and Prentice, N., 1975, Fate of ochratoxin A in brewing, *Appl. Microbiol.* **30**:1048-1049.

NMKL (Nordic Committee on Food Analysis), 1995, No. 152, *Penicillium verrucosum*: determination in foods and feedstuffs (available on www.nmkl.org).

Northolt, M. D., Van Egmond, H. P., and Paulsch, W. E., 1979, Ochratoxin A production by some fungal species in relation to water activity and temperature, *J. Food Prot.* **42**:485-490.

Nychas, G. J. E., 1995, Natural antimicrobials from plants, in: *New Methods of Food Preservation,* G. W. Gould, ed., Blackie Academic and Professional, Glasgow, pp. 58-89.

O'Callaghan, J., Caddick, M. X., and Dobson, A. D. W., 2003, A polyketide synthase gene required for ochratoxin A biosynthesis in *Aspergillus ochraceus, Microbiology* **149**:3485-3491.

Olsen, M., Möller, T., and Åkerstrand, K., 1993, Ochratoxin A: occurrence and intake by Swedish population, in: *Proceedings of the United Kingdom Workshop on the Occurrence and Significance of Mycotoxins*, K. A. Scudamore, ed., Central Science Laboratory, MAFF, Slough, 21-23 April 1993, pp. 96-100.

Osborne, B. G., Ibe, F. I., Brown, G. L., Patagine, F., Scudamore, K. A., Banks, J. N., and Hetmanski, M. T., 1996, The effects of milling and processing on wheat contaminated with ochratoxin A, *Food Addit. Contam.* **13**: 141-153.

Paster, N., Menasherov, M., Ravid, U., and Juen, B., 1995, Antifungal activity of oregano and thyme essential oils applied as fumigants against fungi attacking stored grain, *J. Food Prot.* **58**:81-85.

Pervaiz, S., 2001, Resveratrol: from the bottle to the bedside? *Leukemia Lymphoma* **40**:491-498.

Pitt J. I., 1987, *Penicillium viridicatum, Penicillium verrucosum* and production of ochratoxin A, *Appl. Environ. Microbiol.* **53**:266-269.

Puntaric, D., Bosnir, J., Smit, Z., Skes, I., and Baklaic, Z., 2001, Ochratoxin A in corn and wheat: geographical association with endemic nephropathy, *Croatian Med. J.* **42**:175-180.

Ramakrishna, N., Lacey, J., and Smith, J. E., 1996, Colonization of barley grain by *Penicillium verrucosum* and ochratoxin A formation in the presence of competing fungi, *J. Food Prot.* **59**:1311-1317.

Ramakrishna, N., Lacey, J., Smith, J. E., 1993, Effects of water activity and temperature on the growth of fungi interacting on barley grain, *Mycol. Res.* **97**:1393-1402.

Riaz, M. N., 2000, *Extruders in Food Applications*, Technomic Publishing Co. Inc., Lancaster, PA, 223 pp.

Ricelli, A., Fabbri, A. A., Trionfetti-Nisini, P., Reverberi, M., Zjalic, S., and Fanelli, C., 2002, Inhibiting effect of different edible and medicinal mushrooms on the growth of two ochratoxigenic microfungi, *Int. J. Medicinal Mushrooms* **4**:173-179.

SCF, 1998. Opinion of the Scientific Committee on Food on Ochratoxin A (available at the European Commission web page: http://www.europa.eu.int/comm/food/fs/sc/* scf/out14_en.html.)

Schmidt, H., Ehrmann, M., Vogel, R., Taniwaki, M. H., and Niessen, L., 2003, Molecular typing of *Aspergillus ochraceus* and construction of species specific SCAR-primers based on AFLP, *System. Appl. Microbiol.* **26**:138-146.

SCOOP, 2002, Assessment of dietary intake of ochratoxin A by the population of EU member states, European Commission web page: http://europa.eu.int/comm/food/fs/scoop/3.2.7_en.pdf

Scott, P. M., 1996a, Effects of processing and detoxification treatments on ochratoxin A, *Food Addit. Contam.* **13** (suppl):19-22

Scott, P. M., 1996b, Mycotoxins transmitted into beer from contaminated grains during brewing, *J. AOAC Int.*, **79**:875-882.

Scudamore, K. A., 1996, Ochratoxin A in animal feed-effects of processing, *Food Addit. Contam.* **13** (suppl.): 39-42.

Scudamore, K. A., 2003, Report on questionnaire replies from cereal experts in Europe. Report available at www.slv.se/otaprev

Scudamore, K. A., Banks, J. N., and MacDonald, S. J., 2003, Fate of ochratoxin A in the processing of whole wheat grains during milling and bread production, *Food Addit. Contam.* **20**:1153-1163.

Scudamore, K. A., Banks, J. N., and MacDonald, S. J., 2004, Fate of ochratoxin A in the processing of whole wheat grain during extrusion, *Food Addit. Contam.* **21**:488-497.

Scudamore, K. A., Patel, S., and Breeze, V., 1999, Surveillance of stored grain from the 1997 harvest in the United Kingdom for ochratoxin A, *Food Addit. Contam.* **16**:281-290.

Seitz, L. M., Sauer, D. B., and Mohr, H. E., 1982, Storage of high-moisture corn: fungal growth and dry matter loss, *Cereal Chem.* **59**:100-105.

Smith, E. A., and Sokhansanj, S., 1990, Moisture transport caused by agricultural convection in grain stores, *J. of Agric. Engng Res.* **47**:23-34.

Steel, I. F., Saul, R. A., and Hukill, W.V., 1969, Deterioration rate of shelled corn as measured by carbon dioxide production, *Trans. ASAE* **12**:685-689.

Subirade, I., 1996, Fate of ochratoxin A during breadmaking, *Food Addit. Contam.* **13**(suppl.):25-26.

Tanaka, H., and Yoshida, K., 1984, Heat and mass transfer mechanisms in a grain storage silo, in: *Engineering and Food,* B. M. McKenna, ed., Elsevier, New York, NY.

Tooley, P. W., O'Neill, N. R., Goley, E. D., and Carras, M. M., 2000, Assessment of diversity in *Claviceps africana* and other *Claviceps* species by RAM and AFLP analysis, *Phytopathology* **90**:1126-1130.

Turner N. W., Piletska, E. V., Karim, K., Whitcombe, M., Malecha, M., Magan N., Baggiani C., and Piletsky, S. A., 2004, Effect of the solvent on recognition properties of molecularly imprinted polymer specific for ochratoxin A, *Biosen. Bioelectron.* **20**:1060-1067.

Vanne, L., and Haikara, A., 2001, Mycotoxins in the total chain from barley to beer, Proc. 28th Congr. Eur. Brew. Conv., Budapest 2001, CD-ROM. pp. 839-848.

Vos, P., Hogers, R., Bleeker, M., Reijans, M., van de Lee, T., Hornes., M., Frijters, A., Pot, J., Peleman, J., Kuiper, M., and Zabeau, M., 1995, AFLP: a new technique for DNA fingerprinting, *Nucl. Acids Res.* **23**: 4407-4414.

Wilson, M., and Lindow, S. E., 1994a, Co-existence among epiphytic bacterial populations mediated through nutritional resource partitioning, *Appl. Environ. Microbiol.* **60**:4468-4477.

Wilson, M., and Lindow S E., 1994b, Ecological similarity and coexistence of epiphytic ice-nucleating (Ice +) *Pseudomonas syringae* strains and a non-ice-nucleating (Ice –) biological control agent, *Appl. Environ. Microbiol.* **60**:3128-3137.

White, N. D. G., Sinha, R. N., and Muir, W. E., 1982, Intergranular carbon dioxide as an indicator of biological activity associated with the spoilage of stored wheat, *Can. Agric. Eng.* **24**:35-42.

Wolff, J., 2000, Ochratoxin A in cereals and cereal products, *Arch. Lebensmittelhygiene* **51**:85-88.

Yin, M. C., and Tsao, S. M., 1999, Inhibitory effect of seven *Allium* plants upon three *Aspergillus* species, *Int. J. Food Microbiol.* **49**:49-56.

RECOMMENDED METHODS FOR FOOD MYCOLOGY

In the opinion of the International Commission on Food Mycology (ICFM), the following methods are the most satisfactory currently available for the mycological examination of foods. The methods outlined below are based on those published by Pitt et al. (1992) and have been developed after numerous collaborative studies carried out by ICFM members since its inception. Formulations for all media discussed appear in the Media Appendix.

1. GENERAL PURPOSE ENUMERATION METHODS AND MEDIA

1.1. Methods

1.1.1. Dilution plating

Dilution plating is recommended for liquid foods and powders, and also for particulate foods where the total mycoflora is of importance, as for example in grains intended for manufacture of flour.

Samples: Samples should be as representative as possible. For suitable microbiological sampling procedures, refer to the publication by the International Commission on Microbiological Specifications for Foods (ICMSF, 1986).

The sample size should be as large as possible, consistent with equipment used for homogenisation. If a Stomacher 400 is used, 10 to 40 g samples are suitable.

Diluents and dilution: The recommended diluent for fungi including yeasts is 0.1% aqueous peptone. Dilution should be 1:10 (1+9).

Homogenisation: A Coleworth Stomacher or equivalent Bag Mixer is the preferred type of homogeniser, used for 2 minutes per sample. Blending for 30-60 sec or shaking in a closed bottle with glass beads for 2-5 min are less preferable alternatives.

Plating: For mycological studies, spread plates are recommended; pour plates are less effective. Inocula should be 0.1 ml per plate, spread with a sterile, bent glass rod.

Incubation: For general purpose enumeration, 5 days incubation as 25°C is recommended. Plates should be incubated upright. A higher temperature, e.g. 30°C, is suitable in tropical regions.

1.1.2. Direct plating

Direct plating is considered to be the more effective technique for mycological examination of particulate foods such as grains and nuts. In most situations, surface disinfection (also commonly referred to as surface sterilisation) before direct plating is considered essential, to permit detection and enumeration of fungi actually invading the food. An exception is to be made for cases where surface contaminants become part of the downstream mycoflora, e.g. wheat grains to be used in flour manufacture. In such cases, grains should not be surface disinfected.

Surface disinfection: Surface disinfect food particles by immersion in 0.4% chlorine solution (household bleach, diluted 1:10) for 2 minutes. A minimum of 50 particles should be disinfected and plated. The chlorine solution should be used only once.

Rinse: After pouring off chlorine solution, rinse once in sterilised distilled or deionised water. Note: this step has not been shown to be essential, but is generally recommended.

Surface plate: As quickly as possible, transfer food particles with sterile forceps to previously poured and set plates, at the rate of 5-10 particles per plate, depending on the size of the particles.

Incubation: The standard incubation regimen for general purpose enumeration is 25°C for 5 days. A higher temperature (30°C) may be used in the tropics. Plates should be incubated upright.

Results: Express results as per cent of particles infected by fungi. Differential counting of a variety of genera is possible using a stereomicroscope.

1.2. Media

Dichloran Rose Bengal Chloramphenicol agar (DRBC; King et al., 1979; Pitt and Hocking 1997) and **Dichloran 18% Glycerol agar** (DG18; Hocking and Pitt, 1980; Pitt and Hocking, 1997) (see Media Appendix) are recommended as general purpose isolation and enumeration media.

DRBC is recommended for fresh foods, including fruits and vegetables, meats and dairy products. Note that media containing rose bengal are sensitive to light. Inhibitory compounds are produced after relatively short exposures to light. Prepared media should be stored protected from light until used.

For foods of reduced water activity, i.e. less than 0.95 a_w, DG18 is preferred. Although originally formulated for enumeration of xerophilic fungi, DG18 is now widely used as an effective generally purpose medium, both for foods and for sampling of indoor air (Hoekstra et al., 2000). Its water activity (0.955) reduces interference from both bacteria and rapidly growing fungi.

Chloramphenicol is the antibacterial agent of choice as it is heat stable and can be autoclaved after incorporation into the agar. It should generally be incorporated at a concentration of 100 mg/kg. If a high bacterial population is expected (as in soil or fresh meat), the concentration of chloramphenicol may be doubled, or an equal concentration of a second antibiotic such as oxytetracycline can be added aseptically after autoclaving.

2. SELECTIVE MEDIA

2.1. Media for xerophilic fungi

Dichloran 18% Glycerol agar (DG18; Hocking and Pitt 1980; Pitt and Hocking 1997) is the recommended medium for enumeration of common xerophilic fungi in foods. Incubation at 25°C for up to 7 days is recommended. Growth of *Eurotium* species on DG18 is rather rapid, and colonies do not have discrete margins.

For isolation of extreme xerophiles, e.g. *Xeromyces bisporus, Eremascus* and xerophilic *Chrysosporium* species, **Malt Yeast 50% Glucose agar** (MY50G; Pitt and Hocking, 1997) is recommended. Direct plating should be used. Incubate plates at 25°C and examine after

7 days and, if no growth appears, after longer periods, up to 21 days. Plates should be incubated in polyethylene bags to prevent desiccation.

2.2. Media for *Fusarium* species

The most effective medium for isolation of *Fusarium* species from foods is **Czapek Iprodione Dichloran agar** (CZID; Albidgren et al., 1987). Dichloran Chloramphenicol Peptone agar (DCPA; Andrews and Pitt, 1986) can also be used, but is less selective.

2.3. Media for toxigenic *Penicillium* species

Dichloran Rose bengal Yeast Extract Sucrose agar (DRYES; Frisvad, 1983; Samson et al., 2004) is useful for detecting and distinguishing *Penicillium verrucosum* and *P. viridicatum*, particularly in grains from cool climates.

2.4. Media for species producing aflatoxins

Aspergillus Flavus and Parasiticus Agar (AFPA; Pitt et al., 1983; Pitt and Hocking, 1997) is effective for detecting and enumerating *Aspergillus flavus* and *A. parasiticus* and closely related species. Plates should be incubated at 30°C for 2-3 days to allow development of orange reverse colours in colonies of these species.

3. METHODS FOR YEASTS

3.1. Diluents

For enumeration of yeasts in high a_w foods and beverages, the recommended diluent is 0.1% aqueous peptone. For concentrates, syrups and other low a_w samples, 20 to 30% glucose (w/v) in 0.1% aqueous peptone is recommended.

3.2. General purpose media

For products such as beverages, where yeasts usually predominate, nonselective media such as Malt Extract Agar (Pitt and Hocking, 1997) or Tryptone Glucose Yeast extract agar (TGY; Pitt and

Hocking, 1997) plus chloramphenicol or oxytetracycline (100 mg/kg) are recommended.

For products where yeasts must be enumerated in the presence of moulds, DRBC is recommended.

For products where bacteria may be encountered, or are suspected (e.g. cultured dairy products, fresh meat products) examination of representative colonies under the microscope is important.

3.3. Detection of Low Numbers of Yeasts

For detection of low numbers of yeasts in liquid products, membrane filtration is recommended. If the substrate is not filterable, enrichment techniques are available but are not quantitative.

3.4. Preservative Resistant Yeasts

An effective medium for detection of preservative resistant yeasts is Tryptone Glucose Yeast extract agar with 0.5% acetic acid (TGYA) (Hocking, 1996). Malt Extract Agar with 0.5% acetic acid (MAc) may also be used, but the pH is lower than TGYA, and the medium may be too inhibitory for recovery of some injured cells.

4. PREPARATION OF DRIED SAMPLES FOR DILUTION PLATING

Where dried particulate foods such as cereal grains are to be dilution plated, soaking for 30 minutes in 0.1% aqueous peptone solution before homogenising is recommended.

5. ENUMERATION OF HEAT RESISTANT FUNGI

The method for detection of heat resistant fungi detailed in an earlier chapter of this volume (Houbraken and Samson, 2006) is based on the method previously endorsed by ICFM. Note that very acid samples should be adjusted to pH 3.5-4.0 with NaOH before heating. Care should be taken to avoid contamination during pouring of plates.. Heavily sporulating colonies of, for example, *Penicillium* and *Aspergillus* species generally indicate contamination. Such colonies should be ignored.

6. REFERENCES

Albidgren, M. P., Lunf, F., Thrane, U., and Elmholt, S., 1987, Czapek-Dox agar containing iprodione and dicloran as a selective medium for the isolation of *Fusarium* species, *Lett. Appl. Microbiol.,* **5**:83-86.

Andrews, S., and Pitt, J. I., 1986, Selective medium for isolation of *Fusarium* species and dematiaceous hyphomycetes from cereals, *Appl. Environ. Microbiol.* **51**:1235-1238.

Hocking, A. D., 1996, Media for preservative resistant yeasts: a collaborative study, *Int. J. Food Microbiol.* **29**:167-175.

Hoekstra, E. S., Samson, R. A., and Summerbell, R. C., 2000, Methods for the detections and isolation of fungi in the indoor environment, in: *Introduction to Food- and Airborne Fungi*, 6th edition, R. A. Samson, E. S. Hoekstra, J. C. Frisvad and O. Filtenborg, eds, Centraalbureau voor Schimmelcultures, Utrecht, Netherlands, pp. 298-305.

Houbraken, J., and Samson, R. A., 2006, Standardization of methods for detection of heat resistant fungi, in: *Advances in Food Mycology*, A. D. Hocking, J. I. Pitt, R. A. Samson and U. Thrane, eds, Springer, New York. pp. 107-111.

ICMSF (International Commission on Microbiological Specifications for Foods), 1986, *Sampling for Microbiological Analysis: Principles and Specific Applications,* 2nd edition, University of Toronto Press, Toronto.

King, A. D., Hocking, A. D., and Pitt, J. I., 1979, Dichloran-rose bengal medium for enumeration of molds from foods, *Appl. Environ. Microbiol.* **37**:959-964.

Pitt, J. I., and Hocking, A. D., 1997, *Fungi and Food Spoilage*, 2nd edition, Blackie Academic and Professional, London.

Pitt, J. I., Hocking, A. D., and Glenn, D. R., 1983, An improved medium for the detection of *Aspergillus flavus* and *A. parasiticus*, *J. Appl. Bacteriol.* **54**:109-114.

Pitt, J. I., Hocking, A. D., Samson, R. A., and King, A. D., 1992, Recommended methods for the mycological examination of foods, 1992, in: *Modern Methods in Food Mycology*, R. A. Samson, A. D. Hocking, J. I. Pitt and A. D. King, eds, Elsevier, Amsterdam, pp. 365-368.

Samson, R. A., Hoekstra, E. S., and Frisvad, J. C., eds, 2004, *Introduction to Food- and Airborne Fungi*, 7th edition, Centraalbureau voor Schimmelcultures, Utrecht, Netherlands, 389 pp.

APPENDIX 1 – MEDIA

All media are sterilized by autoclaving at 121°C for 15 min unless otherwise specified. Czapek agars may be made from basic ingredients, or their production can be simplified by the use of Czapek concentrate and Czapek trace metal solution which contain the required additional salts in the specified concentrations. Both options are provided in this appendix.

AFPA: Aspergillus flavus and parasiticus agar (Pitt et al., 1983; Pitt and Hocking 1997)

Peptone	10 g
Yeast extract	20 g
Ferric ammonium citrate	0.5 g
Chloramphenicol	100 mg
Agar	15 g
Dichloran	2 mg
	(0.2% w/v in ethanol, 1 ml)
Water, distilled	1 litre

After addition of all ingredients, sterilise by autoclaving at 121°C for 15 min. Final pH 6.0-6.5.

CMA: Corn Meal agar (Samson et al., 2004).

Add 60 g freshly ground cornmeal to 1 litre water. Heat to boiling and simmer gently for 1 h. Strain through cloth. Make volume up to 1 litre with water, add 15 g agar, heat to dissolve agar, then autoclave for 15 min at 121°C. Also available commercially.

CREA: Creatine Sucrose agar (Samson et al., 2004).

Creatine.1H_2O	3.0 g
Sucrose	30 g
KCl	0.5 g
$MgSO_4.7H_2O$	0.5 g
$FeSO_4.7H_2O$	0.01 g
$K_2HPO_4.3H_2O$	1.3 g
Bromocresol purple	0.05 g
Agar	15 g
Distilled water	1 litre

Final pH 8.0 ± 0.2 (adjust after medium is autoclaved)

Czapek concentrate (Pitt and Hocking, 1997)

$NaNO_3$	30 g
KCl	5 g
$MgSO_4.7H_2O$	5 g
$FeSO_4.7H_2O$	0.1 g
Water, distilled	100 ml

Czapek concentrate and trace metal solution (below) do not require sterilisation. The precipitate of $Fe(OH)_3$ which forms in time can be resuspended by shaking before use.

Czapek trace metal solution (Pitt and Hocking, 1997)

$CuSO_4.5H_2O$	0.5 g
$ZnSO_4.7H_2O$	1 g
Water, distilled	100 ml

Cz: Czapek Dox agar (Samson et al., 2004)

$NaNO_3$	3 g
K_2HPO_4	1 g
KCl	0.5 g
$MgSO_4$ż$7H_2O$	0.5 g
$FeSO_4$ż$7H_2O$	0.01 g
$ZnSO_4$ż$7H_2O$	0.01 g
$CuSO_4$ż$5H_2O$	0.005 g
Sucrose	30 g
Agar	20 g
Water, distilled	1 litre

Final pH 6.3 ± 0.2

CYA (1): Czapek Yeast Autolysate agar (Samson et al., 2004)

$NaNO_3$	3 g
K_2HPO_4	1 g
KCl	0.5 g
$MgSO_4$ċ$7H_2O$	0.5 g
$FeSO_4$ċ$7H_2O$	0.01 g
$ZnSO_4$ċ$7H_2O$	0.01 g
$CuSO_4$ċ$5H_2O$	0.005 g
Yeast extract (Difco)	5 g
Sucrose	30 g
Agar	20 g
Distilled water	1 litre

This formulation should be used to prepare CYA from base ingredients, without the use of Czapek concentrate or trace metal solutions.

CYA (2): Czapek Yeast extract agar (Pitt, 1979; Pitt and Hocking, 1997)

K_2HPO_4	1 g
Czapek concentrate	10 ml
Trace metal solution	1 ml
Yeast extract, powdered	5 g
Sucrose	30 g
Agar	15 g
Water, distilled	1 litre

Prepared using Czapek concentrate and Czapek trace metal solution. Table grade sucrose is satisfactory for use provided it is free from sulphur dioxide. Final pH 6.7.

CZID: Czapek iprodione dichloran agar (Albidgren et al., 1987; Pitt and Hocking 1997)

Sucrose	30 g
Yeast extract	5 g
Chloramphenicol	100 mg
Dichloran	2 mg
	(0.2% in ethanol, 1 ml)
Czapek concentrate	10 ml
Trace metal solution	1 ml
Agar	15 g
Water, distilled	1 litre
Iprodione (suspension)	1 ml

Add iprodione suspension [0.3 g Roval 50WP (Rhone-Poulenc Agro-Chemie, Lyon, France) in 50 ml sterile water, shaken before addition to medium] after autoclaving at 121°C for 15 min.

CY20S: Czapek Yeast Extract agar with 20% Sucrose (Pitt and Hocking, 1997; Samson et al., 2004)

K_2HPO_4	1 g
Czapek concentrate	10 ml
Yeast extract	5 g
Sucrose	200 g
Agar	15 g
Water, distilled	1 litre

Final pH 5.2.

DCPA: Dichloran chloramphenicol peptone agar (Andrews and Pitt, 1986; Pitt and Hocking 1997)

Peptone	15 g
KH_2PO_4	1 g
$MgSO_4.7H_2O$	0.5 g
Chloramphenicol	100 mg
Agar	15 g
Dichloran	2 mg
	(0.2% w/v in ethanol, 1 ml)
Water, distilled	1 litre

After addition of all ingredients, sterilise by autoclaving at 121°C for 15 min. Final pH 5.5.-6.0.

DG18: Dichloran 18% Glycerol agar (Hocking and Pitt, 1980; Pitt and Hocking 1997)

Glucose	10 g
Peptone	5 g
KH_2PO_4	1 g
$MgSO_4.7H_2O$	0.5 g
Glycerol, A.R.	220 g
Agar	15 g
Dichloran	2 mg
	(0.2% w/v in ethanol, 1 ml)
Chloramphenicol	100 mg
Water, distilled	1 litre

Add minor ingredients and agar to *ca* 800 ml distilled water. Steam to dissolve agar, then make to 1 litre with distilled water. Add glycerol:

note that the final concentration is 18% weight in weight, not weight in volume. Sterilise by autoclaving at 121°C for 15 min. Final a_w 0.955, pH 5.5 to 5.8.

DRBC: Dichloran Rose Bengal Chloramphenicol agar (Pitt and Hocking, 1997)

Glucose	10 g
Peptone, bacteriological	5 g
KH_2PO_4	1 g
$MgSO_4.7H_2O$	0.5 g
Agar	15 g
Rose bengal	25 mg
	(5% w/v in water, 0.5 ml)
Dichloran	2 g
	(0.2% w/v in ethanol, 1 ml)
Chloramphenicol	100 mg
Water, distilled	1 litre

Final pH 5.5 – 5.8. Store prepared media away from light; photoproducts of rose bengal are highly inhibitory to some fungi, especially yeasts. In the dark, the medium is stable for at least one month at 1–4°C. The stock solutions of rose bengal and dichloran need no sterilisation, and are also stable for very long periods.

DRYES: Dichloran Rose Bengal Yeast Extract Sucrose agar (Frisvad, 1983; Samson et al., 2004)

Yeast extract	20 g
Sucrose	150 g
Dichloran	2 mg
	(0.2% in ethanol, 1 ml)
Rose bengal	25 mg
	(5% w/v in water, 0.5 ml)
Agar	20 g
Chloramphenicol	0.1 g
Water, distilled	to 1 litre

Final pH 5.6 (adjusted after medium is autoclaved). This medium detects *P. verrucosum* and *P. viridicatum* by production of a purple reverse colour. Can be modified by adding 0.5 g $MgSO_4.7H_2O$.

HAY: Hay Infusion agar (Samson et al., 2004)

Sterilise 50 g hay in one litre of water at 121°C for 30 min. Strain through cloth and make volume up to 1 litre. Adjust pH to 6.2 with K_2HPO_4 and add 15 g agar. Autoclave for 15 min at 121°C.

MEA: Malt extract agar (Pitt and Hocking, 1997; Samson et al., 2004)

Malt extract, powdered	20 g
Peptone	1 g
Glucose	20 g
Agar	20 g
Water, distilled	1 litre

Commercial malt extract used for home brewing is satisfactory for use in MEA, as is bacteriological peptone. Do not sterilise for longer than 15 min at 121°C, as this medium will become soft on prolonged or repeated heating. Final pH 5.6.

MY50G: Malt extract yeast extract 50% glucose agar (Pitt and Hocking, 1997)

Malt extract	10 g
Yeast extract	2.5 g
Agar	10 g
Water, distilled,	to 500 g
Glucose, A.R.	500 g

Add the minor constituents and agar to *ca* 450 ml distilled water and steam to dissolve the agar. Immediately make up to 500 g with distilled water. While the solution is still hot, add the glucose all at once and stir rapidly to prevent the formation of hard lumps of glucose monohydrate. If lumps do form, dissolve them by steaming for a few minutes. Sterilise by steaming for 30 min; note that this medium is of a sufficiently low a_w not to require autoclaving. Food grade glucose monohydrate (dextrose) may be used in this medium instead of analytical reagent grade glucose, but allowance must be made for the additional water present. Use 550 g of $C_6H_{12}0_6.H_20$, and 450 g of the basal medium. Final a_w 0.89, final pH 5.3.

MME: Mercks Malt Extract agar (El-Banna and Leistner, 1988)

Malt extract	30 g
Soy peptone	3 g
$ZnSO_4$ċ$7H_2O$	0.01 g
$CuSO_4$ċ$5H_2O$	0.005 g
Agar	20 g
Distilled water	1 litre

Final pH 5.6

PCA: Potato Carrot Agar (Simmons, 1992; Samson et al., 2004)

Carrots	40 g
Potatoes	40 g

Wash, peel and chop carrots and potatoes. Boil carrots and potatoes separately in 1 litre water each for 5 min then filter off. Sterilise filtrates (121°C, 15 min), add 250 ml potato extract and 250 ml carrot extract to 500 ml distilled water, add 15 g agar and sterilise at 121°C for 15 min.

PDA: Potato Dextrose agar (Pitt and Hocking 1997; Samson et al., 2004),

Potatoes	250 g
Glucose	20 g
Agar	15 g
Water, distilled	to 1 litre

PDA prepared from raw ingredients is superior to commercially prepared media. Wash the potatoes, which should not be of a red skinned variety, and dice or slice, unpeeled, into 500 ml of water. Steam or boil for 30 to 45 min. At the same time, melt the agar in 500 ml of water. Strain the potato through several layers of cheese cloth into the flask containing the melted agar. Squeeze some potato pulp through also. Add the glucose, mix thoroughly, and make up to 1 litre with water if necessary. Autoclave at 121°C for 15 min. Final pH 5.6 ± 0.1.

SNA: Synthetischer nährstoffarmer agar (Nirenberg, 1976; Samson et al., 2004)

KH_2PO_4	1.0 g
KNO_3	1.0 g
$MgSO_4$ċ$7H_2O$	0.5 g
KCl	0.5 g
Glucose	0.2 g
Sucrose	0.2 g
Agar	20 g
Water, distilled	1 litre

Note: pieces of sterile filter paper may be placed on the agar. Recommended for cultivation of *Fusarium*, but also for poorly sporulating Deuteromycetes.

V8: V8 juice agar (Simmons, 1992; Samson et al., 2004)

V-8 juice	200 ml
$CaCO_3$	3 g
Water, distilled	1 litre

V-8 juice is vegetable juice, available commercially (Campbell's Soup Company). Add ingredients, mix well and autoclave at 110°C for 30 min.

YES: Yeast Extract Sucrose agar (YES) (Frisvad and Filtenborg, 1983)

Yeast extract (Difco)	20 g
Sucrose	150 g
$MgSO_4$ċ$7H_2O$	0.5 g
$ZnSO_4$ċ$7H_2O$	0.01 g
$CuSO_4$ċ$5H_2O$	0.005 g
Agar	20 g

Distilled water 1 litre

Recommended for secondary metabolite analysis. Can be modified by adding 0.5 g $MgSO_4.7H_2O$.

REFERENCES

Abildgren, M. P., Lund, F., Thrane, U., and Elmholt, S., 1987, Czapek-Dox agar containing iprodione and dicloran as a selective medium for the isolation of *Fusarium* species, *Lett. Appl. Microbiol.* **5**:83-86.

Andrews, S., and Pitt, J. I., 1986, Selective medium for isolation of *Fusarium* species and dematiaceous hyphomycetes from cereals, *Appl. Environ. Microbiol.* **51**:1235-1238.

El-Banna, A. A., and Leistner, L., 1988, Production of penitrem A by *Penicillium crustosum* isolated from foodstuffs, *Int. J. Food Microbiol.* **7**: 9-17.

Frisvad, J. C., 1983, A selective and indicative medium for groups of *Penicillium viridicatum* producing different mycotoxins on cereals, *J. Appl. Bacteriol.* **54**:409-416.

Frisvad, J. C., and Filtenborg, O., 1983, Classification of terverticillate Penicillia based on profiles of mycotoxins and other secondary metabolites, *Appl. Environ. Microbiol.* **46**: 1301-1310.

Hocking, A. D., and Pitt, J. I., 1980, Dichloran-glycerol medium for enumeration of xerophilic fungi from low moisture foods, *Appl. Environ. Microbiol.* **39**:488-492.

Nirenberg, H., 1976, Untersuchungen über die morphologische und biologische Differenzierung in der *Fusarium*-Sektion *Liseola,* Mitteilungen aus der Biologische Bundesanstalt für Land-und Forstwirtschaft. Berlin-Dahlem **169**:1-117.

Pitt, J. I., 1979, *The Genus* Penicillium *and its Teleomorphic States* Eupenicillium *and* Talaromyces, Academic Press, London.

Pitt, J. I., and Hocking, A. D., 1997, *Fungi and Food Spoilage*, 2nd edition, Blackie Academic and Professional, London.

Pitt, J. I., Hocking, A. D., and Glenn, D. R., 1983, An improved medium for the detection of *Aspergillus flavus* and *A. parasiticus*, *J. Appl. Bacteriol.* **54**:109-114.

Samson, R. A., Hoekstra, E. S., and Frisvad, J. C., eds, 2004, *Introduction to Food- and Airborne Fungi*, 7th edition, Centraalbureau voor Schimmelcultures, Utrecht, Netherlands, 389 pp.

Simmons, E. G., 1992, *Alternaria* taxonomy: current status, viewpoint, challenge, in: *Alternaria Biology, Plant Diseases and Metabolites*, J. Chelkowski and A. Visconti, eds, Elsevier, Amsterdam, pp. 1-35.

APPENDIX 2 – INTERNATIONAL COMMISSION ON FOOD MYCOLOGY

Aims

The aims of the Commission are:

(1) to improve and standardise methods for isolation, enumeration and identification of fungi in foods;

(2) to promote studies of the mycological ecology of foods and commodities;

(3) to interact with regulatory bodies, both national and international, concerning standards for mycological quality in foods and commodities;

(4) to support regional initiatives in this area.

The Commission further aims to extend understanding of the principles and methodology of food mycology in the scientific community by publishing its findings, and by sponsoring meetings, specialist workshops, courses and sessions dealing with aspects of its work.

Members, 2003

Chairman

Dr Ailsa D. Hocking, Food Science Australia, PO Box 52, North Ryde, NSW 1670, Australia

Secretary

Dr John I. Pitt, Food Science Australia, PO Box 52, North Ryde, NSW 1670, Australia

Treasurer

Dr Robert A. Samson, Centraalbureau voor Schimmelcultures, P.O. Box 85167, 3508 AG Utrecht, Netherlands

Members

Dr Larry R. Beuchat, Center for Food Safety, University of Georgia, Griffin, GA 30223-1797, USA

Dr Lloyd B. Bullerman, Dept of Food Science, University of Nebraska, Lincoln, NE 68583-0919, USA

Dr Maribeth A. Cousin, Food Science Department, Purdue University, West Lafayette, IN 47907-1160, USA

Dr Tibor Deak, Dept of Microbiology, University of Horticulture, Somloi ut 14-16, Budapest, Hungary

Dr Ole Filtenborg, Biocentrum-DTU, Technical University of Denmark, 2800 Lyngby, Denmark

Dr Graham H. Fleet, Food Science and Technology, School of Chemical Engineering and Industrial Chemistry, University of New South Wales, Sydney, NSW 2052, Australia

Dr Jens C. Frisvad, Biocentrum-DTU, Technical University of Denmark, 2800 Lyngby, Denmark

Dr Narash Magan, Institute of Bioscience and Technology, Cranfield University, Silsoe, Bedfordshire MK45 4DT, United Kingdom

Dr Ludwig Neissen, Inst. für Technische Mikrobiologie, Technische Univ. München, 85350 Freising, Germany

Dr Monica Olsen, National Food Authority, P.O. Box 622, 751 26 Uppsala, Sweden

Dr Emilia Rico-Munoz, BCN Research Laboratories, P.O. Box 50305, Knoxville, TN 37950, U.S.A.

Dr Johan Schnürer, Department of Microbiology, Swedish University of Agricultural Sciences, 750 07 Uppsala, Sweden

Dr Marta Taniwaki, Instituto de Tecnologia de Alimentos, C.P. 138, Campinas, SP 13073-001, Brazil

Dr Ulf Thrane, Biocentrum-DTU, Technical University of Denmark, 2800 Lyngby, Denmark

Dr Bennie C. Viljoen, Dept Microbiology and Biochemistry, Univ. of the Orange Free State, P.O. Box 339, Bloemfontein 9300, South Africa

Mr Tony P. Williams, Williams and Neaves, 28 Randalls Road, Leatherhead, Surrey KT22 7 TQ, United Kingdom

INDEX